Riding the New York Subway

Infrastructures Series

edited by Geoffrey C. Bowker and Paul N. Edwards

Paul N. Edwards, *A Vast Machine: Computer Models, Climate Data, and the Politics of Global Warming*

Lawrence M. Busch, *Standards: Recipes for Reality*

Lisa Gitelman, ed., *"Raw Data" Is an Oxymoron*

Finn Brunton, *Spam: A Shadow History of the Internet*

Nil Disco and Eda Kranakis, eds., *Cosmopolitan Commons: Sharing Resources and Risks across Borders*

Casper Bruun Jensen and Brit Ross Winthereik, *Monitoring Movements in Development Aid: Recursive Partnerships and Infrastructures*

James Leach and Lee Wilson, eds., *Subversion, Conversion, Development: Cross-Cultural Knowledge Exchange and the Politics of Design*

Olga Kuchinskaya, *The Politics of Invisibility: Public Knowledge about Radiation Health Effects after Chernobyl*

Ashley Carse, *Beyond the Big Ditch: Politics, Ecology, and Infrastructure at the Panama Canal*

Alexander Klose, translated by Charles Marcrum II, *The Container Principle: How a Box Changes the Way We Think*

Eric T. Meyer and Ralph Schroeder, *Knowledge Machines: Digital Transformations of the Sciences and Humanities*

Geoffrey C. Bowker, Stefan Timmermans, Adele E. Clarke, and Ellen Balka, eds., *Boundary Objects and Beyond: Working with Leigh Star*

Clifford Siskin, *System: The Shaping of Modern Knowledge*

Lawrence Busch, *Knowledge for Sale: The Neoliberal Takeover of Higher Education*

Bill Maurer and Lana Swartz, *Paid: Tales of Dongles, Checks, and Other Money Stuff*

Katayoun Shafiee, *Machineries of Oil: An Infrastructural History of BP in Iran*

Megan Finn, *Documenting Aftermath: Information Infrastructures in the Wake of Disasters*

Ann M. Pendleton-Jullian and John Seely Brown, *Design Unbound: Designing for Emergence in a White Water World*, Volume 1: *Designing for Emergence*

Ann M. Pendleton-Jullian and John Seely Brown, *Design Unbound: Designing for Emergence in a White Water World*, Volume 2: *Ecologies of Change*

Jordan Frith, *A Billion Little Pieces: RFID and Infrastructures of Identification*

Morgan G. Ames, *The Charisma Machine: The Life, Death, and Legacy of One Laptop per Child*

Ryan Ellis, *Letters, Power Lines, and Other Dangerous Things: The Politics of Infrastructure Security*

Riding the New York Subway

The Invention of the Modern Passenger

Stefan Höhne

The MIT Press

Cambridge, Massachusetts

London, England

Originally published as *New York City Subway: Die Erfindung des urbanen Passagiers*, by Böhlau Verlag: © 2017 Böhlau Verlag, GmbH & Cie, Köln Weimar Wien Ursulaplatz 1, D-50668 Köln, www.boehlau-verlag.com

The translation of the this work was funded by Geisteswissenschaften International – Translation Funding for Work in Humanities and Social Sciences from Germany, a joint initiative of the Fritz Thyssen Foundation, the German Federal Foreign Office, the collecting society VG Wort and the Börsenverein des Deutschen Buchhandels (German Publishers & Booksellers Association).

This book was set in Bembo Book MT Pro by New Best-set Typesetters Ltd. Printed and bound in the United States of America.

Library of Congress Cataloging-in-Publication Data

Names: Höhne, Stefan, author.
Title: Riding the New York subway : the invention of the modern passenger / Stefan Höhne.
Other titles: New York City subway. English | Infrastructures series.
Description: Cambridge, Massachusetts : The MIT Press, [2021] | Series: Infrastructures series | "Originally published as New York City Subway: Die Erfindung des urbanen Passagiers, by Böhlau Verlag: © 2017 Böhlau Verlag"—title page verso. | Includes bibliographical references and index.
Identifiers: LCCN 2020012668 | ISBN 9780262542012 (paperback)
Subjects: LCSH: Subways—New York (State)—New York—History—20th century. | Urban transportation—New York (State)—New York—History—20th century. | Subways—Social aspects—New York (State)—New York—History—20th century. | Human-machine systems—New York (State)—New York—History—20th century.
Classification: LCC HE4491.N65 H6513 2021 | DDC 388.4/2097471—dc23
LC record available at https://lccn.loc.gov/2020012668

10 9 8 7 6 5 4 3 2 1

New York, New York—A hell of a town,
The Bronx is up and the Battery's down.
The People ride in a hole in the ground.

—*from the musical* On the Town *(1944)*

CONTENTS

As I am finalizing the translated manuscript of this book in early May 2020, the New York City subway and its passengers are once again making headlines throughout the United States and beyond. A few months ago, news emerged of a new and highly infectious coronavirus, spreading rapidly through China and Europe and threatening to severely hit the United States. At the time I am writing, New York City has become the nationwide epicenter of the disease known as COVID-19. With hospitals overwhelmed and casualty numbers soaring, one answer as to how the virus spread so rapidly throughout the city's five boroughs was soon at hand. In mid-April, an economist from the Massachusetts Institute of Technology released a headline-grabbing paper, provocatively titled: "The Subways Seeded the Massive Coronavirus Epidemic in New York City."[1] Linking data on passenger volume from busy train lines to areas of high infection rates, the study claimed that the ramified subway system was the fuse that ignited the outbreak, infecting countless passengers who would then carry the virus into their homes and workplaces. For many politicians and journalists, the paper validated their suspicion that the subway was a determining factor in spreading the deadly disease in the city; one conservative critic even claimed that it was "possibly the single most important means in the entire country."[2]

While this paper has been nearly universally criticized for its flawed methodology and for failing to provide any statistical evidence, it did prove how easily the subway and its passengers can be blamed for crisis in New York City. In early May, after ridership dropped by more than 90 percent, the twenty-four-hour system shut down its overnight service for the first time in its 115-year history to disinfect trains and stations. This not only deprived thousands of essential workers and caretakers of their primary means of transit, it also left many homeless passengers without refuge; they were no longer able to use the subway to avoid crowded shelters, which have become potential breeding grounds for the virus.

This is not the first time that the subway system has been accused of being a public health hazard, nor is it the first time that subway passengers have come under suspicion of being unhygienic and infectious. In the years following the subway's opening in 1904, newspapers and health officials stoked passengers' fears of suffocation due to a supposed lack of fresh air in the tunnels, and warned of severe eye damage from looking out of subway car windows. The uncontrolled spread of dangerous diseases among its riders was an even greater concern. People were warned not to use the subway because it would be brimming with aggressive germs, exposing its passengers to pneumonia or tuberculosis. Politicians and doctors soon declared the allegedly widespread practice of spitting in the subway as a major disseminator—if not the primary transmission vehicle—of the TB pandemic. In the following decades, the unsanitary practices of subway passengers were similarly blamed for outbreaks of influenza or polio in the city. This association testifies to the deep stigma surrounding public transit in the US, which relates to the demographics of a ridership composed largely of people from marginalized groups.

When I began my research in the New York Transit Museum Archives in early 2009, eventually leading to this book, it soon became apparent to me that throughout its history, the subway's passengers—heralded as progressive heroes, scorned as barbaric hordes, or feared as vigilantes and violent offenders— have been at the forefront of many larger societal conflicts and transformations. Even during the three years since the publication of the book in German in 2017, the subway has repeatedly been the site of large protests against fare hikes as well as police brutality and systemic racism, echoing events in the 1960s and 1970s and affirming the long legacy of passengers' struggles.

The fact that such a study on the history of New York subway passengers was originally written and published in German is admittedly a bit unusual. When I started my research as a young graduate student nearly twelve years ago, however, it seemed unthinkable to me to write the book in English. Thanks to the positive reviews following its original release, as well as relentless encouragement of my peers, I eventually reached out to the MIT Press. To my delight, the editors of the acclaimed Infrastructures Series, Geoffrey C. Bowker and Paul N. Edwards, enthusiastically agreed to take on the English edition. With the continuous support of the press, as well as translation funding from the Börsenverein des Deutschen Buchhandels, this book can now appear in English in an updated form.

Despite nearly a decade of research and writing, the book is far from being without shortcomings. While aiming to prioritize the historical reconstruction of the experiences of passengers, it also focuses on the processes of governing such experiences, and the factors involved in people becoming subjects. At times, these dynamics are given analytical preference, and the perspectives of actual subways users shift into the background.

Given the relative scarcity of first-person sources, especially from the first decades of the subway's operation, I frequently had to rely on the perspective "from above" found in scholarly and artistic publications, scientific studies, and media coverage in order to analyze the subjectivity of passengers. As a result, these voices can often be heard more clearly in the discourse than the articulations of urban passengers themselves.

Even more significantly, while going to considerable lengths to hunt down and incorporate historical material allowing for a multidimensional account of diverse subway passenger cultures, I was only partially able to achieve this goal. A large percentage of the available sources from journalists, engineers, planners, politicians, or cultural theorists do privilege white, male, middle-class perspectives. Accounts of how women, people from lower social classes, and people of color have experienced the system are much rarer. While such accounts have been given particular attention whenever possible, these voices remain underrepresented in this study.

One could easily blame these limitations on the availability and partiality of archival sources—a typical historian's excuse, and often a valid one. Nonetheless, this bias should have been reflected and marked much more distinctly throughout the analysis.

At best, my study will inspire further research and archival explorations, allowing for a much more multifaceted picture of what is probably the most heterogeneous crowd ever to travel underground.

As I am writing these lines, however, New York's underground community may be under greater threat than at any point in its history. My deepest hope is that the subway and its passengers will be able to endure and overcome this crisis, as they have done so many times before. Now more than ever, the recovery of the city and the well-being of its inhabitants may depend on the subway system. Ensuring reliable service is critical for doctors, nurses, first responders, and other essential workers, allowing them to get to work and save lives. Continuous accessibility to the subway is also urgently needed to provide refuge—however imperfect—to its most vulnerable and marginalized passengers.

During its turbulent history, the New York subway has seen countless moments of solidarity among passengers, protecting one another from assaults and harassment, helping the elderly and people with disabilities, or offering their local knowledge to newcomers. Subway rides also provide opportunities for a friendly conversation, romance, entertainment, or the simple offer of some change or food. Passengers frequently hold the doors for those who are slower or in a hurry, or help to carry each other's furniture, strollers, or groceries. Even more often, they just try to get out of the way, silently making room for one more passenger in an overcrowded car, or giving each other space for rest, privacy, and quiet on a long ride home at night. Combining anonymity and intimacy, indifference and consideration, these multiple forms of cooperation and encounter among strangers that have become so characteristic of New York's diverse subway culture are now being put to the test. But where better than the subway for passengers to once again demonstrate that they are more than capable of coming together by staying apart?

Stefan Höhne
Berlin, May 2020

ACKNOWLEDGMENTS

A great number of people have contributed to the development of this book, to whom I owe special thanks for critical readings, discussions, advice, and support. First of all, I would like to thank my supervisors Dorothee Brantz and Hartmut Böhme for their tireless support and advice on the dissertation that gave rise to this book. I am also very grateful to the German Society of Urban History (Gesellschaft für Stadtgeschichte und Urbanisierungsforschung, GSU), especially Gisela Mettele and Dieter Schott, for recognizing the thesis with their dissertation prize.

For support and stimulating debates, I am thankful to the transatlantic graduate program "History and Culture of Metropolises in the 20th Century" at the Center for Metropolitan Studies, Technical University of Berlin, and the members of Hartmut Böhme's colloquium in the Department of Cultural History and Theory at the Humboldt University of Berlin. I also want to thank the NYLON research group of Richard Sennett and Craig Calhoun at New York University and the London School of Economics for many inspiring discussions and friendships.

To Volker Berghahn I gratefully owe a research stay at Columbia University, which moved this work a long way forward. Thanks to the knowledgeable and generous support of archivist Carey Stumm during my extended research in the New York Transit Museum Archives between 2008 and 2014, I gained countless insights and made crucial discoveries. I am also grateful to Jonathan Soffer for valuable feedback, and for giving me the opportunity to present the outlines of this book at the Department of Technology, Culture and Society at New York University.

Furthermore, I would like to express my deepest gratitude to my friends, who have supported me during more than a decade of work on this book, be it through scientific advice and inspiration, or by offering sympathy, encouragement, or forbearance: Hillary Angelo, Martin Behnke, Stefan Bünnig, Thomas Bürk, Marian Burchardt, Thomas Deittert, Heike Delitz, Sasha Disco-Schmidt, Christian Driesen, Ray Daniels, Theresa Elze, Hanna Engelmeier, Alexander Friedrich, Alexander Klose, Boris Michel, André Reichert, Jutta Schinscholl, Ulrich Johannes Schneider, Nico Stutzin, René Umlauf, and Boris Vormann.

I would like give special thanks to Janin Zingelmann as well as Cornelia and Frank Höhne. Not only did they read every line of the German manuscript, but without their encouragement and consolation, this work probably never would have found successful completion.

The English translation was generously supported by the Börsenverein des Deutschen Buchhandels. Its release would be unthinkable without the incredible editing and revisions of Sage Anderson, who not only meticulously improved every sentence of this manuscript, but also offered invaluable feedback and advice. I will be forever grateful for her patience, dedication, and inspiration in completing this challenging project. I am also deeply thankful to David O'Neill, who devoted countless hours to finalizing its more than one thousand footnotes, as well as to the anonymous reviewers for their helpful and inspirational comments and recommendations.

This book is dedicated to Marlene, who practiced her first baby steps on the 7 train to Flushing in the summer of 2015. I deeply hope that she will one day find as much inspiration and excitement in wandering the streets of New York and riding its subways as I do.

On October 27, 1904, shortly before one o'clock in the afternoon, New York City notables emerged from an elaborately decorated City Hall.[1] Outside they were greeted by an enormous crowd celebrating a remarkable occasion: the opening of one of the grandest technological wonders of the times, the New York City subway, a system that would eventually alter both the city and the lives of its inhabitants more radically than any infrastructure had ever done before.

Since early that morning, people had been gathering in anticipation of the event (figure I.1).[2] The police soon found themselves roping off routes and forming motorcades to contain the crowds.[3] In the large assembly hall inside, efforts to get overcrowding under control delayed the commencement of the opening ceremony, which was then to be followed by an official maiden voyage from City Hall station to Harlem.

The esteemed delegation had difficulty elbowing its way through the masses down to the underground, where crowding intensified despite strong police presence. As a result, a large number of ordinary New Yorkers managed to board the brand new subway cars along with the official guests. More than a thousand people were aboard the maiden train when it left the station. Already delayed and absolutely packed, the very first subway ride offered passengers a taste of things to come.

Figure I.1 Festively decorated New York City Hall on the morning of the opening day of the subway, October 27, 1904.

When at last New York City Mayor George Brinton McClellan Jr. reached the operator's cab, he started the engine with a silver key specially designed by Tiffany's for the occasion (figure I.2). Under the euphoric applause and cheers of its passengers, the train pulled out of the underground station at 2:35 p.m.[4] Protocol dictated that Mayor McClellan was to hand over the controls to a trained operator. Intoxicated by speed, instead he held on for a large part of the trip, accompanied by the panicked cries of passengers.[5]

McClellan even pulled the emergency brake at one point, causing the first passenger injuries on record for the New York subway. To the relief of everyone on board, an experienced motorman finally took control. Shortly thereafter, the train approached the viaduct over Manhattan Valley between 122nd and 133rd St. and then emerged from the depths. As it shot out of the tunnel into the daylight, passengers were met with an impressive sight. Tens of thousands of cheering spectators were lining the streets or perched on fire escapes and rooftops, having waited hours for this historical moment. The train slowed down and greeted them with a long whistle. Out on the river and in the harbor, boats blew their horns in reply.[6] School and factory sirens added to the din. Bells rang out from all of the churches in Manhattan, fireworks burst in the sky, and bands began to play. Following instructions from the newspapers, countless New Yorkers joined in the chorus of machines with whistles and musical instruments.[7] According to reports of the day, the immense clamor of cheers, music, horns, bells, and whistles continued for more than a half hour after the maiden train had disappeared back into the tunnel.[8]

The next day, the *New York Times* described the jubilation as "the city's glad uproar," a mixture of nostalgic church chimes and modern industrial noise heralding economic growth and prosperity.[9] Combining the sounds of humans and machines, this polyphonic orchestra of the "subway city" marked a historical turning point in the eyes of contemporaries. The thrill of the train emerging from the underground signaled the approach of a new epoch of liberty, civility, and progress. The euphoria of New Yorkers welcoming their subway was an expression of their hopes and aspirations for becoming passengers. For many, this transformation was nothing less than "the only cure to the illnesses of the city."[10]

Figure I.2 Mayor George B. McClellan (center) in the driver's cab of the first subway on October 27, 1904, shortly before the maiden voyage.

The remarkable outburst of pride prompted by the opening of what was at the time the world's largest urban transit system calls for further explanation to be understood in full. The collective ecstasy over the implementation of this element of infrastructure gains greater significance against the backdrop of historic transition taking place around the turn of the twentieth century in New York and other metropolises, such as Berlin, Paris, London, and Cairo. In the decades before and after 1900, residents of these and other cities were witness to profound changes.

While new methods of high-rise construction and the invention of the elevator led to the rapid, dense development of vertical urban space,[11] increased migration kept city populations growing and urban areas expanding at an unforeseen pace. And yet one of the most radical changes took place *below* the pavement. The "rapture of circulation" that metropolitans experienced during this period was in large part due to the new underground infrastructure of mass transit.[12] The subway packed human bodies together and shifted them into accelerated motion in an entirely unfamiliar manner. This was fascinating and exciting, but to many it also felt unnatural and threatening. Traveling underground aroused so much uncertainty and anxiety that it necessitated entirely new behavioral norms, modes of perception, and technologies of control.

According to philosopher Ian Hacking, modern sciences "create kinds of people that in a certain sense did not exist before."[13] The same can be said of modern infrastructures. Like almost any new large technological system, underground transportation not only produced new forms of knowledge and modes of governing, but also altered perceptions and experiences. Moreover, it changed the individuality and collective subjectivity of its users. Upon closer examination, what is commonly referred to in simplistic terms as "the implementation of infrastructure" turns out to be a highly complex and often conflict-laden process. Creating large-scale urban infrastructure requires immense concentration and coordination of labor, skill, resources, and money, as well as people who will step up to use the apparatus. This is an affirmation of Karl Marx's thesis that modern industrialized society "not only creates an object for the subject, but also a subject for the object."[14] The new subject created by the subway was the modern urban passenger.[15]

NEW SUBJECTS FOR A NEW SOCIETY

If we understand infrastructures of urban transit as highly effective machinery fostering social, political, and economic integration, it is remarkable how little attention historians have paid to their user cultures.[16] While tourists, migrants, and modern nomads have been the topic of in-depth research,[17] subjects in transit have remained largely absent from scholarly focus.[18]

This is where my study begins, tracking down the multifaceted subject forms of the urban passenger throughout the twentieth century. I aim to show how the daily transit experience was involved in shaping modern urban life and subjectivity.[19] This means taking a close look at the intricate power relations between those who operated the system and those who used it.[20] When we conceive of underground transit infrastructure as "media of the social,"[21] it becomes clear that such machines have been decisively implicated in the transformation of social subject attributions related to class, race, age, and gender. Underground transportation supported new methods of urban governance, with the aim of fostering smooth, well-ordered circulation of city dwellers. Besides new demands and constraints, this also opened up new possibilities and liberties. As subway authorities sought to transform the often unruly and stubborn passenger masses into predictable and controllable subjects, they simultaneously subsumed countless individuals under legal and administrative categories associated with specific competencies and requirements.

However, the history of subway passenger culture does not follow a simple narrative of continually increasing discipline and control. Authorities were often success in shaping people's behavior in the system by means of institutional regulations, discursive appeals, and material design.[22] But the process was much too complex for smooth, frictionless implementation. Passengers repeatedly resisted these imperatives, or devised different ways to undermine or ignore them. This is true with respect to their movements and bodily practices, for example, which again and again conflicted with the technical demands of the system. From the outset, passengers developed attitudes and interactions that clearly did not conform to subway-imposed norms.

My interest lies precisely in these contested dynamics of "configuring the user" of subway services.[23] How were people subjectified as urban passengers, and why did some accept these strategies while others rejected them? By exploring how these behaviors and forms of knowledge were condensed into historically specific forms of passenger subjectivity, it is possible to bring these forms into clear view for cultural analysis.

In order to reconstruct these dynamics in depth, this study focuses on a single infrastructure system: the New York City subway. Thanks to this system, New York can be regarded as the world capital of passengers in the twentieth century. After opening in 1904, the subway soon became the largest urban transit system ever built. Its sheer volume of riders overshadowed that of all other big cities for nearly a century. Today, within any given twenty-four-hour period, the system transports more than five million passengers along roughly 655 miles of track, with twenty-six lines serving 460 stations.[24]

When New York became a city of passengers and commuters, it was neither the first nor the only metropolis to undergo this transformation. Once the world's first underground railway had taken up operation in London in 1863, the cities of Budapest (1896), Glasgow (1896), Boston (1897), Paris (1900), and Berlin (1902) followed suit. But the impact was nowhere greater than in New York City. The subway transformed the city and urban life in a way that had never been seen before.[25]

Exploring these developments, this study examines roughly a century of subway history. The timeframe stretches from early discursive anticipation of the future passenger in the context of the first plans for the subway around 1860 to its opening in 1904, and from there to the ultimate consolidation of subsystems under the roof of the Metropolitan Transportation Authority (MTA) in 1968. This relatively long time span facilitates an analytical perspective of *longue durée* that allows us to trace subtle historical transformations in the subway's passenger cultures. In order to be effective, the scope of such an analysis must be limited, and some aspects of passenger culture will remain untouched. This study concentrates primarily on the particularly dominant subject forms that emerged during the time period in question. Around 1900 social reformers envisioned the passenger of the future as a messenger of morality and prosperity. When the idea of technological progress was seriously called into question a few decades later, the subway passenger came to be seen as embodying all the evils associated with Fordist mass culture. Passengers then came to be characterized as fragile subjects, violent criminals, denunciatory informers, patriarchal commuters, resolute feminists, consumers, and patriots. This scope in characterization alone shows how many different self-images and external expectations were attributed to the subway's passengers over the years.

WHAT DO WE KNOW ABOUT URBAN PASSENGERS?

Before exploring how people became passengers of the New York City subway, and situating them within larger historical dynamics, a few words are in order regarding the guiding concepts, methods, and sources used in this book. Additionally, it is necessary to address the question of what, if anything, we already know about passengers.

Encyclopedias and lexica offer little assistance. Although the entire world was in motion, major twentieth-century reference works rarely mentioned the passenger per se.[26] The *Oxford Universal Dictionary* from 1959 does offer a few lines on the subject: "one who travels in some vessel or vehicle, esp. on board ship or in a ferry or passage boat; later also applied to travelers by any public conveyance entered by fare or contract (the prevailing sense)."[27] By the 1970s, definitions of what it meant to be a passenger disappeared completely from reference works. Being a passenger was apparently such a self-evident condition at this point that it needed no further definition.

Some scholars in history, literature, and cultural theory have taken a closer look at the passenger as such, most prominently Wolfgang Schivelbusch with his influential study on nineteenth-century railway culture.[28] We can also find inspiration and insights in works by Michel de Certeau,[29] Leo Marx,[30] Bernhard Siegert,[31] and Paul Virilio.[32] They all bring the passenger into perspective, but not specifically in the urban context.[33] This is all the more remarkable considering that there has been a higher volume of transit within cities than between cities since the beginning of the twentieth century at the latest.[34]

While the history of urban passenger culture has remained largely unexplored, much has been written on subway history. Probably no other urban transit system in the world has been the subject of as many historical studies as the subway in New York. Over ten monographs and dozens of essays deal intensively with the design of the New York subway, its technical details, and its management.[35] But while they provide precise accounts of how the subway system was planned, built, and operated, these works have a tendency to marginalize the users of the infrastructure.[36]

A handful of chroniclers of the New York subway have also examined the experience of transit and its effects on life, art, politics, and society in general. In particular, Tracy Fitzpatrick's work on the history of subway motifs in art and Michael Brooks's writings on the historical and cultural significance of this infrastructure have been indispensable for my own study.[37] The same is true for Clifton Hood and Brian Cudahy's standard works on the history of the New York subway.[38] But if we want to understand the multifaceted historical manifestations and transformations of its user cultures, we will have to write the history of the New York subway once more, this time as a history of its passengers.

Many brief, disconnected, almost fleeting traces of passenger cultures wait largely unanalyzed in the New York Transit Museum Archive (NYTMA). The repository examined in the course of this study includes newspaper and magazine articles, photographs, bureaucratic evaluations, internal studies, protocols, and much more. In particular, thousands of previously unevaluated complaint letters from 1954 to 1968 have much to tell us about conflict and social order among passengers.

To explore images and imaginations of the subway and its users, we will turn to painting, film, photography, music, and literature from the period in question. We will also have a closer look at material artefacts like turnstiles, car interiors, and sign systems, asking what role they played in passenger perception and behavior. Additionally, we will have recourse to works of cultural theory and social philosophy that were relevant and specific to certain places and times; they hold valuable clues regarding the contemporary theorization of metropolitan experience.[39]

However, analyzing these materials in order to historically reconstruct passengers' multifaceted experiences and actions clearly highlights the limits and boundaries of such an endeavor. In large part, the twentieth-century sources from archives and elsewhere examined here stem from authorities, planning and engineering departments, administrations, prominent intellectuals, or the media. While offering countless invaluable insights for this study, these sources are far from unbiased or all encompassing. Instead, most of them offer white, male, middle-class perspectives, often neglecting or marginalizing the experiences of other passenger groups. Aiming to tackle this

9

bias, I have worked to identify and incorporate material that provides information on how women, African Americans, or Latinx passengers, among others, have perceived and made use of the system. Nonetheless, these perspectives unfortunately often remain underrepresented, not only in the archives and historical records, but also in this study. In addition, many insights into the gendered and racialized ways of ordering passenger interactions stem from scientific studies and media coverage, as well as intellectual and artistic works, sources that often tend to impose a generalizing viewpoint. This also holds true for the historical reconstruction of processes and strategies of controlling passengers' experiences and behaviors. For the sake of this analysis, the perspectives of subway users themselves sometimes have to take a back seat, at least temporarily.

The fact that the positions and perceptions of the powerful and privileged often feature much more prominently in historical discourses than the actual experiences of many members of the diverse passenger population has crucial implications for the scope of this study. First of all, it means that we have to keep in mind the limitations and biases outlined above, while engaging in critical reflection on the sources analyzed. We also have to pay close attention to how passenger experiences have differed along familiar lines of dominant social order. Furthermore, we have to establish an analytical model that sheds light on the dynamics of infrastructural subjectivity that figure into the metamorphoses of the passenger over the decades.

This approach also raises some fundamental theoretical and methodological questions: How much can we know about past passenger experiences and interactions? How can we understand the relationship between urban infrastructure and its users at a given moment? What role does the function and state of materiality and technology play in the social sphere? In order to address these questions, we must first develop a conceptual apparatus that accounts for the historicity of different passenger subject forms.

SUBJECTS, SCRIPTS, AND INFRASTRUCTURES

One of the perils of speaking of the subject as such is that this term has often been used with very different and sometimes contradictory meanings in philosophy, history, and cultural studies.[40] The relevant meaning for the purposes of this study is not the meaning found in classical philosophy of the subject, which usually considers the subject as an irreducible, autonomous, or even transcendental entity.[41] What we are focused on here is how an individual, at a given time, considers him- or herself as a socially integrated and acknowledged subject who also acknowledges others.

Like any contemporary historical analysis of the subject, this study takes into account Michel Foucault's pathbreaking work on the genealogies and transformations of Western subjectivity.[42] Especially in his later writings, Foucault examines the role that the self (subject) and mechanisms of making selves (subjectivation) play in maintaining or transforming social order.[43] He explores how culture and society shape people into selves through knowledge and discourse, governmental strategies, and technologies of the self.[44] As Foucault has shown, such processes of subjectivation involve powerful, conflictual procedures of discipline and control that are bound up in the daily activities and practices of ordinary people: "This form of power applies itself to immediate everyday life which categorizes the individual, marks him by his own individuality, attaches him to his own identity, imposes a law of truth on him which he must recognize and which others have to recognize in him. It is a form of power which makes individuals subjects."[45]

Understanding everyday life as a powerful arena of subjectivation is especially useful for the purposes of this investigation into the daily use of urban infrastructures.[46] As users of mobility infrastructures, passengers' habits, interactions, and experiences are significantly shaped by the technical environment and logistical requirements of transit.[47] Thus, in order to examine large-scale socio-technical systems as apparatuses of subjectivation, we need to operate with an understanding of infrastructure that encompasses more than the material elements of the system, such as rails, tunnels, and cables.

Upon closer examination, the term "infrastructure" turns out to be much more ambiguous than it might seem. As historian Dirk van Laak has shown, the term was first used within the context of railway development in nineteenth-century France, where it initially meant only railroad tracks, but then came to include additional immobile elements that contribute to mobility.[48] Used only sporadically for many years, "infrastructure" became more familiar after 1950, finding its way into military logistics, economics, and foreign aid.[49] The term was then picked up by economics, political science, and policy planning, eventually entering general public discourse.[50]

For the purposes of this study, we will define urban infrastructures as the socio-technical apparatuses that enable the circulation of people, things, and information within a city, following urban scholars Christopher Boone and Ali Modarres.[51] Here, circulation refers to more than mere spatial mobility: it is a sophisticated process that produces certain "cultures of circulation."[52] An analysis of such infrastructural cultures calls for an examination of technical modalities, bureaucratic regimes, and legal regulations. It also requires attention to questions of how protocols of engineering and technology have been translated into ordinary habits, and how normative ideas, fantasies, and collective imagination have altered these protocols in turn. Based on these various considerations, it makes sense to characterize infrastructure as what Foucault describes as a dispositif (or an apparatus):

> What I am trying to pick out with this term, is [. . .] a thoroughly hetero-geneous entity consisting of discourses, institutions, architectural forms, regulatory decisions, laws, administrative measures, scientific statements, and philosophical, moral, and philanthropic propositions—in short—the said as much as the unsaid. Such are the elements of the apparatus.[53]

For Foucault, a dispositif—or apparatus—fulfills a decisive purpose: it structures the self-images and subjective experiences of the people it addresses.[54] A subjectifying apparatus develops its power by linking heterogeneous elements. These include orders of knowledge, artefacts, and bureaucracies, as well as certain cultural techniques, territories, and bodies.[55] When we consider infrastructure as a kind of subjectifying dispositif, we find that it cannot be understood merely as a conglomeration of technical objects. Instead, infrastructure must be seen as a complex form of organization that follows the logic of technology as much as the logics of culture, politics, and economics.[56]

In the 1980s, historian of technology Thomas P. Hughes developed a groundbreaking model for the role and function of technology and infrastructures in modern societies, focusing on what he called large technological systems.[57] Technology and society, he found, should not be understood as two distinct and separate spheres; we must view technical artefacts as inherent to the social. These artefacts are inseparably woven into the fabrics of discourse, habits, forms of knowledge, and methods of governing, and they must be analyzed accordingly.

However, if we understand modern society and culture as inherently *technomorph*, we must be careful not to frame technology as deterministic, nor to treat subject cultures merely as a subordinate effect or product of a superior apparatus.[58] Neither should we adopt the stance of social constructivism that sees technology primarily as a variable dependent on cultural or economic trends.[59] Instead, I suggest that we understand technical artefacts—including infrastructures—as *co-constituents* of the social.[60]

Cultural historian Hartmut Böhme stresses the significance of the relationship between elements of infrastructure and their users, emphasizing that "every device object and every technical system include codes of how to interact with them."[61] Actor-network theory (ANT) understands these codes as *scripts* or *inscriptions*. Developed primarily by Madeleine Akrich and Bruno Latour, this approach has proved highly useful for the purposes of describing how designers and engineers attempt to predict and control how users interact with certain technical artefacts.[62] The aim is to incorporate desired product/user interactions right into the object itself, while simultaneously discouraging interactions deemed inappropriate or dangerous.[63] "Like a film script," successfully designed technical objects "define a framework of action together with the actors and the space in which they are supposed to act."[64]

While this approach can help us to understand how technical design anticipates the skills, motives, and behavior of users, it does not imply behavioral or technical determinism. In most cases, the actual handling of an object may deviate considerably from what designers and engineers believe an ideal user will do. Such deviations, or uses that change or circumvent the originally intended script, can be understood as *deinscriptions*.[65] Particularly pronounced discrepancies between intended use and actual use can result in disruption or even crisis, especially when it comes to networked objects or technological systems, such as the subway.

Beyond controlling the social and material conditions of transit, such scripts also function as a set of highly normative tools, designed to regulate the behavior of an ideal user with respect to the political, social, and moral codes of the times. At first glance, infrastructures might present themselves as apolitical, neutral technical systems. Yet

---

in their function as media of social and political integration, they are in fact crucial elements in structuring urban inclusion and exclusion.[66] As we will see in the course of this book, operators and engineers of the New York subway implemented scripts not only to control the interactions of passengers with the system, but also to educate and discipline them in matters of hygiene, manners, and civic duty.

In a historical study such as this, any isolated subject can only represent an individual case. As a result, I will use the concept of "subject forms" to mediate between the self and the cultural order that both presupposes and produces that self. Often, subject forms do not gel in a way that yields a clearly distinct, stable picture.[67] As sociologist Andreas Reckwitz emphasizes, these forms are better understood in terms of "a correlate of shifting patterns of subjectivation."[68]

For the subject forms of the passenger under examination here, the subway is the primary site of subjectivation. However, the mode of operation of this subjectivation is intertwined with other subjectifying factors. These factors include, for example, a political economy based on the spatial separation of home and work, and habits of consumption, sociability, and leisure that depend on urban mobility. Adopting the subject form of the passenger has ramifications that go far beyond the site of subjectivation. I will show that this is particularly true for New York City, where the subway plays a constitutive role in so many aspects of daily life.

At the same time, it is important to keep in mind that successfully transforming people into passengers requires more than merely subjecting them to the power of the infrastructural dispositif of the subway. Subjects themselves must acknowledge and adopt the codes and models presented to them. The successful establishment of any socio-technical order depends on stabilizing the relationship between individual practices and overarching technical dispositifs. The concept of "infrastructured subject forms" aims to mark the relay between infrastructure and individual autonomy.[69] These forms can be understood as generalizations that include behavioral models, modes of perception and experience, and horizons of meaning that provide subjects with orientation and self-understanding.[70]

In order to describe the many historical subject forms of the New York subway passenger, we must reconstruct cultural and perceptual habits, as well as codes and mechanisms inherent to the infrastructure and technical artefacts. We must also explore the imaginations, forms of knowledge, and administrative procedures that circulated in discourses surrounding the passenger. Finally, we must combine these various levels of analysis in such a way that the passenger stands out clearly from these complex interconnections.

STRUCTURE OF THE BOOK

In reconstructing the historical dynamics of passenger subject forms, this study basically proceeds chronologically, with an emphasis on events that took place between 1904 and 1968. But because it is often impossible to precisely pinpoint the emergence or demise of a particular subject form, and because many codes developed over long periods of time, we will sometimes diverge from strict chronological order.

Chapter 1 begins with the anticipation of future passengers in the course of the subway's planning and construction. By the 1860s, New York was in urgent need of mass mobility as a solution to the many problems arising from overcrowding. As the imperative of circulation gained momentum, the idea of a gigantic underground transit system transformed from an absurdity into a necessity. The subway buoyed hope for a glorious future: New York would grow and prosper as a city of underground passengers. As the analysis of contemporary sources will show, elites in particular expressed great expectations for the system. They imagined passengers as heroic subjects who would embody the blessings of circulation. With the turbulent and spectacular events of opening day, however, these high expectations and hopes were punctured the very first time people set foot in the subway.

The next three chapters cover passenger culture from 1904, when the first line went into operation, until 1953, when three formerly independent subsystems were consolidated under one roof with the New York City Transit Authority. During this period, New York—along with industrialized societies in general—went through various processes of mechanization associated with Taylorism, Fordism, and mass culture. In retrospect, the transformation of US society in particular appears so radical and consequential that many historians refer to this as the *machine age*.

Chapter 2 shows how the beginning of this age was marked by a profound crisis of traditional models of subjectivity, along with new modes of mass production and consumption, new technology-driven experiences, and the spread of machines throughout society. At the same time, the concept of *the masses* gained significance. This emerging form of collective subjectivity initially evoked associations of an impulsive and threatening mob that needed to be disciplined and controlled. The perceived need for control applied particularly to crowds of subway passengers, whose unruly behavior seemed to pose a threat not only to public moral order, but also to the technical workings of the system itself.

In order to bring passenger masses into line with the technical requirements of the subway, a dispositif of the machine emerged. Through methods of rationalization and discipline, passengers were transformed into predictable units. Documents from the New York Transit Museum Archives show that this was largely achieved by applying biopolitical methods of statistical standardization, and by building scripts and behavioral codes into the material elements of the system, such as cars, turnstiles, and station interiors. In this way, passengers were coded as standardized, rational *container subjects* with a kind of *container ethics* aligning social and technical norms.[71]

Chapter 3 shows how the spread of machines throughout society changed passenger perceptions and interactions. An analysis of different sources, including photographs, medical reports, and studies in design, provides clues as to how subjects adapted their sensory techniques to cope with the excessive demands of subway travel. We discuss how visual elements such as maps and signs shaped passenger perception and practices. As we will see, the subway's first passengers in particular found underground transportation frightening and overwhelming in many respects.

In reaction to the many unfamiliar sights, sounds, and smells of the subway, passengers developed habits for modifying sensory perceptions. Contemporaries described it in terms of protection against stimuli or industrial awareness, indicating that this was a process of containerization. Passengers practiced techniques of isolation, shutting themselves off from their surroundings to reduce emotional strain. Reinforcing the boundary between their inner lives and the machine environment, passengers simultaneously generated new forms of social interaction, particularly when it came to looking.

The priority of sight in passengers' perceptual techniques was reflected by the wide variety of signs, symbols, maps, and advertisements in the subway, implemented to shape their interactions, perceptions, and moral convictions. The subway's visual regimes also aimed to ensure proper circulation within the system, and to discourage undesirable behavior like smoking, spitting, and shoving. Advertisements and posters addressed passengers as consumers, patriots, taxpayers, and civilized citizens, highlighting the ways in which the subjectivation of passengers went well beyond merely compelling them to interact properly with the system's equipment and routines. Visual elements played a role in the efforts of subway operators to propagate certain norms. These efforts did not always go over well with the public, as we know for example from the angry reactions that followed the introduction of advertisements in the subway. Such protest and outrage can be seen as a refusal to meet the imperatives of conformist, consumerist mass culture.

Chapter 4 discusses the critique of mass culture in general, and the lonely crowds of the subway in particular. In the early 1950s, as the machine age was coming to an end, attitudes toward the once so enthusiastically welcomed transportation system began to shift. The subway came to be seen as an instrument of alienation, exploitation, and oppression. The system had fulfilled the expectations of enhanced mobility and new urban cultural experiences, but it could not make the dream of a better life into reality. As the automobile began to promise more individual freedom and renewed modernization, public opinion of subway transportation deteriorated. The subway passenger now stood as a deterrent example of what mass culture could do to a person. New York's commuters in particular became emblematic of how industrialized society standardized, isolated, and exploited its subjects. Works from the times by E. B. White, Fortunato Depero, George Tooker, and Walker Evans all portray subway passengers as exhausted, fragile, defenseless subjects at the mercy of an inhumane bureaucratic regime of labor and mobility.

At the same time, intellectuals increasingly voiced their criticism of the subject orders of mass culture and the impact of technologized society on personal character, especially for middle-class employees. Sociologist David Riesman coined the term "lonely crowd," and C. Wright Mills spoke of "cheerful robots." While many contemporary sociologists and philosophers lamented the pathologies of the masses, Lewis Mumford and Sigfried Giedion began looking at how the spread of machines affected all areas of life. Mumford even referred to the subway as a "megamachine" of novel exploitative dimensions, similar in its dangerous capacity to the nuclear bombs dropped on Hiroshima.

Intellectuals and artists were not alone in their criticism of the subway in New York City. Policy makers, members of the political and economic elites, and urban planners also increasingly regarded it as an outdated, expensive apparatus not worth the price of maintenance. Finally, New Yorkers themselves began abandoning the subway, if not the city itself. In the wake of exodus to the suburbs, passenger culture began eroding in the 1950s, and by the 1960s the rate of crime in the neglected system had risen substantially.

Chapter 5 looks at how this crisis was experienced by analyzing passenger complaint letters written to the Transit Authority between 1954 and 1968. This fascinating corpus includes accusations, denunciations, petitions, replies, and other material, explored for the first time by this study. This material offers intimate insights into passenger experience, showing how subjects strategically presented themselves when communicating with authorities. At the same time, these documents attest to a dramatic rise in conflicts in the subway, often along the lines of gender, class, age, and race. Reporting their experiences of violence and fear, passengers frequently demanded that certain groups of passengers be barred from using the system, or complained of rampant vandalism and material neglect. Especially as of early 1960s, there were also frequent reports of arbitrary and brutal police conduct. Through their letters of complaint, subway passengers speak to us as fragile, insecure subjects, outraged denunciators, or concerned parents.

Authorities followed up with those who had written complaints, revealing these routines as a mode of negotiation between the powerful dispositif of the system and its users. Analyzing these dynamics, as well as the tricks of those who submitted complaints in order to evade the grasp of bureaucracy, it is possible to bring the process of subjectivation into sharper focus. As we will see, this is not only a matter of subjugation, it also involves moments of self-empowerment. Highlighting the inherent ambivalence of complaints, these letters expressed discontent and criticism while simultaneously acknowledging authority, aiming to uphold social order. Yet the fact that passengers called on authorities, questioned their legitimacy, and demanded to be recognized as subjects demonstrates that this instrument had the potential for transformation and emancipation as well.

The archiving of complaint letters ends with December 1968, after the New York subway system as well as all busses and ferries were consolidated into the Metropolitan Transportation Authority. But this consolidation did not mean an end to the crisis of the subway system, nor to the erosion of its passenger culture. Thus, this study closes with an account of further developments in the subway and the emergence of new subject forms, particularly during the 1970s and 1980s, and after September 11, 2001. Finally, we discuss the processes of containerization, as well as the desires for individuality and autonomy that remained unsuppressed throughout the course of twentieth-century subjectivation.

For now, let us return to the beginning of this history by exploring the final decades of the nineteenth century, and the utopian expectations built up around the new subway and the future passenger. In Foucault's view, a dispositif presents a "formation which has as its major function at a given historical moment that of responding to an *urgent need*."[72] This certainly holds true for the socio-technical dispositif that constitutes the New York subway. What are the needs, demands, and crises that made the introduction of this gigantic, expensive element of infrastructure seem inevitable? What are the subject forms and subject orders produced and transformed by this dispositif?

To answer these questions, we begin by reconstructing the events and discourses that led to the call for an underground transit system for New York City, and ultimately to its realization. How was the future passenger of this system anticipated by these dynamics? What was the passenger, before emerging underground?

1
_____

Utopian Passengers

New Yorkers will never go into a hole in the ground to ride.[1]

—*Russell Sage (1900)*

The Subway [. . .] will give them more time, more ease. It will carry them
to better and cheaper and healthier homes. It will change their very lives.[2]

—*John McDonald (1904)*

In the late nineteenth century, as New York politicians, entrepreneurs, and residents
discussed the technical details of the coming subway and debated how to fund it, they
also thought about what effects the great machine might have on its users. Portraying the subway as a universal remedy for all urban problems, various camps proposed
discursive formulations attributing concrete properties to the still utopian passenger-
subject. The idea of the passenger grew from an increasing awareness that municipal policies for social, political, and economic order had become dysfunctional and
something had to be done about it. At the same time, the passenger also appeared in
the discourses of new scientific and scholarly findings, and new paradigms of social
organization and government. The reshaping of urban society would alter traditional
aspirations. Future subway passengers were to be the exact opposite of pedestrians and

slum dwellers, subject forms perceived as precarious and dangerous, along with the earlier passengers of horse-drawn omnibuses and even elevated trains. The passenger of the future was to be a savior figure who would play a key role in overcoming economic stagnation.

The coding of passengers in this way followed from a specific idea of circulation, found in ideologies of hygiene as well as in areas of economics and strategies of urban governmentality. Turning citizens into passengers was decisive in the effort of the urban elites to convince people that regulated circulation meant salvation. As the gigantic underground system became part of the imagination of the city population, the passenger appeared as an eagerly desired collective ideal.

Visions for the passenger community of the future encompassed a number of elements: urban order, civilization, and control; debates on medical and social hygiene; and infrastructure as an element of social regulation. Circumstances were indeed critical. The city was overpopulated and the economy sluggish. Political elites, social reformers, and entrepreneurs all had an eye on the passengers who would be. Heroes in motion, they were going to introduce a new era.

In order to explore the passenger's discursive formation, in this chapter we will look at the time from around 1860 until the day the subway opened on October 27, 1904. By the time the London Underground took up operation in 1863, it had become impossible to ignore the need for a new system of urban transportation in New York. In 1866 the senate advisory board decided that the construction of a subway system was indispensable for urban growth.[3] The first planning committees were set up and projects were proposed, but bureaucratic conflicts and political turbulence hindered progress time and again.

Some forty years later, opening day became a key event in the metamorphosis of New York into a city of passengers. The start-up of subway operation represented a risky moment, especially for investors, politicians, and operators. From the very beginning, proposals and plans for the underground system had met with objections and doubt. Opening day would show whether the huge effort to popularize the subway would pay off. In that moment, discursive imagination and prognostication of the ideal passenger would meet the actual experience of people in the system. For residents, opening day marked a significant turning point in the history of New York.

It was the beginning of a new era that would permanently change the face of the city and the daily practices of those who lived there.

How did the subway passenger come into the world? The spectacular events of opening day offer some clues. Twelve hours are of particular interest, from noon, when the opening ceremony began, until midnight, when the last exhausted passenger finally returned home. To account for the significance of these hours and what they meant to the people of the city, it is necessary to consider the context of events and some of the historical developments that preceded them. Throughout this inquiry, we will look back at discourses and processes that led to the realization of the twentieth century's largest urban transit system. It started with an extraordinary economic boom and unexpected population growth beginning in the 1820s, which led to serious problems for the remainder of the nineteenth century.[4] Let us begin, however, with the opening day and the great expectations that the people of New York had for their new subway.

OPENING DAY

On the morning of October 27, 1904, New York's city hall presented itself in exuberant festivity.[5] Flags and banners hung in every window to welcome guests and onlookers who had come to witness the epochal event of the opening ceremony. Crowds gathered so quickly that it became difficult to maintain order in the rush on the great hall. Police officers held people back as ushers tried to get the situation under control. Some of the official guests had to wait outside. City Mayor George Brinton McClellan (1865–1940) led a delegation of honor into the jam-packed hall to the sound of thundering applause (figure 1.1).[6] New York Bishop David H. Greer opened the ceremony with a prayer, blessing the new infrastructure and invoking its power to lead the city and its inhabitants to moral, economic, and spiritual prosperity. He further insisted that the new machinery was nothing short of pleasing to God.

After an act of consecration and a collective prayer for passenger safety, there were brief speeches by a few distinguished individuals who had helped to make the subway a reality. Along with William Parsons (chief engineer), August Belmont (financer), and Alexander E. Orr (president of the Board of Rapid Transit Commissioners), general

Figure 1.1   The opening ceremony of the subway system at New York City Hall on October 27, 1904. Courtesy of New York Transit Museum.

contractor John B. McDonald stressed the superhuman effort of everyone involved in the execution of this huge infrastructure project. In addition to engineers, bankers, and workers, he emphasized the crucial role of the inhabitants of New York City. They had weathered the strain of construction with remarkable dignity and stoicism, and now they were to have their reward: "To the citizens of New York, the men who have borne almost without complaint the inconvenience which the construction of the Subway has necessitated, all praise is due. I scarcely believe that their patience and forbearance have been or will be equaled elsewhere, but I trust that the result will amply repay you all."[7]

The mayor then addressed the tightly packed guests of honor, highlighting the system's importance for the people and city of New York. Opening the subway represented a huge step toward unity for the city, in both material and imaginary terms. McClellan effusively promised that the citizens of New York, a territorial unit that had been created just eight years earlier by incorporating the districts of Manhattan, Brooklyn, Queens, the Bronx, and Staten Island, would now live in harmony as never before.

At that time, what had been completed of the system consisted of only twenty-eight stations along a single nine-mile line that ran from City Hall to 145th street in the Bronx. Extensions were already feverishly underway.[8] The mayor declared that the subway would not only grow rapidly in the years to come, but it would eventually reach all parts of the city. He claimed that this would allow New Yorkers to forget which part of town they came from, realizing that they were all children of the greatest metropolis in the history of mankind, inseparably bound together by common hope and shared destiny. Transform New York into an urban society of passengers meant nothing less than fulfilling the city's historical purpose.[9]

McClellan's vision demonstrates the strong contemporary coding of the subway as a utopian promise of progress and prosperity. It was meant to allow people to reach a higher level of human development. The mayor's emphasis on historical destiny echoed a recurrent theme in late nineteenth-century semantics, namely, the ideology of progress linking technical innovation to societal change.[10] Yet despite attempts by the elites to present the subway as progressive and modern, it did not emerge without

25

contradictions, as we will see again and again. Like most large technical systems, it mobilized fear and suspicion along with fascination and euphoria.

After a standing ovation for the mayor's prophetic words, investor August Belmont presented McClellan with an ignition key of pure silver, made and engraved by Tiffany's for the occasion. "I give you this controller, Mr. Mayor, with the request that you put in operation this great road, and start it on its course of success and, I hope, of safety."[11] The bishop blessed the system once again, and as the crowds cheered, the subway was declared open for operation.

Underlying the pathos of progress and destiny invoked at the opening ceremony, along with the repeated assurances of safety, was a discourse of crisis widely debated in the last decades of the nineteenth century. This discourse revolved around overpopulation and lack of circulation in the world's fastest growing metropolises.[12] Everyday life in New York City came to be viewed as a kind of "modern martyrdom."[13] Increasingly, the atmosphere of crisis was associated with the subject form of pedestrians, flaneurs, and the early passengers of horse-drawn streetcars, omnibuses, and elevated trains.

METROPOLITAN MARTYRDOM

In 1800 New York was a largely insignificant city with little manufacturing and trade. London and Tokyo already had populations reaching over one million, while New York had fewer than 80,000 inhabitants.[14] Other US cities such as Philadelphia and Boston were not only larger, but they also had greater economic and cultural power. With the expansion New York's harbor, things changed rapidly. In the first decades of the nineteenth century, the city rose to become one of the world's leading metropolises. As a steady influx of migrants from Europe and the southern United States came to New York in search of manufacturing and dock jobs, the city population began to increase at an unprecedented pace. The population doubled once between 1820 and 1840, doubled again between 1840 and 1860, and doubled a third time between 1860 and 1880.[15] At the turn of the twentieth century, with almost 3.5 million inhabitants, New York had become the second largest city on the planet.[16]

A serious problem that had already become clear in the early nineteenth century was further exacerbated as the population grew over the following decades: the geographic formation of Manhattan as a long but narrow island. Surrounded by the great Hudson and East Rivers, its geography was ideal for trade and ports, but not at all suited to such a heavy influx of people. By the middle of the nineteenth century, it was clear that the scarce landmass of southern Manhattan had reached its capacity for accommodating newcomers. The only option was to expand to the north, where a street grid introduced in 1811 had laid out the pattern of future settlement.[17] Yet the capitalist real estate market had little interest in developing this area so far removed from the commercial centers at Manhattan's southern tip.[18] As a result, around 1900 more than 1.4 million New Yorkers lived packed together on the southern part of the island, in conditions that were catastrophic even for the times.[19] Historians estimate that around 1900, people in the Lower East Side lived in the greatest density in the entire history of settlement.[20] The expansion of factories and warehouses along with the construction of the first skyscrapers for banks and insurance companies led to an overwhelming concentration of people and goods. Overcrowded tenement housing was one major problem, and mobility on the city's narrow streets was another.

The crisis of overcrowding is particularly evident in the increasing precarity and stigmatization of the subject form of pedestrians and early passengers. Their mobility practices were criticized by contemporaries as uncivilized, dangerous, and unhygienic, based on assumptions of economic, social, and physical stagnation. Until the late nineteenth century, New York was essentially a city of pedestrians. There were some early carriage and bus networks, but most people could afford neither horses nor fares. They walked to work. The average city dweller had a daily radius of about two miles, or the distance that could be walked in about a half hour. It was difficult for pedestrians to navigate streets crammed with countless coaches, wagons, and busses all competing for space. And overcrowding was not the only factor behind the lack of circulation on the streets. Traffic flow had only very little regulation or rhythm. Central twentieth-century innovations in traffic logistics such as the separation of directions of travel and the creation of two opposite traffic streams had yet to be introduced. The same goes for other regulatory elements such as traffic lights or crosswalks.[21] It was extremely challenging and dangerous for pedestrians to maneuver through all the commotion.

Not only in New York but also in other expanding American and European cities, the crisis of the pedestrian city crystallized in descriptions of the flaneur, whose practices and forms of perception were quickly becoming impossible due to the transformation of the modern metropolis.

THE DEMISE OF THE FLANEUR

As a social type, the flaneur is defined as a gentleman of leisure, strolling alone through the streets and surrendering himself to impressions of the city and its inhabitants. Walter Benjamin famously described the flaneur as a subject form that was already starting to disappear in the Paris of the second half of the nineteenth century. More and more, this figure appeared as the relict of a time in which the processes of acceleration and transformation had not yet reached their peak.[22] As American literary scholar Dana Brand has shown, the flaneur as a subject can be located not only in European cities like Berlin, Paris, and London.[23] Literary examples are also situated in major cities of the United States, especially New York, as in the writings of Walt Whitman (1819–1892). In his 1860 poem "A Broadway Pageant," however, Whitman also compared the crowds on the streets of New York to regiments of soldiers barging through and leaving havoc in their wake.[24]

Flaneurs always made up a marginalized group of masculine city dwellers, and the massive overcrowding of urban space posed an existential threat to their practices. The crisis in flanerie also reflected the increasing crackdown on loitering, a central factor in the governmentality of major cities as of the middle of the nineteenth century.[25] The flaneur was painted as a bohemian, a dandy, and a person carried along aimlessly by the crowd, with associations of dubious morality and potential danger attached to these interrelated subject forms.[26] Coded as a subject refusing the capitalist logic of exploitation, the flaneur was increasingly seen as a parasite. According to Benjamin, the growing economization of the urban environment led to heightened pressure on the flaneur to participate in the labor market through work. As commodification gained influence over both public space and social relationships, bohemians, loiterers, and people of leisure were sucked into the cycle of utility.[27] Not limited to flaneurs, the effects of economization and stigmatization were also felt by nineteenth-century

subjects such as street urchins, loafers, and other members of the so-called "street corner societies" in metropolises.[28]

The diagnosis of crisis in different urban subject cultures, especially that of the
flaneur, testifies to a more general crisis in the pedestrian city. The streets of New
York were recoded as dangerous and hazardous to one's health, with countless reports
describing a walk through crowds of pedestrians, coaches, omnibuses, and horse-drawn
streetcars as potentially life threatening. In 1837 columnist Asa Greene wrote under the
title "Traffic, Dirt and Cholera": "To perform the feat with any degree of safety, you
must button your coat tight about you, see that your shoes are secure at the heels, settle
your hat firmly on your head, look up street and down street, at the self-same moment,
to see what carts and carriages are upon you, and then run for your life."[29]

Horse-Drawn Streetcars and Omnibuses

After around 1830 there were rapid developments in the transit systems of horse-drawn
streetcars and omnibuses, but this did not reduce massive chaos and congestion on the
streets.[30] The world's first horse-drawn streetcar system began operation in New York
in 1832. Simple wooden wagons, each accommodating twelve to fifteen passengers,
were drawn along iron rails in the ground. Known as trams in Europe and streetcars
in the United States, this technology was seen as a triumphant step for major cities.
John Stephenson, who was just twenty-three years old when he invented the vehicle,
built more than 25,000 of them during his lifetime and exported them worldwide, to
France, New Zealand, and India, among other places.

Horse-drawn omnibuses created a furor in New York. Although they were slower
and less comfortable than the railways, they quickly became a common sight, giving
New York its reputation as a "City of Omnibuses." These vehicles radically changed
the look of streets in large US cities from Boston to New Orleans. They followed
schedules on fixed routes and stopped at regular intervals. They were reliable. They
permitted limited urban expansion and the development of neighborhoods in the less
densely populated outskirts of the city, at least for residents who could afford real estate
as well as the price of transit, and who were willing to cope with hours of uncomfortable riding back and forth.

Thanks to the omnibus, New York boasted the world's largest offering of urban transit long before the subway opened. But even contemporaries found the system's organization rather backward.[31] The menagerie of horse-drawn omnibuses, streetcars, and cable cars caused traffic mayhem. A number of companies, some of which ran just a single line, competed for the same riders in busy districts.[32] Harlem and the Bronx remained difficult to reach, and none of the transportation options reduced urban density as hoped. Additionally, constant traffic jams all but eliminated the would-be advantage of greater speed. Omnibus transportation was anything but safe and comfortable. Chaotic conditions and a lack of regulation on the streets led to frequent accidents, collisions, and passenger injuries. Theft, fistfights, and assault on transit users were commonplace.[33] Even more decisive in terms of public opinion was the fact that fares were prohibitively high for most city residents, at five cents for a ride on the omnibus and ten cents for the streetcar. The first urban passenger experiences were thus limited to New Yorkers from the middle and upper classes who could afford to use the horse-drawn vehicles, while this new form of transportation remained unavailable to the broader masses of workers and immigrants.

According to reports from the passengers of the day, the experience of confinement and stagnation on the street was intensified in the wagons of the omnibuses and streetcars. This description by a *New York Herald* journalist in 1864 is particularly drastic:

> People are packed into them like sardines in a box, with perspiration for oil. The seats being more than filled, the passengers are placed in rows down the middle, where they hang on by the straps, like smoked hams in a corner grocery. To enter or exit is exceedingly difficult. Silks and broadcloth are ruined in the attempt. As in the omnibuses pickpockets take advantage of the confusion to ply their vocation. Handkerchiefs, pocketbooks, watches and breastpins disappear most mysteriously. The foul, close, heated air is poisonous. A healthy person cannot ride a dozen blocks without a headache. For these reasons most ladies and gentlemen prefer to ride in the stages, which cannot be crowded so outrageously, and which are pretty decently ventilated by the cracks in the window frames. The omnibus fare

is nearly double the car fare, however, and so the majority of the people are compelled to ride in the cars, although they lose in health what they save in money.[34]

In just a few lines, we find all of the pathologies associated with life in nineteenth-century metropolises on the verge of collapse: congestion, immobility, crime, lack of circulation, and the danger of infection and disease. Transportation shortcomings were visible more than half a century before the subway opened. Contrary to predictions, omnibus and streetcar passengers developed techniques of aggression to cope with the transit experience. The form of togetherness that emerged for passengers on the crowded vehicles was anything but civilizing, to the horror of public transit advocates.[35] Aside from constant conflict and strife, passengers found one another antisocial, ignorant, and rude. Transportation systems appeared to reflect the epoch's heightened contradictions and social conflicts.[36] The *New York Times* noted in May of 1860:

> There is something irresistibly comic in an omnibus full of utter strangers. Face to face they sit, and their muscles are perfectly rigid; they nod forward, like an old dame dozing, at every jolt of the omnibus, and threaten each other with their noses; they stare vacantly at the brims of each other's hats or bonnets; their hands are crossed upon the handles of their sticks or umbrellas; and they speak never a word unless the gentleman at the top should still have sufficient confidence in the kindliness of human nature left to ask the gentleman near the door to tell the conductor "Fleet-street." . . . You there see man in his primitive state, not of nudity, of course, but of incivility; and you may form a notion, from the behavior of the passengers, of the probable manners in the days of acorns. Nobody, you will observe, dreams of voluntarily making room for a new-comer.[37]

From the early image of the urban passenger offered by such contemporary sources, it is crystal clear that this was not an aspirational identity in the eyes of New Yorkers. At the same time, as we have already seen, there was already an awareness that the forms of subjective experience and social interaction brought about by urban transport

31

systems were something entirely new. Public debate from the time shows the need for a new kind of infrastructure to solve the problems of overpopulation. Some solution was needed not only to overcome the stagnation of the city streets, but also to change passenger behavior to something more acceptable. Discursively, the person riding the omnibus or horse-drawn streetcar was characterized as uncivilized, dangerous, and a hazard to public health, in part in order to emphasize the contrast between this subject form and that of the future subway passenger associated with projections of greater circulation and civilization. The discourses surrounding plans for new urban mobility thus combined two separate issues, with the search for new technological modalities linked to the attempt to define a new subject culture that would establish itself through innovative transit systems.

ELEVATED TRAINS

Initially, a different kind of apparatus had been expected to establish just such a subject culture: the Els, or elevated trains. These were steam-powered cable trains on elevated tracks. It soon became apparent, however, that elevated railways were not a perfect solution.[38] In 1868 in New York, the world's first elevated line in met with widespread skepticism. At about 20 mph it offered comparatively speedy travel, but the spindly and often makeshift rails were not reassuring to potential passengers.[39] Accidents, defects, and financial difficulties impeded the further development of elevated trains during the last decades of the nineteenth century.[40]

But the Els did leave a lasting impression. They opened up a new verticality in urban space, allowing passengers to see the city and its inhabitants from an unaccustomed perspective.[41] On the one hand, some of these new sights were perceived as picturesque and spectacular. On the other hand, they offered an often unsettling glimpse into the lives of the poor and the marginalized.[42] Several articles in the *New York Times* characterized the experience of riding high as dizzying, frightening, and a strain on the eyes.[43] Complaining that elevated railway capacity was insufficient and compartments were uncomfortable and overcrowded, such articles described the experience as "far from one of pleasure or security," emphasizing that "it requires some little nerve to sit at the car window and look steadily down upon the street."[44]

More significantly, the newly erected elevated lines drastically reduced the quality of life for the people below them. The structure blocked the sun, leaving streets with a permanent atmosphere of twilight. The train was noisy and the steam engine puffed exhaust, leading to massive criticism. Doctors warned that riding the system was not conducive to health.[45] Above all, elevated trains did nothing to straighten out the city's tangled settlement patterns. They provided access to areas in the north of the city for more housing development, but neither their capacity nor speed could keep pace with population growth.

Even during construction, contemporaries saw elevated trains as only a temporary solution for a city growing so fast. In reality there was a need for a far more extensive and innovative form of transit. Following a blizzard in 1888 that shut down public life in the entire city for several days, the call for a new system could no longer be dismissed.[46] This dramatic disruption in transportation demonstrated that the lack of circulation was not only inconvenient for the city's residents, but indeed a threat to New York's basic economic, social, and political structures.

In the second half of the nineteenth century, concrete ideas for a new infrastructure of mass transit began to emerge against this backdrop of crisis, facilitated by new technological prospects for urban transformation. Yet while there was general agreement that new technology was necessary to transform the city into a sphere of circulation, it was not at all apparent what this would look like. For a long time, there was no decision as to where in Manhattan a new system of transit might be implemented, and no clear sense of how it should be organized or how it would function. Underground tracks were far from the only possibility.

VISIONS OF TRANSIT

During the last decades of the nineteenth century, there were a number of visionary ideas for new kinds of circulation systems that might put an end to "modern martyrdom." These ideas were often accompanied by spectacular plans and illustrations that were circulated in newspapers and magazines in order to arouse public enthusiasm.[47] Carried away by inventiveness and the belief in progress, some visionaries proposed models that resembled pneumatic tube systems or vertically doubled streets (figures 1.2 and 1.3).

Figure 1.2 Dr. Rufus Gilbert's covered atmospheric railway, 1874, from *Frank Leslie's Illustrated Newspaper* (March 18, 1874).

Figure 1.3 John M. August Will, proposed arcade railway under Broadway. View near Wall Street, 1869.

Inventor and entrepreneur Alfred E. Beach actually followed through with his plan for an underground train driven by air compression. In 1869 he secretly had a pneumatic tube laid under Broadway that could shoot a wagon carrying a few passengers about three hundred feet between two stations. The project was a sensation upon public demonstration in 1870, but bureaucratic and technical challenges soon put an end to it.[48]

One of the most spectacular proposals came in 1871 from inventor and entrepreneur Alfred Speer (1823–1910). He suggested erecting a raised system, like the elevated train, but with no rails (figure 1.4).[49] Instead, the structure would support a huge conveyor belt driven by underground steam engines. Gliding through the entire city at about 11 mph, this moving street would make a wonderful promenade. There would be also smoking cars for the gentlemen and powder rooms for the ladies offering weather protection.

Many of these proposals helped to spark interest among the people of New York for a future transit system. While differing widely in terms of design and technological solutions, all of them presented public transportation as comfortable, safe, fast, and affordable, a true alternative to the existing jumble and confusion of city traffic. The illustrations only show affluent, well-dressed men and women in pleasant company. The streets look clean, spacious, and inviting, in stark contrast to the real conditions of the overcrowded city.

With the exception of Beach's short-lived pneumatic experiment, none of these utopian ideas were ever realized. Either no investor could be found for a given system, or it was impossible to actually construct, or it involved too many legal issues.[50] Instead, the idea of an underground transit system slowly gained acceptance. Advantages were clear: such a system would utilize the newly harnessed technology of electricity, it would be protected from the weather, and it could be built using new methods for comparatively simple and cost-efficient construction under the street. Plus, while any ground-level construction would have a negative impact on the quality of life and the real estate market, this was not the case for a subterranean system.

By 1880 at the latest, there was general consensus regarding the technological modalities of a future transit system in New York. At the same time, it was becoming clear that the realization of such a subterranean infrastructure would be an ambitious

Figure 1.4   Railway plan by Alfred Speer from *Frank Leslie's Illustrated Newspaper* (March 21, 1874).

endeavor requiring a huge amount of resources. Conditions in the city continued to deteriorate as debate dragged on. Population growth continued to increase demand for already scarce living space. Congested streets impeded reliable, timely deliveries of goods to shops and manufacturers, and it took workers and employees much too long to get to work. Observers agreed that if the transit problem were not solved soon, the city's social order and economy might well collapse, with unpredictable consequences.

The historical dynamics that finally led to the realization of the subway were primarily the result of a growing awareness of crisis in the noncirculation of goods and labor. But for political and economic elites, as well as for social reformers, the idea of a future city of passengers was about more than this. They saw it as the only way to eliminate poverty, amorality, the deterioration of social norms, and cultural stagnation. The demand for a new infrastructure that would radically change the distribution of society revolved around a new vision for urban subject culture.

PROMISING CIRCULATION

The poor diffusion of goods, materials, and labor was a primary concern for New York's entrepreneurs. Local economies could not thrive under the prevailing conditions. Manufacturers, banks, merchants, and realtors claimed that the lack of circulation was already a threat to the growth of their businesses, and they saw worse yet to come: sooner or later, existing circumstances would destroy the basic structures of New York's economy.[51]

Since the beginning of the nineteenth century, economic theory had stressed that modern capitalist economic systems depended on the effective circulation of goods and capital, in addition to production.[52] Defenders of capitalism never tired of emphasizing that the ever faster, more frictionless, and more reliable circulation of capital would contribute to growth and general well-being.[53]

For urban economies that relied on trade, the circulation of capital was decisive, along with the circulation of goods, materials, and labor. New York's development depended on increasing the volume of maritime trade, expanding ports, and building larger factories. While the city's geography facilitated access to transatlantic routes and domestic rivers, its port capacity could not sustain the growing volume of goods.[54] At the same time, the city aimed to surpass other significant East Coast port cities such as Philadelphia and Baltimore. To do so, it was necessary to straighten out internal circulation routes and multiply networks while increasing speed all around. The fact that the city was situated on an island, a major advantage in its development from a sixteenth-century trading post, now posed more and more of a problem.[55]

Given their stake in this situation, it is no surprise that entrepreneurs and manufacturers became central actors in the demand for a new transportation system. The Chamber of Commerce, the most powerful business organization in New York at the time, firmly supported construction: "the future of this city as the commercial metropolis of the United States [. . .] requires the very best system of rapid transit."[56] The chamber negotiated major building contracts and took over a large part of the planning.[57] The Rapid Transit Commission, founded in the 1890s for the purpose of supervising subway construction, was led by the city's wealthiest, most influential entrepreneurs. The business elite hoped that employees and workers would be able to reach offices and factories with ease, which would in turn reduce the number of pedestrians and omnibuses, creating more space on the streets for the unencumbered circulation of goods. They also believed that this would lead to an unimaginable rise in levels of production and consumption. Last but not least, the subway would trigger expansion in urban construction, which would solve the housing problem while generating immense profit for speculators and builders. The transformation of New York from a city of pedestrians into a city of passengers carried with it the promise of untold prosperity.

The belief in the beneficial effects of circulation expressed by entrepreneurs was not limited to the economic sphere. If the subject form of the subway passenger can be seen emerging as the heroic embodiment of an imperative for circulation, this must be understood within the context of a wider field of scholarly discourse related to ideas of circulation throughout the eighteenth and nineteenth centuries.[58] The notion of circulation gained in significance beginning in the sixteenth century, and by the nineteenth century, it had become a guiding scientific principle.[59] While knowledge of circulation was initially concentrated in biological inquiry, especially dealing with the circulation of blood, by the eighteenth century the concept had become detached from a fixed notion of circular processes.[60] Circulation came to designate the reciprocal dependence of separate elements, and it was used as a general concept to describe systemic interaction within biological systems.[61] This concept found application in many other disciplines as well, and the nascent social sciences found it particularly useful as an explanatory model.[62] With connotations of vitality, health, and prosperity, in the course of the nineteenth century circulation became a paradigm for organizing modern society.[63]

Simultaneously, the concept of circulation turned up on the horizon of disciplines including urban development, education, chemistry, and engineering. Although these disciplines put the term to different uses, they all viewed well-ordered circulation as productive, profitable, and conducive to development.[64] During the second half of the nineteenth century, models and ideas of circulation provided for valuable transfer between scholarship and social policy, generating links between various academic fields and political, social, and economic debates of the day. By extension, the subject form of the subway passenger emerged as a point of convergence between the discursive fields of hygiene, urban planning, economics, and policy.

New York's entrepreneurs, investors, and speculators were not the only groups lobbying for a new form of industrial transit: the town's political elite called for modernization as well, along with the social reformers, or progressivists. By analyzing the crisis diagnoses and suggested solutions of these two groups, we will see how the passenger appeared as a subject circulating on the horizon of a new governmental technology. Using a term from the work of Michel Foucault, this process can be described as a security dispositif, meaning that efforts were primarily focused on "organizing circulation, eliminating its dangerous elements, making a division between good and bad circulation, and maximizing the good circulation by diminishing the bad."[65]

The imperative of circulation took effect on many levels, responding to issues of individual hygiene and cultural assimilation, and feeding into debates on the social and technical design of urban space. Underlying these debates was the belief that stagnation meant corruption and degeneration, along with faith in circulation as a cure-all that would make the city safer, healthier, and more liberal.

Governmental Technologies

In the second half of the nineteenth century, New York's political elites went from dismissing the idea of a gigantic underground transit system as ludicrous to embracing it as an urgent necessity. There were two main reasons for this shift. For one thing, the prospect of economic prosperity attached to the future subway promised a massive increase in city tax revenue. For another, politicians saw it as a powerful tool for reestablishing social and political order in the crisis-shaken city. After initial resistance from within New York's Democratic Party, by the turn of the century all of the city's political parties, interest groups, and members of the city council had agreed to give top priority to the construction of a new transit system.[66] This would entail major challenges and a massive mobilization of resources, but ultimately the subway would allow for the realization of a new form of urban governmentality, generating social order through controlled circulation.

These ideas reflect a profound change in the notion of good and rightful sovereignty. By the nineteenth century, many traditional governmental technologies had reached their breaking points. Frictionless circulation was becoming a central premise not only in economics, but also in the function of bureaucratic apparatuses of control, which were increasingly dependent on precision and speed when it came to orders and regulations as well as knowledge, ideas, and laws. With the urbanization of North America, the need for a new form of social control and authority came into sharp focus. Throughout the nineteenth century, issues of social and moral order in the cities and industrial centers of the United States became more and more explosive.[67]

In 1800 the United States was still largely rural, with only a handful of big cities. Within a century, it was transformed into an urban industrial nation with a shrinking agricultural sector. Arriving in waves throughout the nineteenth century, immigrants landed mostly in the urban centers, where they had a deep impact on both social and architectural structures. At the beginning of the twentieth century, between sixty and eighty percent of the population in US metropolises were of foreign origin. Faced with large numbers of immigrants from very heterogeneous cultural backgrounds, political leaders and authorities perceived a threat to the existing social order that had already been so difficult to establish.[68] In New York in particular, where many different groups had to share scarce living space—from Italian Catholic farmers to Orthodox Jews from Eastern Europe—the political elite saw the city at the "ragged edge of anarchy."[69] Confronted with unprecedented levels of destitution, gambling, prostitution, and violence, the chronically understaffed police force struggled to keep problems under control. The elites feared that poor moral and hygienic conditions in the overcrowded slums would tear apart the city's social fabric.

It soon became evident that repressive governmental technologies based on discipline and punishment were insufficient for controlling and integrating new parts of the population. Leaders looked for ways to transform the subjectivity of individuals, changing their moral values, cultural norms, and social practices. During the first half of the nineteenth century, they tried to achieve this by reverting to rural models of order and community, and promoting traditional values. Charities, Bible societies, and Sunday schools joined in the effort along with politicians and journalists.[70] But with the new political, economic, and cultural challenges emerging in US cities, the idea of restoring traditional rural society proved futile and misguided by the mid-nineteenth century at the latest. Alternative strategies for achieving moral and political order arose, based primarily on class, heritage, and race.[71] These different strategies all responded to the question of how to integrate the growing number of immigrants into the subject culture of the emerging American middle class in order to pacify them and cement their morals.[72]

Such integration efforts brought urban populations into focus in a whole new way. Since the seventeenth-century, the concept of population control had been based on biopower and discipline as technologies of rule. Over the course of the nineteenth century, a fundamental shift in these principles emerged in Western metropolises. Losing some of its rigidity as a monolithic concept, social order became a more open and flexible governmental technology that involved continuous regulation. Rather than ruling through prohibition and punishment, the emphasis was on governing subjects by mobilizing and directing them. As Michel Foucault has shown, "the territorial sovereign became an architect of the disciplined space, but also, and almost at the same time, the regulator of a milieu, which involved not so much establishing limits and frontiers, or fixing locations, as, above all and essentially, making possible, guaranteeing, and ensuring circulations: the circulation of people, merchandise, and air, etcetera."[73]

Good and efficient governance came to distinguish itself by whether it controlled circulation in a way that could promote social order and achieve integration.

Historian Patrick Joyce also describes this paradigm shift in *The Rule of Freedom: Liberalism and the Modern City*.[74] Inspired by Foucault's studies on the history of governmentality, he traces the emergence of liberal rule in large Western cities of the nineteenth century. Like Foucault, Joyce characterizes this new governmental technology in terms of the importance it placed on the freedom and mobility of subjects. Requiring less direct disciplining of the masses, the new form of liberal rule was subtle and indirect, transforming the physical structure of the city as a means of improving the morals of its inhabitants.[75] Along with political provisions, the emphasis lay on social technologies and infrastructures that both enabled and regulated the freedom of subjects. One element of the strategy, for example, was to extend the system of public lighting. More streetlights made the city seem cleaner and safer, giving residents a certain kind of liberty.[76] Other infrastructures such as streets and improved sewage systems did the same, linking what Joyce has called the triad of "sanitary city," "moral city," and "social city."[77] Like the new forms of government, these social technological remedies for the city's problems were based on the paradigm of circulation. Joyce shows how liberal policy focused on mobilizing urban structures and the behavior of individual subjects: "this freedom was realized around the city and the person as both now themselves sites of free movement, free association, with the person now freely choosing, responsible and therefore self-monitoring."[78]

Coding the subject as liberal, the task for good and rightful governance was to create a city that could guarantee well-ordered circulation for its population.[79] This task was complicated by the fact that there were other forms of circulation beyond what was state organized and legal, such as smuggling, vagrancy, and the spread of infectious diseases. Prevention of these dangerous forms of circulation went hand in hand with efforts to make the city tidy, peaceful, and prosperous.

While political elites and law enforcement authorities saw the need to bring about the regulated freedom of subjects through orderly circulation, a large-scale sociotechnical apparatus was required make this possible. The introduction of sewage systems and gas and electricity lines had already proved successful in generating an orderly and hygienic urban environment, and it seemed like an obvious solution to create a similar system of circulation for the population itself.[80] A subway would transform the chaos on the streets into an orderly, channeled flow of integrated passenger-subjects. As predicted by William McAdoo, New York's police commissioner at the time, the subway also promised to be an effective tool for law enforcement: "In case of great local disturbances or riots, great bodies of police can be transported quickly underground from one part of the city to another, and therefore, from a practicable and strategic standpoint, the tunnel will be an important factor in the police life of New York."[81]

There was no time to lose. The city's wealthier residents had already begun moving to less crowded, more suburban neighborhoods in Brooklyn and New Jersey. In order to keep financially powerful, well-educated classes in town, it would be necessary to transform these subjects into passengers along with the residents of slums. Soon entrepreneurs and investors were not alone in calling for a subway; city leaders agreed that the fast construction of an underground mass transit system could no longer wait.

SOCIAL REFORMERS: HYGIENE AND MORALS

Dirt offends against order.[82]

—*Mary Douglas*

Along with the political and economic elites, a third group also actively advocated the transit system: social reformers, or so-called progressivists. Primarily from the aspiring middle class, this heterogeneous group included physicians and lawyers as well as journalists and engineers.[83] They were politically active and had a great deal of influence, playing a key role in the discursive formation of the future urban passenger-subject. In the construction of a new subway, they saw beyond potential economic value to the utopian promise of a new kind of urban society. While politicians and entrepreneurs tended to think of New York residents in a totalizing manner as the population or the masses, social reformers saw their fellow townspeople as individual residents. They wanted to transform individual practices, ways of life, and value systems.

In their advocacy for a new infrastructure of circulation, reformers took up a debate on social hygiene that had already been playing out between metropolitan politicians on both sides of the Atlantic for more than fifty years.[84] The progressive movement became widespread at the beginning of the twentieth century, and their political influence grew with respect to social issues.[85] Many reformers were active in charities and political committees, or wrote articles in reputable journals such as *McClure* and *Municipal Reform*. They addressed a wide range of problems, from corruption, anomy, poor education, and moral decay to neglect, low wages, and child labor. This was primarily a secular movement with a political agenda that can be seen as a reaction to some of the rapid social, cultural, and economic processes of modernization that had resulted in poverty for large parts of the population. In New York in particular, reformers sought to improve the lives of immigrants and poor people by making changes to their environment.[86]

Danish immigrant and journalist Jacob A. Riis (1849–1914) was a central figure of the social reform movement in New York.[87] In a famous report from 1890 titled "How the Other Half Lives," he vividly documented miserable living conditions in the tenements of the Lower East Side and Five Points.[88] Despite his propensity for prejudice and ethnic stereotyping, his photographs and descriptions provide valuable insight into how space was organized in the city's poorer districts (figures 1.5 and 1.6).

Riis describes how heavy foreign immigration and domestic migration to the cities forced the lower classes to settle in inferior, cramped housing structures, which grew in number as ever more newcomers arrived.[89] The resulting living conditions were catastrophic even for the day, as Riis emphasizes in his discussion of a report from 1857: "Their large rooms were portioned into several smaller ones, without regard to light and ventilation, the rate of rent being lower in proportion to space and or height from the street; and they soon became filled from cellar to garret with a class of tenantry living from hand to mouth, loose in morals, improvident in habits, degraded, and squalid as beggary itself."[90]

The strong link between morality, hygiene, and social order was by no means limited to discourses surrounding New York's slums. As anthropologist Mary Douglas argues in her study *Purity and Danger*, almost all societies have concepts of impurity and strategies of cleansing. Poor hygiene is often associated with deviant behavior, moral decay, and a dismissal of social norms.[91] According to Douglas, the perception

Figures 1.5, 1.6 Photos of tenements in New York (1890), taken by Riis to illustrate his reportage.

of impurity leads to the development of cleansing practices and rituals in order to maintain societal order.

Faith in orderly circulation as a means of achieving cleanliness permeated Western discourses of public health dating back to the eighteenth century. This conviction spread in calls to introduce sanitary infrastructure in the living quarters of the underprivileged, in the hope that it would have an ordering and civilizing effect. At the heart of such efforts was a very specific notion of hygiene, propagated as a government responsibility as well as an individual bodily practice. The priority was to maintain proper circulation and prevent congestion in city pipes and people's bodies. These ideas were closely associated with ideas of spatial transmission and contagion. Poverty-stricken areas were portrayed as breeding grounds for all kinds of disease.[92] These spaces were coded as infectious not only in medical terms (as in epidemics of typhus, cholera, tuberculosis), but also in terms of the moral attitudes of the poor.

The discursive entanglement of filth and disease in poor areas with their inhabitants' unhealthy way of life meant that it became the task of good government to fight both. In the implementation of hygiene regimens, social reformers began discussing how to change slum dwellers' beliefs and practices, an effort that became just as urgent as the modernization of municipal infrastructure. One way to achieve this was to integrate the ideology of hygienic circulation into the curricula of the public schools (figure 1.7).[93]

Riis gives us one example of an attempt to educate children in hygiene. After visiting a school in the poor Jewish district around New York's Allen Street, he reports:

> The question is asked daily from the teacher's desk: "What must we do to be healthy?" and the whole school responds:
>
> "I must keep my skin clean,
> Wear clean clothes,
> Breathe pure air,
> And live in the sunlight."
>
> It seems little less than biting sarcasm to hear them say it, for to not a few of them all these things are known only by name.[94]

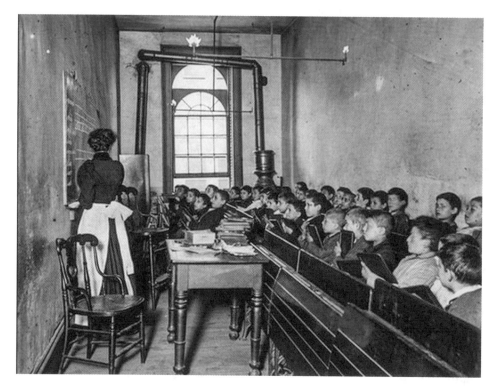

Figure 1.7 School lessons on the Lower East Side, ca. 1888. Photo: Jacob Riis.

Access to fresh air, sunlight, and clean water for all New Yorkers soon became a central public demand. In an early phase of social engineering, questions of discrimination and unequal access to resources were reformulated as organizational problems that could be solved by technology.[95] Urban planning became a matter of hygiene, and the elimination of stagnation in poorer districts a matter of utmost urgency. As these ideas spread throughout the late nineteenth century, New York and a number of other metropolises began introducing extensive programs that would reform sanitary conditions.[96] One of the goals was to connect poor districts to existing urban infrastructure, and another was to create incentives for investors to build new, less dense housing structures. These programs did succeed in somewhat lowering the high mortality rate in poor districts, but they also led to new forms of marginalization and exclusion. As medical discourse and debate on social hygiene continued to stress the importance of sunlight, fresh air, and clean water, the capitalist real estate market of New York began to commodify these resources.

Soon the poorest residents could no longer afford to live in renovated housing with improved connections to circulation. Many tenement buildings deteriorated as owners could not procure the capital needed for modernization, and conditions got even worse. In the late nineteenth century, even though cities like New York, Paris, London, and Berlin officially prohibited the rental of dilapidated housing, poor migrants continued to be jammed into overcrowded, unsafe buildings that were seen as death-traps. Police raids and eviction orders were common.

In light of these developments, social reformers realized that they would have to take more drastic measures to improve living conditions in the slums. For circulation to solve problems on a grand scale, it would take nothing less than the fundamental infrastructural transformation of the entire city. Various efforts at reform converged in the political call for a vast transit system. One of the most prominent reformers of the times, economist Adna F. Weber (1870–1968), offered scientific findings in support of the demand. Having studied the distribution of populations in Western metropolises, Weber came to the following conclusion in his groundbreaking 1899 study, *The Growth of Cities in the Nineteenth Century:*

It is now clear that the growth of cities must be studied as a part of the question of distribution of population, which is always dependent upon the economic organization of society—upon the constant striving to maintain as many people as possible upon a given area. The ever-present problem is so to distribute and organize the masses of men that they can render such services as favor the maintenance of the nation and thereby accomplish their own preservation.[97]

Radically reorganizing and redistributing the population came to be seen as a crucial means of securing economic growth and social cohesion for the cities of the future. Weber saw the only solution in the intensification of suburbanization, which had already been taking place at the periphery of large cities since the end of the nineteenth century. In his view, suburbs promised an escape from the misery in Western cities that were collapsing under the pressure of millions of inhabitants. More than merely improving distribution, suburbanization presented an entirely new kind of community. Yet resettling people at the periphery also meant rethinking structures of life and work from the ground up. In Weber's words: "Though population must be concentrated, it does not follow that population must be congested unless we assure that a man's abode cannot be separated from his workplace."[98]

The creation of a far-reaching transit system that would allow people to live at some distance from their places of work promised to disentangle urban tenements and slums by keeping people in circulation.[99] Once successfully connected to the goods and capital flowing through the city, people on the move would be culturally integrated at last. They could be physically and socially mobilized while finally escaping the sickness, misery, and hopelessness of the slums. For social reformers, getting New York's residents to circulate did not necessarily imply changing society's systems of rule and power structures; reformers wanted to bring the poor and the disadvantaged into existing political, social, and economic systems. Anticipating the subject form of the commuter who circulates between home and work, reformers were also able to justify their call to reduce the number of hours worked daily based on the projected increase in transit time.

Weber, who along with Riis and other social reformers had considerable say in debates on urban development, maintained that creating suburbs at the metropolitan periphery would simultaneously create a new kind of urban culture, a new form of life lived by a new kind of subject. Suburbs would unite the advantages of the city with the benefits of village life.[100] For Weber as for most reformers of the day, in contrast to opinions from the first half of the nineteenth century, the return to rural life was no longer a viable option. Many reformers had come from the country and made careers for themselves in the city; they were convinced of the value of opportunities offered by an urban environment. For them, creating suburbs meant combining urban opportunities for education and earning with the stable social relationships and moral integrity associated with rural life. But before urbanites could enjoy the privileges of the suburbs along with the benefits of downtown, they first had to be transformed into passengers.

Hero of the Liberal City

As a circulating subject primed for integration into a new form of urban society, the future subway passenger was discursively equipped with a variety of hopes and expectations. For politicians, entrepreneurs, and reformers, the passenger carried the promise of a morally sound, politically adjusted, economically successful life. The antithesis of stigmatized slum dwellers, pedestrians, and earlier passengers, the future passenger emerged with the image of a savior for the city in crisis.

The political elite believed that including the working class in the infrastructure of circulation would promote assimilation and peace. Reformers such as Weber and Riis believed that connecting people to the city's socio-economic structures would also help tenants to become property owners. Despite reform efforts and new building regulations, tenements remained impoverished, and reformers saw this as a dead end for urban development.[101] Perceiving a destructive lack of community and excess of individualized anonymity in the slums, reformers saw the future of urban communal life in the solidarity of property owners. The elites hoped that the isolation of the tenements would give way to the pride of home ownership, leading to the development of new personal traits such as abstinence, morality, responsibility, and financial prudence.[102]

51

The future subway passenger was expected to entirely assimilate to urban culture. Reformers viewed the tenements of the Lower East Side, Greenwich Village, and Five Points as stagnant places, where immigrants spoke their old-world languages and upheld obsolete customs, religions, and other cultural practices.[103] The transformation of slum dwellers into home owners and circulating passengers fit into the reformers' plans for radical social and economic integration. In this respect, becoming a passenger represented a kind of amnesia. Once immigrants started moving to the rhythm of urban transit, they would forget the past, adopting the cultural patters and value systems of their new home.

The elites also believed that making workers into commuters would strengthen family ties.[104] Slum dwellers were branded as promiscuous individuals with unstable private lives. Drawing poor families into the fold of urban economic utility would turn them into close-knit, stable units. Leaders were convinced that clear work and transit schedules would promote reliable personal relationships. Commuting would encourage husbands to pursue employment and adhere to a strict time regime. Wives would be encouraged to cultivate domestic instincts that had supposedly been suppressed in the tenements.[105] And in the suburban environment, no longer confronted with the corruption of inner city street life or the exploitation of child labor, children would grow up to become respectable, law-abiding citizens. They, too, would become passengers, commuting to schools and training centers, building their moral character as urbanites. In short, for entrepreneurs, politicians, and reformers, facilitating new underground mobility went hand in hand with eliminating personal and economic standstill. Becoming a passenger would mean climbing from the bottom rungs of the social ladder up to the middle class, with the support of stable family relations, steady employment, and moral integrity. As Weber pointed out: "All that is needed is cheap and rapid transit between home and workplace."[106]

This meant that the subway of the future had to be fast, reliable, clean, and affordable to all. Some European transit systems had low-fare cars reserved for workers, but American reformers rejected that idea outright. While certainly not free of prejudices based on class and race, they emphasized that America was to be a classless society. At least in their rhetoric, there was an egalitarian aspiration for people of all classes and all races to become passengers.[107]

The future passenger was envisioned as a utopian hero, a new form of individual who would also initiate a new form of urban community. This vision of unity for the city and its residents, presented at the subway opening ceremony on October 27, 1904, had its roots in the first efforts to create such a system forty years earlier. As early as 1866, a senate committee investigating New York's mobility crisis came to the following conclusion: "commercial, moral, and hygienic considerations all demand an immediate and large addition to the means of travel in the city of New York."[108] Through decades of proposals and plans, the elites never tired of claiming that the coming subway would be much more than simply a banal element of public infrastructure.[109] They saw it as a collective achievement, a monument to free civil society. As Mayor McClellan underscored in the opening ceremony, New York's subway would be a matchless accomplishment of the "sons of the mightiest metropolis the world has ever seen."[110]

According to the intentions of its planners, the subway would constitute a democratic technology, available to all New Yorkers, irrespective of class, race, and origin. Low fares were crucial in this. The first tickets could be purchased for half the price of a ride on the elevated train or horse-drawn omnibus. At five cents each, they were affordable for workers and slum dwellers alike.[111] Leaders took every opportunity to point out that such inclusion should be a source of pride for all New Yorkers, "actuated by a united hope and united in a common destiny."[112]

The pathos frequently expressed in this context shows the immense ideological effort that went into making the underground transit system appeal to everyone.[113] Support came from theologians and missionaries, physicians and engineers, authors and publishers, all of whom contributed to the portrayal of the subway as a liberal infrastructure that would give its users new freedom.

While London, Paris, Boston, and Berlin already had underground railways, there are many reasons why the New York subway was not built until the early twentieth century: political intrigues and power struggles; complicated and contradictory legislative regulations and judicial issues; as well as the stock market crash in 1893 and subsequent recession.[114] One high hurdle was an administrative restriction limiting public investment in urban infrastructure to not more than fifty million dollars per year, an insufficient sum for such an enormous project. Due to all of these factors, thirty years passed between the senate committee recommendation for fast action and the ground-breaking for the subway in 1900.

MAKING UTOPIA REALITY

Entrepreneurs from New York's Chamber of Commerce eventually reached a break-through by devising an entirely new financing model for the subway.[115] The subway would become city property, but construction and operations would be carried out by a private company. In 1899 "Contract I" was given to John B. McDonald (1844–1911) and wealthy investor August Belmont Jr. (1853–1924) for the construction and operation of the subway. They promised to build the first sections of the subway for thirty-five million dollars—the largest contract sum ever paid to an investor out of the public coffer up to that point.[116] The newly founded company was named the Inter-borough Rapid Transit Company (IRT). Chief engineer was the relatively young William Barclay Parsons (1859–1932). Parsons had been to Europe to acquaint himself with various modes of electric railway transit, and he would now use his expertise to create the world's largest and most modern underground railway. After much nego-tiation and many contentious planning sessions, in 1897 a decision was finally made on where to lay the first line. It would begin at the station City Hall and then pro-ceed along Manhattan's west side up to the Bronx.[117] Plans for further expansion were expected.

New electric motors made it possible to cover the large distances and extensive net-work that today characterize New York's subway. Electricity provided an even flow of energy and made tight scheduling possible. While the world's first underground railway in London ran on steam, New York was the world's first metropolis to install a large network that ran on electricity. An additional innovation was the four-track system: one for a local train and one for an express train in each direction. Compared to other cities such as London and Budapest, New York's system had further range and much higher speeds. There was some doubt about the profitability of express trains, but as they were associated with progress, planners firmly upheld the idea.[118]

Tens of thousands of people, kept under control by more than a thousand police offers, attended the groundbreaking ceremony for the first stations on March 24,

1900. According to the *New York Times*, the excitement for "one of the most important events in the history of the city" was indescribable.[119] Only the subway's maiden voyage in 1904 would create more of a stir. Public enthusiasm was vital, because it would take four long years of inconveniences to New Yorkers to complete just the first stretch. Instead of the tunneling technique used for the London Underground, subway construction in New York was done using the cut-and-cover method, which meant that the entire street was torn up so that tunnels and stations could be installed from above (figure 1.8).[120]

Engineers saw a long-term advantage in cut-and-cover, which allowed for the construction of stations closer to the surface, eliminating the need for complex stairways and elevators.[121] The method turned out to be more challenging and expensive than anticipated, however. Planners had underestimated Manhattan's complex geology: granite rock and drift sand exasperated engineers and "subway miners" again and again.[122] At the same time, it was necessary to rope off large sections of the already crowded streets for long periods, even years. This led to angry protests and further construction delays. In addition, New York already had a multifaceted underground network of sewage tunnels, gas pipes, and water mains that had to be redirected during construction. Rarely a month passed without New Yorkers reading negative news about the project in the papers: frequent blasting accidents, a cave-in that killed sixteen people, damage to the foundations of surrounding buildings.[123] Subway advocates nonetheless continued to praise the project, reciting its blessings and appealing to residents to persevere. Their new lives as passengers would more than compensate for these inconveniences.[124]

While the characteristics of the ideal passenger had already been shaped through discourse, when it came to how people would actually experience the system, the subject form of the passenger remained to be determined. What kinds of practices and interactions would emerge? Anticipation of the subway's effects on the subject cultures of the population had to include the behavior of passengers in transit. As a result, entrepreneurs and engineers focused their efforts on clarifying and controlling the affects of the first passengers.

Figure 1.8 View of the 1902 construction of the subway at Broadway station and 104th Street to the south. Shows the last streetcars that were soon to be replaced by the subway, as well as construction activity on the adjoining properties, illustrating the immense real estate boom that the subway brought with it.

Emotional Controls

As opening day drew near, it became clear to investors and politicians just how much was on the line. After decades of planning and conflicts, and long years of construction, the moment had finally come when all sides would know whether their expectations would be met. How would people react to the system? Would they acknowledge it as the feat of civilization heralded by its creators? Would they overcome their fears of technology? And most importantly: Were they willing to pay for it?

The weeks leading up to October 27, 1904, marked a new phase in the system's implementation. After gaining political endorsement, finding a way to fund the project, and actually building the first segment, the time had finally come to take up operation. It was the most critical moment of the entire turbulent construction period. Never before had such a gigantic urban infrastructure been opened all at once. Neither London nor Budapest had done anything comparable. Not only were their subway systems much smaller, they also opened for the public one step at a time. The London Underground, for example, was only five kilometers long when it opened in 1863, and it had an entirely different mode of operation.[125]

Everyone realized that the opening of the New York subway was going to be an extraordinary event, and that failure would mean a major crisis for the city's elites. Technical reliability was one risk factor, and the unpredictable behavior of the passengers themselves was another. Although the opening ceremony had been planned as a dignified, composed occasion, the event itself posed a serious threat to public order. Riots, chaos, or accidents would decimate social acceptance and trust in the system. It was crucial for everything to go smoothly, and for the people of New York to be convinced of the advantages of becoming passengers. Investors and politicians had thrown their entire political and economic weight into the enterprise. Failure would spell ruin.

Despite the massive campaign to present the subway to the people as a technology that would pull the city out of crisis and pave the way to a new kind of community, subway construction had in fact been laden with risk at every step. There was some uncertainty as to whether people would accept the idea, and subway supporters faced resistance and skepticism from the start. In the early phases especially, many members of the political and economic elites were deeply wary of the project, partly because they found the idea of underground transit strange and unnatural.[126] Influential public figures such as railroad financier Russell Sage also warned of the dangers of such a system.[127] According to Sage, "people would go below ground only once in their lifetime—and that was after death."[128]

Faced with such strong reservations, the elites mobilized a variety of strategies in order to assuage future passengers' mistrust of the system and encourage calculable behavior. While the use of the subway was coded as heroic behavior, there was also a need to address people's fears, deflecting consumer concerns with scientific authority. The strategies that were employed were based on three particularly effective elements: recoding underground urban space as safe and healthy; introducing regulations and behavioral scripts that would standardize and discipline the practices of passengers; and providing a massive array of fire, police, and medical services to restore public order in the event of riots or emergencies. Taken together, these three elements aimed to situate passengers within a dense network of subjectifying channels that would make their first contact with the subway a compliant, calculable experience.

Such strategies dealt not only with the bodies of passengers, but also with their affective states, especially deep-rooted anxieties regarding the city's underground spaces. Affect control played a key role in the subjectivation of the first subway passengers. Norbert Elias has explored how controlling and regulating affect is vital to maintaining social order: "No society can survive without a channeling of individual drives and affects, without a very specific control of individual behavior. No such control is possible unless people exert constraints on one another, and all constraint is converted in the person on whom it is imposed into fear if one kind or another."[129] In the case of the subway passenger, the implementation of affect regimes was part of the creation of a subject who would face the system with a minimum of emotional arousal, bowing to the dictates of technical rationality and convinced of their rightness.

SUBTERRANEAN FEARS

Many future passenger concerns about subway safety stemmed from the simple fact that the system was almost entirely underground.[130] Some found the mere thought of subterranean transit ominous and frightening. The thought of being confined in a dark, unnatural territory with strangers and moving at great speed below the city evoked defensive reactions. Many of the dangers associated with subterranean transit were linked to more general cultural connotations of the underground, with deep roots in Western traditions.

For thousands of years, the underground has played a prominent role in human thought and fantasy. Collective ideas and emotional codes associated with the underworld change over time and with the dynamics of different societies. Yet within Western culture, the underworld has been consistently coded as a place of crisis experience and hidden truth, as well as horror and vice.[131] In the symbolic geography of antiquity, the underworld was a realm of death and decay, a beyond, a counterpart to the world of the living. The world below the ground was seen as having the power to reverse all that counted as normal above. In Christian symbolism, the underworld has often represented the opposite of the divine region of heaven, a place for the dead, demons, and the devil. In early modern times, the regions under the earth were increasingly secularized and industrialized, for example through mining. And yet these spaces have never lost their ambivalent character. They continue to be seen as spheres that are neither completely part of our world nor completely of some other world. They appear to be just as real as the world we know on the surface, but much more disturbing, threatening, and unfamiliar.[132]

As both an imaginary realm and a real place, the underground unites profoundly unfamiliar things with things that are trivial and ordinary. In the symbolic world orders of modernity, the underground is neither a simple reflection of the world above the surface, nor its opposite. It serves as a place to which we ban things and people when they become obsolete and useless, and a place of refuge for outcasts.[133] We find this idea in phenomenology, for instance in works by Gaston Bachelard, who understands the psychology of the built environment as coded primarily in vertical terms, with the cellar and the attic as opposite poles.[134] According to Bachelard, human imagination is marked by two extremes: the rationality of the upper floor and roof, and the irrationality of the cellar and other subterranean rooms. Underground darkness provides a perfect setting for projections of the uncertainties and monstrosities of modern life. As a seismograph and crisis indicator for the world above, the underground continues to mark a problematic and fascinating topos of collective fantasies and projections.

Particularly in cities, the underground is where society's fissures and margins come to light.

Before the eighteenth century, the urban underground was not an object of much administrative control and regulation. This began to change as matters of public hygiene and a lack of urban space compelled those seeking to exert public control in Western metropolises to focus on subterranean territory. They reformed cemetery regulations, prohibited the mining of natural resources below settled areas, and installed the first sewage systems. As more and more infrastructure systems were placed underneath the streets, the urban underground became a highly engineered space.[135] Lanced with pipes, cables, and channels, the underground was transformed from a hidden, private territory into a public sphere of circulation. As historian David L. Pike has shown, nineteenth-century discourses related to the construction of subterranean transit systems revived a number of disconcerting concepts and images from both ancient mythology and Christian teachings.[136] This was true not only in New York. When the London Underground opened, one skeptic invoked theology to make his argument: "Why not build an overhead railway? . . . It's better to wait for the devil than to make roads down into hell."[137]

Beyond these symbolic demons, the fears and threats associated with the underworld found rational expression in questions of subway safety and public health. The first passengers were less wary of harassment by strangers on the train than they were of the technical apparatus itself. In order to dispel qualms about underground transit in order to convince the masses of its value, the urban underground had to be recoded as a territory that was as safe as it was healthy.

Thus, presenting the first plans for the underground train, planners and engineers stressed that the new spheres of the New York subway would be anything but dark, cramped, and dank.[138] The system would create an entirely new form of underground. Its first architects, George Lewis Heins (1860–1907) and Christopher Grant LaFarge (1862–1938), had made a name for themselves erecting the sacred space of cathedrals and chapels. They emphasized that this subway would be a monument to the technological and artistic achievements of the times.[139] They sought to give the subway an aura of sublimity and safety, drawing architectural inspiration from nineteenth-century urban train stations that had been met with enthusiasm as "modern cathedrals."[140] French author Theóphile Gautier (1811–1872) saw such transit infrastructures as "palaces of modern industry exhibiting the religion of the age: the railways. These cathedrals of the new mankind are the points where nations meet, the center where all converges, the nucleus of gigantic iron-rayed stars that stretch to the ends of the earth."[141]

The technological and utilitarian sacrality of nineteenth-century train stations was to be showcased in subway stations as well. Signaling comfort and safety, mosaics, light shafts, and other elements of modern architecture contributed to a new underground aesthetic that was spacious, functional, and elegant.[142] There was a conscious decision to leave elements like steel girders, rails, and control units exposed in order to demonstrate the structure's sturdiness and technical perfection.[143]

Alongside concerns regarding the stability and security of the subway, skeptics worried about potential health risks for passengers. Fear of suffocation due to a lack of fresh air in the tunnels, trains, and stations was a major issue. Fear of infection and the uncontrolled spread of disease was another. According to historian Christopher Cumo, a widely read article in the *New York Medical Journal* warned people not to use the subway because it would be brimming with aggressive germs and viruses, exposing passengers to tuberculosis or pneumonia.[144] This scenario found support among the same scholars who had made reference to "germ theory" to warn of the dangers of infection spreading through city streets. The medical discourse that had once coded ground-level urban space as a danger to public health now threatened to spread to the space of the subway as well.

Efforts to refute these claims required incontrovertible evidence that the system was not hazardous. Once again, the principle of circulation was mobilized as an indication of health and safety. Leaders needed to demonstrate that clean, healthy air was able to circulate smoothly in underground tunnels and stations, in order to show that passengers could do the same. To offer such proof, they had to find scientific methods that would make the circulation of air visible and measurable. This was a fairly new idea, as it had only been a few decades since laboratory techniques had first allowed scientists to determine the properties and chemical composition of air.[145] Health officials and subway operators turned to Charles Frederick Chandler (1836–1925), a renowned chemist and professor at Columbia University, and asked him to analyze air in the subway. They hoped that a thorough analysis by a recognized scientific authority, employing the "mechanical objectivity" of the laboratory, would produce an unbiased result that even system skeptics would have to accept.[146]

Professor Chandler began to put the air in the subway system to the test. He took samples of air from every station platform and compared them to samples from the city above. After meticulous laboratory analysis, the result was clear: in terms of oxygen content and carbon dioxide levels, air from above and below ground was almost identical.[147] Chandler's findings represented a milestone for subway operators in the effort to promote acceptance. They now had objective, unassailable proof that the system was not dangerous. Additionally, the study served as a crucial political signal. The subway now had science on its side. Chandler's study can be seen as the first scientific investigation of the New York subway, earning him a place in the history books and appreciative recognition in obituaries: "He was the first to come forward as a defender of subway air and to back up his statement with proof."[148]

Proof that the quality of the air above and below the streets of New York was practically the same was taken as further evidence of the power of circulation to create a healthy and safe urban environment. Chandler's research not only testified to the achievement of planners and engineers, it also indicated permeability between underground space and the city above. Whether on the train, at home, or at work, one always inhaled the same air. Chandler's study served as retrospective scientific support for the widespread nineteenth-century idea that a city is a kind of organic entity. According to this idea, the large modern city formed a complex and continuous metabolism, whose separate areas above and below ground infused and stabilized one another through circulation. This organicist perspective on the city was derived in part from new developments in medical discourse.[149] The spatial organization of the urban sphere appeared to correspond with the arrangement of human bodily organs. Such ideas linked Kant's notion of the power of the self-organization of organic matter to new medical knowledge about human blood circulation and the system of arteries and organs.[150] While the semantics of circulation persisted as a combinatory motif, the dynamics of circulation required further scientific articulation and political promotion. Even before the subway was put into use, the knowledge that urban air was flowing freely between the city above ground and the tunnels and stations below reflected the system's successful integration into the urban metabolism of New York.[151]

Despite this success, just forty-eight hours before the official opening, there remained one more obstacle in the process of dispelling passenger concerns: the subway system had to pass inspection by the city board of health. To the great relief of operators, politicians, and investors, the inspection went off without a hitch. They quickly disseminated the reassuring findings in a series of newspaper articles and pamphlets. The *New York Times* proudly reported that inspectors were impressed.[152] From a hygienic perspective, the system had no deficiencies, the air quality was excellent, there was very little dust, and the risk of danger to future passengers was low. It was clean and not too noisy. To make sure this message reached all passengers, operators printed Professor Chandler's clearance right on subway schedules. The first passengers descending into the stations were handed a schedule with every ticket, with the following messages in bold print on the back: "SUBWAY AIR AS PURE AS YOUR OWN HOME" and "35 MILLION CUBIC FEET OF AIR SPACE."[153] As demonstrated by these extensive efforts to disperse fears, passenger concerns were taken extremely seriously. This was part of a wider attempt to code the subway system as an everyday sphere of public life, not a site of strain or threat. Over the course of the history of the passenger, it would become clear time and again that this attempt was not entirely successful.

BEHAVIORAL SCRIPTS AND POLICE FORCE

While people were meant to experience the subway as fast, reliable, safe, comfortable, and affordable, such an experience could not be guaranteed by business acumen and technological expertise alone. It was imperative to instruct future passengers in the correct use of the infrastructure. Many New Yorkers, especially from the middle and upper classes, were familiar with travel on elevated trains and horse-drawn omnibuses; they found these forms of transit unsafe and uncomfortable, and their fellow passengers uncivilized and antisocial. It was crucial for subway operators to show that the practices and experiences of passengers travelling underground would be very different. Riding the subway was an act worthy of dignity and pride, and passengers were to behave accordingly. Operators saw a need to discipline passengers, homogenizing their practices and synchronizing them with the subway's technical modalities. This

63

required the introduction of rules and instructions.[154] Planners and engineers conceived an initially concise rulebook of prohibitions and commandments, which passengers would ideally internalize before their first contact with the subway.

To this end, in the weeks before the grand opening, the rules were printed in newspapers and posted at the yet unused subway entrances. Under the caption "Some Ifs and Don'ts," the *New York Times* published some of the most frequently asked questions on the eve of the opening, laying out the ten most important prohibitions.[155] Along with questions and answers regarding technical details and train schedules, the primary focus was on safety precautions. The article attempted to alleviate all manner of concerns, from fire, to flooding, to the sudden death of the conductor. Assuring readers that the new system was well-protected from all sorts of possible dangers, the article then went on to list instructions beginning with "Don't" that revealed many potential threats to life and limb. These were things like: "Don't try to stick your head out of the window of a subway train. [. . .] Don't walk across the tracks between station platforms. [. . .] Don't move from your seat if there is an accident. [. . .] Don't deface the stations or trains. If you do, you are likely to be arrested."[156] It is important to note that these rules addressed passenger interaction with the technical apparatus, but not interaction between passengers. Compared to the complex rules of later decades, these early instructions were very rudimentary in their formulation, but they had a strong appellative character from the start.

The safety of the system was to be guaranteed by technology on the one hand, and by passenger behavior on the other. The system was safe as long as passengers followed the rules. But why should they? Because deviation from the rules would challenge the smooth functioning of the infrastructure, and it could be fatal. In this scenario, the passenger gained ambivalent agency as the subway's consumer as well as the producer of its safety. Divergence from codes of behavior threatened the physical integrity of the individual as well as the functionality of the system as a whole. As a result, the unconditional sanctioning of divergent practices through police measures appeared legitimate and necessary.

Maintaining public order on the opening day of October 27, 1904, was critical for the authorities. In the eyes of the police, crowds of euphoric subway pioneers presented a serious risk, as they could quickly transform into a violent mob. In the weeks before the opening, authorities devised plans to actively anticipate overcrowding, accidents, and riots. The public was informed ahead of time about safety precautions. New York City Police Commissioner William McAdoo announced in the newspapers that on October 27, heavy police presence would be in place to ensure public order: "Every policeman in the city will be on duty that day."[157] This was no exaggeration. Police patrolled the streets and trains. Two police officers stood guard at every stairway of the entire system, with up to fifteen officers posted on every platform. Five hundred policemen were on duty at City Hall station alone.[158] Prepared for anything, all fire departments and hospitals were on alert. Additional trains and subway personnel stood by in case the situation escalated. In order to play it extra safe, the requests of various bar owners for permission to serve drinks the entire night of the opening were denied.[159]

All of this goes to show that public acceptance by future passengers was not the only element of uncertainty in the subway's opening. The massive mobilization of resources and personnel demonstrates that this represented an exceptional situation for the people of New York. The opening was purposely scheduled for a Thursday in order to reduce crowd size by making it difficult for port and factory shift workers to attend.[160] Calculations were made for carrying twenty thousand passengers per hour—the upper limit of what was deemed manageable by IRT. Operators anticipated slight disturbances and delays, but in general they were confident that within forty-eight hours, the subway would run as smoothly as could be expected of such a gigantic railway system.[161]

During the last few days before the opening, while the final details of the system were being finished in a hurry, local media eagerly reported on technical details and hectic preparations for the event. For weeks the press had worked at getting the public into an appropriate mood for the special day. While the operators ran last trials and people could already hear the noise and feel the vibrations beneath the asphalt, the city above ground was adorned with decorations normally reserved for elections or Independence Day. Suspense was great throughout New York. Everyone awaited the moment when the subway would finally open its doors. But they were not prepared for what actually happened next.

OPENING NIGHT: SUBWAY MADNESS

Just hours after the grand ceremony at City Hall and the afternoon ride for officials and special guests, the opening of the subway for the general public would show whether the strategies and efforts of operators, politicians, and law enforcement would work out as planned. While the system would become accessible to the general public at 7 p.m., it would open at 3 p.m. for around fifteen thousand special guests with exclusive tickets. The distribution of these tickets had already caused some commotion.[162] All day, excited people had gathered at the station entrances, awaiting the moment in the evening when the subway would be open to them. The *New York Times* later reported that the afternoon sight of privileged subway pioneers pouring out of the stations onto the streets was an impressive spectacle for those who had no previous experience with underground transit: "Of this sight New York seemed never to tire, and no matter how often it was seen there was always the shock of the unaccustomed about it. All the afternoon the crowds hung around the curious-looking little stations, waiting for heads and shoulders to appear at their feet and grow into bodies. Much as the Subway has been talked about, New York was not prepared for this scene and did not seem able to grow used to it."[163]

According to this report, despite intense propagandistic preparation on the part of the elites and the operators, people were still shocked by the subway's actual manifestation. Anticipation of what lay ahead of them that evening grew and grew. An hour before the system was to open for the general public, the rides for special guests came to an end and the entire system was shut down. The tunnel was controlled one last time, stations and trains were manned, last construction tasks were completed, and the system was made ready for the sale of tickets beginning at 7 p.m. The gates opened on schedule and people ran down the steps to finally see the city's new acquisition for themselves.[164] Almost instantly, people were scrambling to get the first tickets, which were already considered valuable souvenirs and traded at high prices. Fare vendors and policemen could barely contain the rough-and-tumble of it all. They attempted to form lines and loudly announced that passengers should have exact change ready to make sales go faster. Contrary to operators' predictions, in less than one hour the system had already exceeded maximum capacity. Subway platforms and car doors were too narrow to accommodate the rush. Additional law enforcement officers were called in, but the police soon reached their limits. Trains quickly became so overcrowded that passengers had to stand on the connecting platforms between cars, a practice that was strictly forbidden. The express trains from City Hall to the terminal on 145th Street were in particularly high demand and overcrowded within minutes.

The overloading of the trains caused delays in departures, and in less than an hour the schedules had become obsolete. Panicked conductors left out stops to make up for the delay, much to the outrage of the newly minted passengers. The entire system was so packed that the flow of passengers through the stations soon came to a halt. Suddenly thousands of people were backed up at each end terminal in long lines along the sidewalks. At the north terminal, the situation began to escalate.[165] Growing crowds blocked all entrances and exits, panic broke out, and the police could not drive back the masses. Only when reinforcements arrived from the nearest station was it possible to control the mob and restore order, and not without the use of batons.

Yet despite all of this, newspapers reported that passengers had been euphoric.[166] People had been throwing "subway parties" for weeks leading up to the event to get into the mood.[167] When the day finally came, many residents donned their best clothes, went out for a festive dinner, and then descended into the subway station to celebrate their first experience as passengers of the underground. Countless passengers rode just for the pleasure of it, remaining in the underground until well past midnight. Despite calls for prudence and order, mayhem broke out more and more frequently as the evening wore on. The transformation of people into passengers was supposed to transpire with dignity and pride, but the whole event took a carnivalesque turn. The exciting novelty of underground transit and the incredible speed that had captivated the mayor during his own first ride now seized the crowds as well. Local observers were reminded of New Year's celebrations; New Yorkers had gone "subway mad."[168] Groups of thrilled teenagers occupied entire cars, singing, flirting, and celebrating. Many other people were completely overwhelmed. Scores of people stood in line, bought tickets, and entered the station but then did not dare step into a car: "All they could do was stand on the platform and gawk."[169]

While the subway had been emphatically pitched as a system for New Yorkers of all classes, genders, and races, many members of high society used it just once, on the day it opened. Some not even then. The *New York Times* wrote of the millionaire's wife Virginia Fair Vanderbilt (1985–1935): "Mrs. Vanderbilt came down from Grand Central station, and some of the Subway attendants recognized her at once. The station agent, M. F. Maddigan brought chairs, which were placed on the platform overlooking the express tracks. After the first special went through, Mrs. Vanderbilt, with her party, went to the automobiles which were in waiting for them above. They did not ride on the cars."[170]

Statistics show that on the evening of the opening, 111,881 people bought tickets for a ride on the subway. All proceeds were donated to New York hospitals.[171] While guests at an opening banquet had been assured that the subway would never reach maximum capacity,[172] the overcrowding of the first evening only increased in the days to come. Delays were frequent, as were riots. Police had to close individual stations several times when the situation got out of control. According to an official estimate, the subway carried about 350,000 passengers on its first full day of operation—many more than had been expected.[173]

The rush on the system escalated even more two days later. Many New Yorkers from the lower classes had to work six days a week; Sunday was the only day that remained for their first experience as subterranean passengers. Early in the morning, people from all over the city streamed from bridges and ferries onto the island of Manhattan to admire the technical marvel for themselves.[174] But since the IRT had been instructed to admit only 350,000 passengers, many were turned away or had to wait for hours at the gates. That day almost a million people came to try out the subway, and the system simply could not handle it. Once again panic broke out at the Bronx terminal. People who had come a long way to experience the adventure of the subway aired their complaints in angry protest. Again it took policemen deployment to calm the crowds and restore order.

Despite all of these adverse incidents, the system was nothing short of a sensation. New Yorkers were so thrilled at the experience of being passengers that they invented the phrase: "doing the subway."[175] Their enthusiastic reception of the system demonstrates that anticipatory attempts to dispel fears and concerns had been largely successful. It then remained for people's new existence as passengers to become part of everyday life, and this happened remarkably quickly.

Becoming Routine

In two days it will seem to New York as if it had never ridden anywhere but in the subway.[176]

—New York Times, *October 28, 1904*

Despite the chaos and euphoria of the opening, the subway became a normal part of routine for the people of New York practically overnight. The very next day, a *New York Times* journalist wrote with amazement: "On every hand there were evidences that the novelty was gone soon and the time is not many hours distant when few, save the oldest inhabitant, would be prepared to admit that they had more than a vague reminiscence of the days before the Subway began to run."[177]

The excitement and thrill of being a passenger for the first time soon gave way to normalcy. Taking the subway became a trivial affair, and New Yorkers left it to tourists to marvel at the system as an attraction.[178] Residents soon came to think of the passenger experience as something that would be "the daily routine for the rest of their lives," as one commentator laconically remarked.[179]

All of these factors indicate the successful routinization of the subway as an infrastructure and the subjectivation of passengers that accompanied it. The subway was a success story, and people soon came to find it indispensable for their daily lives. Newspapers proudly reported that after taking a first ride, even the famous subway skeptic Russell Sage was convinced, predicting a glorious future for the new "Subway City."[180] Almost immediately, efforts were made to place the event within historical context. The *New York Times* reported on debates over who could be called the very first passenger: "as now men dispute who was the first man to answer Lincoln's call to arms, and who was the first man to enter Richmond in 1865. This was a historic event, and nobody seemed to doubt it."[181] People felt certain that someday the opening of New York's subway would be seen as a milestone in American history. Papers continued to note remarkable firsts: the first man to offer a woman his seat; the first passenger to buy a transfer ticket. The ease with which the subway had turned New Yorkers into a population of passengers overnight surprised even contemporary observers: "It was astonishing, though, how easily the passengers fell into the habit of regarding the Subway as a regular thing. While the crowds above were still eagerly watching the entrance to see men emerge, were still enthralled by the strangeness of it all, the men on the trains were quietly getting out at their regular stations and going home. [. . .] It is hard to surprise New York permanently."[182]

A person's first subway ride can be seen as a liminal rite of initiation, in which the moment of becoming a passenger represents a kind of coming of age.[183] A single passage through the urban underground seemed to suffice to convert the city's residents into passengers. In light of the novelty of the New York subway in particular and the novelty of underground transit in general, it is remarkable how quickly and successfully people adapted to it. Today's cities can profit from knowledge gained through the implementation of existing systems, but at the turn of the century, cities were often on their own when it came to implementing new kinds of infrastructure. Just a few decades before, people had gotten used to railroads. Now people had to learn to cope with the new experiences of the subway in order to integrate these experiences into their lives.[184]

The effort to make underground transit part of daily routine began the day the system went into operation. According to Georg Simmel, a blasé attitude is central to metropolitan subjectivity.[185] The first subway passengers employed this attitude as a cultural technique, coping with sensory overload by demonstratively exhibiting an unaffected response. Some New Yorkers even reacted to the impressive subway experience with a bit of shame: "The Manhattanites boarded the trains with the sneaking air of men who were ashamed to admit that they were doing something new, and attempting to cover up the disgraceful fact."[186]

Before the opening, the passenger had been heralded as the hero of a better society to come. These hopes were dashed in the very moment of the first actual passenger experiences. The disruption of these hopes and expectations demonstrates the clash between different ideas regarding the subject of this infrastructure. The confusion that followed from system overload suggested that the scripts devised to shape and predict passenger behavior were not functioning as planned. Calling on the police to restore order in many situation, the elites were confronted with the limits of their own power of appeal. As Ulrich Bröckling emphasizes, adopting new forms of subjectivation also means changing them: "The individuality of human behavior insinuates itself in the form of opposing movements, inertia and techniques of neutralization. Regimes of forming the self and others do not provide a blueprint that needs to be followed. They require continual trial and error, invention, correction, criticism and adaption."[187]

The very first day of subway operation showed some evidence of appropriation and adaption, but these processes would take decades to fully unfold. Anticipated as a hero of free circulation, the passenger was quickly revealed to be an ordinary and rather tiresome everyday figure. No one denied that the subway had ushered in a new epoch of New York's history, but there was still no sign of the morally improved, civilized individuals that this machine was supposed to produce. It quickly became clear that the system devised to untangle the city's huge problems with overcrowding actually compounded these problems instead. As we will see in chapter 2, it was not just a matter of altering the city's architectural structures. The increased density of bodies passing through the subway brought new forms of interaction to light, some of which were perceived as transgressive, especially in the eyes of the elites. In the new subject cultures of the subway, the masses that had emerged around 1900 appeared in a particularly dramatic form. Incessant floods of passengers disrupted smooth subway operation and presented a serious threat to public order.

Controlling undisciplined crowds in the subway called for the fast development of new governmental technologies as well as new individual practices. While subway operators and law enforcement authorities learned how to regulate the flow of people through complex spatial passages, the first generation of passengers developed techniques to cope with their fears and adjust to the mechanical workings of the subway. As subjects, they had to mobilize maximum affect control to manage the confusing underground territories of the subway and its unfamiliar speed, along with claustrophobia and newfound bodily proximity to people of different classes and races. As the next chapter will show, establishing standards for rational, efficient behavior turned out to be far more difficult than anticipated.

MACHINES AND THE MASSES

Imperial New York.
Plenty of time yet.
Men and machines.
We are all so young yet.
Wait and see.
Wait and see what New York will do.[1]
—*Sherwood Anderson (1907)*

The opening of the New York City subway at the beginning of the twentieth century coincided with a paradigm shift in societal organization across Western industrialized nations. In light of the increasingly ubiquitous massification and technologization of everyday life, people were experiencing profound disillusionment with the bourgeois principles of the outgoing nineteenth century. As of 1900, we can see evidence of the development of Tayloristic work structures and a revolution in the material world of artefacts, as well as in the experiential worlds of consumer culture and mass media. Traditional notions of community began to appear anachronistic, along with the associated subject codes of morality and interiority.[2] A new guiding principle emerged under the force of these radical changes, revolving around the idea of the machine.

Central to the organization of subway passengers, this principle advanced to become such a dominant social model that historians of the United States refer to the next fifty years as *the machine age*.[3]

The idea of the machine shaped social order and economic organization, extending into different spheres of culture. Revolutionary effects were felt first in big US cities on the East and West Coasts. With the spread of new technologies for mass communication and mass mobility, along with new modes of production, consumption, and the use of electricity, this new machine culture quickly reached every corner of the country.[4]

The motif of the machine also became a dominant aesthetic paradigm in the arts. While early twentieth-century visual arts, architecture, design, film, and music were marked by a wide variety of styles and methods, there is a common fascination with the novel possibilities opened up by machines.[5] This motif not only advanced new forms of perception and imagination, it also produced new forms of community, expanding the reach of machines beyond their material omnipresence. As we will explore in this chapter, the machine functioned as a new ideal, a model of society manifested most clearly in the movements of Taylorism, Fordism, New Objectivity, and mass culture.

Especially in cities, the proliferation of machines in society extended into almost all areas of life, from work and mobility to the home, leisure activities, and consumption. Technological artefacts such as telephones, typewriters, radios, kitchen appliances, automobiles, and elevators became increasingly prevalent in daily life after the First World War. Underground railways further contributed to the experience of an entirely new kind of artificial environment.

At first, the machine age looked like a genuinely North American phenomenon. While following this development closely, Europeans initially viewed it as an "Americanism."[6] However, it did not take long for Fordist labor culture to spread throughout Europe and beyond via new modes of production and mass media.[7] Differently than the association with machines might suggest, this era was characterized by ruptures and contradictions, with the First and Second World Wars as well as the economic boom in the 1920s and recession in the 1930s. At the same time, the mechanization and automation of everyday life became more and more reciprocally intertwined. Mottos of efficiency and rationalization drove engineering, systems development, politics, and business alike.

Belief in technological progress, decisive in the second half of the nineteenth century, gained momentum as the machine age progressed. The revolution in everyday artefacts was met with near messianic enthusiasm, as was the explosion of knowledge in natural and technical sciences; people were excited about the rapid rise in living standards brought about by new urban infrastructures. Factories and administrative offices became more productive than it had previously been deemed possible. Along with the new worlds of experience opened up by the mass media and mass consumption, all of these developments gave people the feeling that they were on the threshold of a new age of self-determination and prosperity. As historian of technology Thomas P. Hughes puts it, the people of the early twentieth century saw themselves "involved in a second creation of the world."[8] This is also due to the fact that traditional value systems and models of behavior no longer provided effective orientation in a quickly mechanizing society.

The new forms of interaction and techniques of perception brought about by mass culture and the machine age can be accurately described as *post-bourgeois*.[9] While planning for the subway was still underway, the passenger was anticipated as a bourgeois hero. Once the system was actually implemented, this image no longer seemed plausible against the shifted backdrop of metropolitan experience. The decline of bourgeois aspirations in a social and material environment undergoing radical change left room for the emergence of a new subject culture in Europe and North America, which sociologist Andreas Reckwitz has called *organized modernity*.[10] Both of these developments are relevant for analysis of the machine age in several respects.[11] Reckwitz describes how traditional nineteenth-century ideals came to be seen as antiquated and incompatible with the interrelated demands of masses and machines. He also points to an increased tolerance for transgression: "Technology advanced to become an excellent opportunity for the material decentering of the subject through human-machine configurations that made subjective humanism and Romanticism obsolete."[12]

Drawing on bourgeois modernity, the subject culture shaped by machines radically recoded inherited cultural patterns in light of new forms of knowledge. Around 1900, alongside the machine, the collective subject of "the masses" emerged into the spotlight of Western metropolises. On both sides of the Atlantic, scholars began to analyze the characteristics of this new form of semi-anonymous collectivization.[13] For example, philosopher José Ortega y Gasset wrote: "Towns are full of people, houses full of tenants, hotels full of guests, trains full of travelers, cafés full of customers, parks full of promenaders, consulting-rooms of famous doctors full of patients, theatres full of spectators, and beaches full of bathers. What previously was, in general, no problem, now begins to be an everyday one, namely, to find room."[14] Many other prominent intellectuals of the machine age—including Sigmund Freud, Siegfried Kracauer, and Robert Ezra Park—also discussed the properties of the masses, often critically.[15] They described forms of social organization connected with the unprecedented crowding of bodies in offices and factories as well as working-class housing, cinemas, and department stores. The experience of being part of the anonymous masses, not least on the subway, also became a central motif in contemporary visual arts, literature, and film, for example in King Vidor's *The Crowd* (1928), John Steinbeck's *Grapes of Wrath* (1939), and Ralph Ellison's *Invisible Man* (1952).[16]

Linked to the rise of mass media, mass culture took shape as a new social form.[17] While bourgeois culture had been primarily embodied by a specific class, mass culture included different segments of the population. Inseparably tied to the rise of mass production and consumption, the mass culture of the machine age spread to encompass different spheres of industrial societies, both capitalist and socialist. Mass culture was a hegemonic and egalitarian way of life introduced by Taylorism and Fordism, disseminated by commerce, and adopted through consumer practices.

The formation of mass culture went hand in hand with the establishment of new forms of interaction that the upper classes often found uncivilized, obscene, and vulgar. For New Yorkers, the vulgarity of the masses was nowhere more evident than in the subway. Regulating the flow of passengers through the technical workings of subway cars and stations urgently called for the development of new means of subjectivation. It quickly became clear that the social norms evolving in underground urban territory could not be modeled on Victorian bourgeois subjectivity. Establishing "subway etiquette" meant abandoning traditional public etiquette, as new codes of conduct were needed to negotiate different kinds of conflict. In order to discipline and standardize passengers as part of the ensemble of the machine, authorities turned to the fields of engineering and logistics. Engineers and technocrats applied principles of standardization, operationalization, improved efficiency, and political neutrality in order to comprehend the mass behavior of passengers. The application of these principles was supposed to reduce the mob to a predictable entity whose distribution in the system could be regulated with precision. Simultaneously, the very same principles were inscribed into the subway's technical workings and architectural design as powerful instruments of control and subjectivation.

Before examining these phenomena more closely, let us consider how the advance of machines contributed to a massive transformation of New York's urban structures. Along with the rise of skyscrapers, the process of implementing new infrastructure and new technology into the city was most visible in the rapid expansion of the subway. We will explore how the subway system evolved from its opening up to the consolidation of its various subsystems in 1953. Specific pivotal moments will be discussed in further detail in later chapters. For now, a cursory outline allows us to trace the historical dynamics of the subjectivation of passengers.

Around 1900 there were already signs that the disciplines of engineering and logistics would be central to technologies of government in the following decades. As we saw in chapter 1, the idea of circulation came out of the nineteenth century as a central concept for explaining or describing various aspects of society. Institutions concerned with political, economic, and social control leveraged this concept to optimize the flow of capital, commodities, and people. The paradigm of circulation was still based on a specific ideology of hygiene, focused on the bodily practices of individuals as well as the organization of urban structures. As the machine developed as a model for mass culture, the premises of this ideology began to change substantially. Promotion and control of circulation remained key issues for public health and urban logistics, but in contrast to the previous program of social reformers, the aim was no longer to transform the moral attitudes of New Yorkers by integrating them into the city's dynamics of circulation. Instead, all efforts went into organizing the masses in a way that would optimize the performance of the system. This utilitarian adaptation in the instrumental rationality of logistics inherited many traits from nineteenth-century notions of utopian hygiene. While nineteenth-century urbanists saw the city primarily as an organism, in the twentieth century the efficient machine replaced the healthy metabolism as an ideal for social organization, as circulation came to be understood in terms of technology rather than organicism.[18]

New York didn't build the subways. The subways built New York.[19]

—*Fiorello LaGuardia (1945)*

As the machine age wore on, it seemed ever more plausible to think of the city as a complex ensemble of machines, in terms of theoretical discourse as well as structural design. In response to immense population increase, New York leaders sought ways to mechanize city life. The subway was principle to their efforts, facilitating the unprecedented circulation of workers. Quickly becoming fundamental to the everyday experience of mass culture, the subway altered the structure of the city more profoundly than any prior infrastructural element. As the elites had predicted, there was a subway-driven surge in economic growth. The city rose to become the nation's leading center of trade and industry. New York's harbors soon handled one-quarter of the entire volume of transshipments in the United States.[20] The city also became a hub for the developing world of finance. During the machine age, New York became a prototype for the global cities that would make up the networks of post-Fordist finance capitalism toward the end of the twentieth century.[21]

Rapid economic growth and an increasing demand for factory and office workers led to an entirely different type of construction in the city.[22] Skyscrapers added a new dimension to urban living; as the city spread horizontally, it also grew vertically.[23] The invention of the elevator was a major factor in the proliferation of skyscrapers and high-rises.[24] While the subway allowed for a new concentration of workers, the elevator opened up a new vertical frontier, ever expanding in the race to build the tallest building.[25]

Newfound verticality altered the symbolic and social orders of New York City. Prior to the introduction of elevators, buildings were not only lower, but also differently organized. Upper floors had been considered the most dangerous and least amenable places in the house, where poor people and domestic workers lived. Elevators took away the stigma of inaccessibility, so that top floors were suddenly attractive and desirable.[26] Vertical expansion increased office space and the number of luxury apartments on the market. At the same time, taller buildings also contributed to new levels of population density. To the dismay of political leaders and social reformers alike, within years after the opening of the subway, Manhattan had become the most densely populated place on the planet. In 1910 more than 2.3 million people lived on the island. One-sixth of all New Yorkers lived on the tip of land south of 14th Street. Contrary to forecasts, densification in the poor districts on Manhattan's Lower East Side broke historical records.[27] During the first decades of the twentieth century, tens of thousands of the island's poorest inhabitants were resettled in the Bronx, Queens, and Brooklyn, but waves of new immigrants kept replacing those who had moved on. Many of the people who had moved to the outskirts returned daily by subway to work in Manhattan. An apparatus for accelerating circulation, the subway also began to propel population growth. Between 1905 and 1920 alone, the population of Manhattan in the area north of 125th Street grew by more than 265 percent.[28]

Central to the development of New York's urban culture as a whole, the subway also played a major role in the formation of Harlem, the world's most famous black "ghetto." Connection to the elevated railway in 1880 had generated a first wave of urbanization and dense construction in this area in the north of Manhattan. When it became clear from plans around 1900 that the first subway line would reach Harlem, investors began building apartment blocks in hope of lucrative profits. Delays in subway construction, however, quickly led to an excess supply of housing. To minimize their losses, owners were forced to rent to African Americans. This marked the beginning of a wave of domestic migration of African Americans to Harlem, especially from southern states. As a result, in the 1920s Harlem became a social and economic center for New York's African American population, sparking the Harlem Renaissance and the blossoming of a rich culture that flourishes to this day.[29]

In the course of the city's expansion, the subway had become an elementary necessity, as demonstrated by ever growing demand. During the first decade of operation, the average number of annual rides per New Yorker rose to 343.[30] Floods of passengers constantly overloaded the system, causing breakdowns, delays, and overcrowding, which in turn frequently led to angry protests.[31]

Under mounting pressure, political leaders and subway operating companies worked feverishly at expanding the system. The first line was quickly extended to the Bronx and Brooklyn, but as opening day had demonstrated, far more extensive expansion was already required. New York's particular political and economic regulations often stood in the way. New infrastructure could be built with financial contributions from the public sector, but construction and operation were reserved for private enterprises. Power struggles among potential investors further complicated the process. August Belmont, who had financed the IRT, refused to expand the system, and in 1905 he bought up the only company that could seriously compete with him.[32] As a result, most of the lines were limited to areas that were already densely populated and thus especially lucrative.

The second phase of subway construction began only in March 1913 with the negotiation of dual contracts for the creation of a separate underground system in Brooklyn, to be operated by the Brooklyn Rapid Transit Company (BRT).[33] Connecting the two systems created a network that more than doubled the existing number of lines, soon comprising around six hundred miles of track. Previously inaccessible meadows and fields in the Bronx, Queens, and Brooklyn were newly developed as "subway suburbs."[34] Gigantic housing complexes shot up everywhere at the edges of the city, creating settlements for the passengers who relied on daily subway transportation to get to and from their places of work and education. Passenger numbers continued to rise rapidly, perpetuating the problems of the overloaded system.

After the First World War, inflation and the economic crises of the 1920s and 1930s delayed further system expansion. Although subway companies had initially turned a considerable profit, the operation of an underground transit system proved to be a financial disaster as time passed. The cost of operation tripled between 1915 and 1925. A lack of material and labor further complicated the construction of new lines. Additionally, due to a contract clause, operating companies were not permitted to raise the price of tickets. Despite several phases of heavy inflation, the original fare of five cents remained in place until 1948. This made it possible for people from almost every class to use the subway regularly, while the operating companies were in constant financial difficulty. For example, in 1918 the BRT had to file for bankruptcy, reconstituting itself as the Brooklyn-Manhattan Transit Corporation (BMT) in 1923.[35]

John Francis Hylan (1868–1936), mayor of New York City from 1918 to 1925, realized that intervention from the public sector would be necessary in order to extend the subway network.[36] Splitting up the system among separate operating companies had proved to be ineffective and expensive, so now the aim was to combine them under a single institution. The economic crisis of the 1920s made this impossible at first. As a result, the public administration decided to create its own network, not only financed but also operated by the city. In 1925 construction began for the Independent City-Owned Subway System (IND), which went into operation seven years later. Whereas the opening of the first line in 1904 and various subsequent expansions had been celebrated with much pomp, grand ceremonies no longer seemed appropriate by the time IND trains started rolling at one minute past midnight on September 10, 1932.[37]

Subway euphoria was a thing of the past. For passengers, the overcrowded subway was now symptomatic of the stresses and strains of everyday urban life, along with the corrupt and incompetent municipal government that wasted time on political intrigues instead of tackling the city's problems. By the 1920s at the latest, the automobile had replaced the subway as a symbol of progress and prosperity.[38] Public transit no longer held as much fascination for engineers and urban planners, who set to work making the city suitable for the cars that now carried the promise of a better way of life. Automobiles made it possible to expand residential areas even further, creating new prospects for the real estate market.

During the first term of Mayor Fiorello LaGuardia (1882–1947) from 1934 to 1937, the IND opened additional lines, but the system's expansion was effectively finished by 1940. With more than 745 miles of track and an annual passenger volume of 2.3 billion, for many decades the New York subway remained the world's largest and most-used underground railway.[39] After long preparation, the three systems were finally consolidated and made public. Buying back facilities from private operators was a massive investment that promised to ultimately reduce costs and achieve greater efficiency for a system that was already heavily subsidized. In June 1940 the New York City Board of Transportation, the largest fusion of railways in the history of the United States, united nearly 35,000 employees under one umbrella organization.[40] Created in

June 1953, the New York City Transit Authority (NYCTA) constituted an even more comprehensive management organization that included subways, buses, and streetcars. The founding of the NYCTA brought almost fifty years of subway standardization and expansion to a close.

The subway had undeniably accelerated the mass circulation of subjects, contributing to the emergence of a new form of urban culture. Within just decades, it had also radically changed the architectural structure of New York City. Yet the utopian society that the subway was meant to produce had failed to materialize. After world wars and severe recessions, people were disillusioned with the promise of salvation through total mechanization. For historians of technology focused on the subway, this is a threshold moment representing the beginning of the system's decline.[41] In 1953, despite shortcomings in maintenance and cleaning, the New York subway was still considered modern and safe. In the following decades, it came to be seen as the most dangerous and ramshackle underground railway in the world. A turning point for the subway in both technological and economic terms, the 1950s also marked a peak in the subjectivation of passengers. In the first half of the twentieth century, the image of the ideal passenger was derived from the subject form of a Fordist worker or employee. This model fell apart in the 1960s, when the city faced economic and social decline. Around 1950 the system carried more passengers than ever before, but a massive exodus commenced soon thereafter. People began to avoid the subway, as more and more New Yorkers left the city for proliferating suburban areas that could only be reached by car.[42]

Historically speaking, it is no surprise that the end of the standardization and expansion of New York's subway system coincided with the end of the machine age. After the Second World War, a growing number of philosophers, journalists, and sociologists voiced concern about the social control and destructiveness generated by a machine-like culture that alienated its subjects and forced them into new dependencies.[43] By the time countercultures were gaining speed toward the second half of the century, these concerns had boiled down to drastic criticism of how thoroughly technology and rationalization had taken over society. For thinkers like Lewis Mumford and Herbert Marcuse, this was particularly evident in the passenger culture of the subway. We will return to critical reckoning with the machine age as well as the erosion of passenger culture in a later chapter. For now, let us take a closer look at the standardization of passenger masses in the first half of the twentieth century as a hegemonic governmental technology and mode of subjectivation.

The new culture that began taking shape across Western metropolises around 1900 was marked not only by the mechanization of social spheres, but also by the unprecedented manner in which bodies were densely assembled. The new face of mass society emerging in overcrowded cities everywhere became particularly distinct in transit machines. Nowhere else was the erosion of traditional social relations and bourgeois interactions more obvious. With their supposedly irrational behavior and transgressive practices, masses of subway passengers gave rise to great concern.

MOBS AND MASSES

The mechanical workings and technical equipment of subway stations and cars presented newly minted subway riders with a host of challenges, from new sights and smells to new experiences of acceleration and deterritorialization. We will explore these sensory regimes later, first focusing on the new forms of social interaction that resulted from the massive concentration of bodies in subway spaces. The collective experience of rapid underground transit disrupted the established subject codes of the city above. As we have already seen, the first subway passengers were faced with a situation that was at once overwhelming and underdetermined, differing from life on the surface to such an extent that it seemed unclear whether existing cultural patterns would still apply. Passengers began behaving in ways that not only ran against the machine logic of the system, but also violated bourgeois-era norms. Yet these new forms of collective behavior did not become apparent from the practices of individuals. The mechanical workings of the overloaded subway processed passengers as a continuous flow, or as one *New York Times* journalist put it: "Endless legions of passengers in mass formations."[44]

Enormous numbers of bodies gathered tightly in one place posed a problem for the subway as well as many other early twentieth-century institutions, and the concept of "the masses" was politically charged. Discourses and theories of mass culture in the machine age became heavily laden with ideology. We will focus on how the concept of the masses was used in a specific historical context to characterize new forms of organization and interaction in subterranean transit. Many contemporary descriptions were marked by tension, presenting masses of passengers as a homogeneous entity on the one hand, and segmenting those same masses according to class, race, age, and gender on the other. Underlying these contradictory characterizations were questions such as: Do the masses form a singular unit, or a heterogeneous and fragmented array of individuals? What new collective practices and forms of behavior do the masses produce? What forms of regulation and control must be established in order to integrate them into the logic of the subway as a machine?

In the course of the machine age, the features and functions of passenger masses came into greater focus as a problem area. We will look at three dimensions in particular: the coding of the masses as a barbaric mob, transgressive practices of the mass subject, and social conflicts among passengers in masses. In the second half of the nineteenth century, the nascent disciplines of sociology and psychology sought to describe and explain the collective subjectivity of people in masses.[45] Scholars had different views on the particular form of social cohesion observed along with this phenomenon. While Marxist theory saw in the masses a potential revolutionary subject, most turn-of-the-century descriptions of the masses were shaped by critical perspectives on contemporary culture.[46]

Particularly influential was an 1895 study by French social psychologist Gustave Le Bon, *Psychologie des Foules*. Translated as *The Crowd: A Study of the Popular Mind*, it was an immediate bestseller when it appeared in America one year later.[47] The study did not paint the masses in a positive light. Based on his interpretation of the events of the French Revolution and the Paris Commune, Le Bon portrayed the masses as an irrational, impulsive, and potentially threatening mob. In his view, while individual subjects may otherwise be thoroughly rational and predictable, they suddenly become the opposite when they come together in masses. As violent as they are impulsive, the easily swayed masses are a threat to rational, liberal social order.[48]

In retrospect, Le Bon's study reads like a reflection of bourgeois anxiety, and contempt for the working class on the part of educated elites. Nonetheless, it was very effective. Gabriel Tarde, Max Weber, and Robert Ezra Park adopted many of Le Bon's ideas, along with Sigmund Freud. In his 1921 book *Group Psychology and the Analysis of the Ego*, Freud complained that Western culture was now literally producing "heaps of people [*Menschenhaufen*]."[49] According to Freud, being among the masses inhibits self-control and encourages emotional outbursts: "A group is impulsive, changeable and irritable."[50]

For contemporary New Yorkers, the description of the masses as an instinct-driven mob was fitting for subway passengers. The chaos of the first days of operation was taken as proof that masses of passengers were prone to unpredictable behavior and panic. Coded as uncivilized, practices like pushing and shoving were cause for concern. There were regular complaints about smoking and spitting, along with harassment and physical altercations. All of this was taken as evidence that the anonymous concentration of bodies in subway cars and stations led to a kind of regression. Le Bon, who had invoked the phantasm of the barbarian masses, went on to say that "by the mere fact that he forms part of an organized crowd, a man descends several rungs in the ladder of civilization. Isolated, he may be a cultivated individual; in a crowd, he is a barbarian—that is, a creature acting by instinct."[51] The impact of this interpretation of urban subjects in the context of the subway is demonstrated by the persistence of this motif in many caricatures from the time (as in figures 2.1 and 2.2).[52]

A drawing published in the *Evening Sun* provides another example of this motif from March 1913, the same month that Duchamp's now iconic painting *Nude Descending a Staircase, No. 2* was creating an uproar at New York's Armory Show (figure 2.3).

Depictions of passengers in barbaric masses highlighted the necessity of developing entirely new everyday practices in response to the new forms of social togetherness shaped by modern machines. The subway's concentration of anonymous bodies encouraged experiments in transgressing the bourgeois order of composure, morality, and shame.

TRANSGRESSION AND INFRASTRUCTURAL EROTICS

In the early years of the New York subway, the transgressive practices of passengers found expression in music, film, and dance. Popular "subway songs" probed underground territory as a site for the transgression of established relationships between genders. The machine age was romanticized, with the interiors of subways, automobiles, and elevators recast as suitable places for erotic fantasy. The first forms of an "eroticism

"The Paleozoic smash is on."

"10,000 years of civilization vanishes."

Figures 2.1, 2.2 Subway passengers as ape-men and members of the primal horde. These drawings accompanied a detailed essay in the *New York Times* titled "Darwin Defied in Our Subways," which attempts to refute the theory of evolution on the basis of the regression of New York passengers.

The Rude Descending a Staircase
(Rush Hour at the Subway)

Figure 2.3 J. F. Griswold, *The Rude Descending a Staircase* (1913).

of the machine" had already emerged in the late nineteenth century, linked to the sensations and mechanical aesthetics of railways and bicycles, and focused on stimulation through the technical apparatus.[53] In contrast, subway songs were more about social changes and the kinds of romantic interaction produced by new technology that brought many bodies close together.

After the subway opened in October 1904, the first sheet music for subway songs sold very well (figure 2.4). Songs like "Come Take a Ride Underground" (1904) and "Subway Glide" (1907) were targeted primarily at the white middle class. They combined catchy waltz rhythms and colloquial lyrics with subtly salacious love stories.[54]

These songs were mostly about the chance meeting of two passengers stuck in a crowd on the subway, unfolding as a romantic and potentially erotic encounter. "The Subway Express" from 1907 is a good example, the story of romance between a young woman and a young man on an overcrowded train traveling from the southern tip of Manhattan to the Bronx:[55]

Boy   It was in no sheltered nook
       It was by no babbling brook
       When romantic'lly we met.

Girl   Ah, the scene I can't forget
       We were thrown together in the Subway Express.

Boy   You were clearly all at sea
       As you wildly clutched at me
       When around that curve we swung.

Girl   Yes, and though I'd lost my tongue
       I made a hit with you, you must confess.

Boy   Yes, you hit me in the back
       And as around and around you flew
       I inquired if I could tender a supporting arm to you.

Girl   To which I answered "No, Sir!"
       When the guard yelled "Move up closer!"
       And clearly there was nothing else to do.

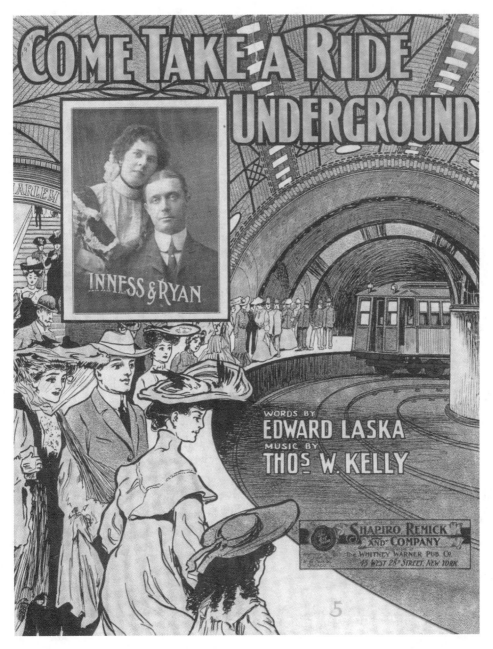

Figure 2.4 Cover of the sheet music for the subway hit "Come Take a Ride Underground," Shapiro, Remnick and Company, New York, 1904.

Boy   When we first met down at Spring Street
      And then upon my word.

Girl I felt I'd known you all my life
      When we reached Twenty Third.

Boy   You won my heart at Harlem

Girl At the Bronx I murmured yes,

Boy   We lost no time in that hour sublime
      On the Subway Express.

Love at first sight between two urbanites who meet by chance is not a new story, of course. This was already a common motif in nineteenth-century descriptions of flanerie. Yet subway encounters differed greatly from those fleeting encounters with the opposite sex experienced in a crowd of pedestrians on the boulevard, as captured by Charles Baudelaire in his 1860 poem "To a Passerby." Passengers pressed against one another in a small space did not experience modernity as excitingly ephemeral, fugitive, or contingent. Instead, the subway presented a very intimate and often oppressive situation without escape.[56]

Like crowding in elevators, anonymous crowding on the subway seemed particularly conducive to erotic energy.[57] The technical and artificial setting of the subway had its appeal, as evidenced by songs like "I Lost my Heart in the Subway (When I Gave My Seat to You)" from 1935, with the lines: "There was no bench, no park / There was no moon above, / An unromantic thoroughfare was where I found my love."[58]

Such descriptions of the subway demonstrate the liminal quality of this terrain in terms of behavioral norms.[59] Passengers saw transit spaces as strange, disconnected realms, neither distinctly private nor distinctly public. The ambivalence and anonymity experienced among masses of people allowed passengers to experiment with their relations to one another, permitting moments of transgression and intimacy. The first subway passengers found confirmation of Freud's claim that when individuals are immersed in a crowd, human libido and its "liability to affect becomes extraordinarily intensified."[60]

Indistinct codes of conduct in the machine environment made room for frivolous behavior. The very first film about the New York subway shows the improper flirtations of a group of passengers (*2 A.M. in the Subway* from 1905; see figure 2.5).[61] In a sequence lasting less than a minute, two police officers take notice and then watch as one of the women gathers up her skirt to show off her striped silk stockings to her fellow passengers. At the end of the scene, police intervention puts a stop to this behavior, highly indecent for the time.

By highlighting various forms of frivolous, romantic, or chivalrous behavior among subway passengers, pop culture served as a normative instrument for negotiating social codes. Remarkably, however, such sources neither glorified nor stigmatized the behavior of individual subjects in their new technological environment. The songs often convey a nuanced understanding of both the impositions and the freedoms belonging to the subject culture of passengers. Similarly, the celebration of new romantic behaviors in subway songs and films indicates how traditional patterns of intimacy and relationships were changing dramatically around 1900.

Among other factors, these changes bore the impact of migration movements in Western metropolises, where young people met and married in ways that were no longer primarily regulated by their parents.[62] Underground territory provided the young protagonists from the songs with a new, unregulated space, free of supervision by traditional authorities. Nonetheless, the songs upheld accepted cultural patterns; romantic narratives may have had a new setting and much faster pace, but they rarely violated basic norms of social interaction. Subway songs demonstratively perpetuated heteronormative role models like that of the gentleman or cavalier, obscuring other possible forms of transgressive sexual behavior.

Subway passengers were subject to yet another feature of infrastructural eroticism: the pleasurable yet uncomfortable libidinous sensations caused by the mechanical vibrations of rail transit. In *Three Essays on the Theory of Sexuality* from 1904/1905, Freud claimed that there was a connection between sensations of motion in the train and sexual arousal.[63] The exciting sensations caused by the machine's rumblings also gave rise to one of the most remarkable pop cultural products of the day, the Subway Express Two-Step.[64] The choreography of this partner dance, wildly popular in the months after the subway opened, imitated the physical swinging motion of standing

91

on the subway. Incorporating moments of physical contact such as that caused unintentionally by the shaking of the train, the dance helped to normalize this phenomenon by presenting it as an incidental, harmless aspect of passenger life. The Subway Express Two-Step also showed that the kind of tacit cooperation passengers had to master was not unlike the complex movement and interaction of a dance.

By defining and popularizing transgressive passenger experiences, subway songs and dances provided New Yorkers with important subject codes, teaching them the bodily techniques and forms of interaction necessary to fulfill their new role as passengers. Such sources also portrayed chance encounters among passengers from various backgrounds as exciting, not threatening. Films, songs, and dances did not explore the ways in which packing bodies into underground spaces could lead to frightening and sometimes violent situations. Masking violence and fear, the dominant discourses of the day focused on specific ideological connotations of the masses, highlighting the homogeneity of subjects in masses and disregarding the tensions, contradictions, and exclusions at work in this social form.[65] Early theorists of the masses such as Freud and Le Bon reinforced their critical arguments by stressing conformity and the collective subordination of individuals.[66] At the same time, intense debates surrounding the New York subway show that the question of who belonged to the masses was in fact up for constant negotiation. As we will see, contrary to the assumptions of Le Bon and Freud, becoming part of the masses also held emancipatory potential.

INCLUSION AND EXCLUSION

During the machine age, the social composition of passenger masses changed, along with the ways in which certain groups were included or excluded from subway transportation. Changes and conflicts were primarily related to the categories of gender, race, and economic standing. In the years after the subway first opened, many people had to fight for the right to join the masses, women first and foremost.

While the subject culture that emerged in Western metropolises around 1900 opened up new liberties for women, they were also confronted with countless incidents of discrimination and violence, from aggressive glances, to insults, to sexual assault.[67] Subway historian Michael Brooks points out the ambivalence of female experience in big cities, drawn between transgression and exclusion: "Women were not so much welcomed into the modern city as channeled into particular parts of it."[68] In many respects, crowded subway stations and cars intensified the ambivalence of gender-specific experience and subject order.[69] With its anonymous masses of bodies concentrated by technology, the subway also constituted a privileged setting for unwanted physical contact, exhibitionism, and other forms of sexual violence.

In response to such incidents, in the few weeks following the subway's opening there was a demand for cars reserved for women, protected areas that would shield them from sexual harassment by men. But in 1909, when the IRT actually began reserving the last car of each train for women, activists from women's rights groups rejected the idea outright (figure 2.6).[70] The Equality League of Self-Supporting Women protested by pointing out that female passengers frequently behaved just as badly as men.[71]

According to historian Clifton Hood, author of one of the authoritative studies on the New York subway, one representative of the movement put it like this: "Get into a suffragette car? Never! I am no better than the men, and the cars that are good enough for them are good enough for me."[72] For suffragettes, being treated differently from other passengers in any way was nothing less than discrimination. The subway was supposed to be a democratic space, as manifested in the inclusion and equal treatment of all New Yorkers.[73] Female passengers' strong insistence on being part of mass culture shows just how fragile and contested their subject position was, in the subway and elsewhere.

Once they had become part of the crowd, many female passengers developed techniques to hold intruders at bay among the tightly packed bodies. According to subway lore, some women used hairpins to prick improper hands reaching out for licentious touch.[74] Many of the city's new spaces of possibility were also contested spaces, with questions of behavior and access under constant renegotiation. This was particularly true of the subway.[75]

Whereas contemporary discourses primarily framed subway passengers as male, statistics show that about half of the passengers were women. The percentage of women among the working population of New York City increased steadily during the machine age. In 1920 more than two million women already worked in Manhattan, for example as secretaries, switchboard operators, or saleswomen.[76] Despite the

Figure 2.5 Film stills from *2 A.M. in the Subway* (dir. G. W. Bitzer, American Mutoscope and Biograph Company, 1905). As an important cultural document, this film is now part of the collections of the Library of Congress.

Figure 2.6 An IRT subway car reserved for female passengers with male supervisory staff (1909).

strain, they depended on the subway as part of the workforce and constituted an integral segment of the passenger masses.

The subway system was a contested space in terms of race as well as gender. Transit machines often served to amplify the social tensions at work in New York and society at large, as evidenced by informal kinds of discrimination against African Americans and Latinxs in the subway.[77] In contrast to the southern states, where Jim Crow laws imposed racial segregation on all public transportation, New York's transit system resembled more of a tightly packed melting pot for people of different skin colors.[78] While racist insults and attacks were commonplace in the subway, people of color were not refused entrance to stations and cars. This may have had less to do with the egalitarian convictions of New York's elite than with the fact that every paying passenger counted; the dramatic financial plight of the enterprise made it urgently necessary to welcome all passengers. Access to the system was certainly also determined by the fact that its major purpose was to keep the working population moving. As this population included people of all races, segregation appeared plainly impractical. Yet this only applied to passengers. For a long time, subway companies did not employ African Americans, Latinxs, or women.[79]

The fact that people of different races were part of the passenger masses from the outset does indicate that the subway system as an apparatus of circulation represented a liberal achievement for the times. The new form of mass collectivity generated by the subway served to level differences of race, class, and gender, at least to a certain extent. Brooks underscores this point: "The paradox of the subway was that it was a relatively unsegregated place in a society characterized by a high degree of segregation."[80]

The heterogeneous composition of the masses of passengers boarding the subway when it opened in 1904 became even more diverse as the system expanded. Especially after the First World War, the percentage of passengers of color and passengers who were underprivileged increased considerably.[81] Contrary to the predictions of social reformers, the number of migrants moving into poor districts continued to rise, yet subway transportation did make it possible for many families to move out of overcrowded slums and into the suburbs. Living conditions were more comfortable there, but residents still had to commute daily to the city center. These new members of New York's commuter culture were generally second- and third-generation immigrants whose parents had once made their way through the city's congested streets by foot to get to their places of work.[82]

In its first decades of operation, the subway was often packed, dirty, and loud, but it did provide reliable service around the clock. During the machine age, the very affordable fare of just five cents and a very low crime rate made subway service available to even the city's poorest inhabitants. The 1920s heralded a golden age for subway passengers, as Hood describes.[83] As the underground system spread out further and further, people began to see their city from a new perspective, extending their own radius of activity. This was truly liberating in terms of urban experience, opening up new places to go and things to do. The subway had been designed primarily to increase the circulation of working people, but it also promoted new cultures of mass entertainment, from theaters and variety shows on Broadway to amusement parks on the beaches of Brooklyn. Along with the city above ground, people from all walks of life found the underground transit experience to be full of the "excitement and urban color that embodied both New York's grit and its endless opportunities."[84]

Nonetheless, the mass culture of passengers also included many forms of exclusion. Material elements such as long stairways and crowded passages made it extremely difficult to navigate for people with physical disabilities as well as people who were elderly or frail. A thorn in the side of the operating companies and other passengers, beggars and homeless people sought alms and protection in the subway, particularly in times of inflation and economic crisis; there was little public protest when a law was implemented in 1933 to ban them from the system. Many passengers deemed the decision humane because it reduced the strain of transit.[85]

As the subway became part of normal daily experience for millions of passengers, its symbolism of progress, speed, and freedom began to fade. The image of the passenger masses changed, too. While at the beginning of the machine age, passengers were seen as ominous, uncivilized mobs, they soon became a symbol of the thoroughly rationalized, conformist culture of employees.[86] Artists and cultural theorists still occasionally raised the subway passenger to the status of a hero, however. Fascinated by the spectacle of heterogeneous masses in subway stations and cars, American author Christopher Morley wrote in 1923: "Someday a great poet will be born in the subway—spiritually speaking; one great enough to show us the terrific and savage beauty of this multitudinous miracle. As one watches each of those passengers, riding with some inscrutable purpose of his own (or an even more inscrutable lack of purpose) toward duty or liberation, he may be touched with anger and contempt toward individuals; but he must admit the majesty of the spectacle in the mass."[87]

While Morley describes the masses of passengers with respect as well as suspicion, during the system's first years of operation they were primarily coded as irrational, violent, and antisocial. Subway operators saw a highly explosive combination in the massive overloading of the system with the overwhelming experience of transit and concentration of passengers from diverse backgrounds. Subway engineers were convinced that if the situation were not diffused, sooner or later the system would collapse. They laid out two prevention strategies: expand the system and discipline the passenger masses.

The chaos on opening day had made it clear from the start that smooth subway operation depended on rational, compliant passenger behavior. Undisciplined behavior threatened to damage the already overloaded system. Chaotic conditions just getting on and off the subway regularly caused schedule delays. Since the maximum number of trains were already on the rails, such delays often caused cascading backups that disrupted the entire system.[88]

The undisciplined mob became a nightmare for engineers and subway operators as well as politicians and policemen. They could only succeed in generating smooth circulation if passengers were made to see themselves as part of the structure of the machine, submitting willingly to the logic of subway operation despite the strain of transit. This meant radically altering subjectivity and surrendering traditional bourgeois models of individualism. Instead of honoring the romantic ideals of virtuous autonomy, it was a matter of producing subjects who would conceive of themselves as controllable, efficient, emotionless agents, compatible with the technical modalities of infrastructure. In other words, passengers would be compelled to transform themselves into cargo. This transformation was to be achieved with methods from engineering and logistics that made it possible to regard individual subjects as functional units of the higher-level structure of a machine.

MACHINE CODE AND LOGISTICS

The human body is studied to discover how far it can be transformed into a mechanism.[89]

—*Sigfried Giedion*

In order to understand how the machine became a dominant model for industrial society in the early twentieth century, it is necessary to consider developments in the field of engineering in the last decades of the nineteenth century. A specific kind of knowledge began to take shape that we will call the "machine code."[90] Drawing on experience in designing, constructing, and implementing new machines in electrical engineering, chemistry, and motor science, this successful dispositif of knowledge was based on five principles: (1) efficiency as an economical form of organization; (2) political neutrality as a feature of technological rationality; (3) systematicity as a property of technical apparatuses; (4) standardization; and (5) modularization. Taken together, these principles cumulated in the idea of a machine as a functional structure made up of interchangeable subelements that must operate according to specific rules in order to achieve efficient functionality.

Extremely practical for building complex technological systems, this paradigm had neither political nor social connotations at first. Around 1900, however, such models began to shift from fields related to technology to discourses of social organization. Pivotal transformations in material culture contributed to this shift, especially with respect to technologies of circulation in transportation, production, communication, and urban planning.[91] Telegraph systems and railroads in particular restructured ordering patterns, generating the experience of "time-space compression."[92] When underground railways were introduced in Western metropolises, this kind of transformation became an everyday experience for millions of people.

The principle of the machine underlying the spread of new technology spread into the social spheres of mobility and employment. Implementation of the machine code led to radically restructured processes in factories and administrative offices. Advocates of the so-called efficiency movement preached the ideals of machine logic, with ideas that later achieved global influence in the form of Taylorism and scientific management.

A major step toward asserting the logic of machines in the organization of work was the 1911 publication of *Principles of Scientific Management* by the American engineer Frederick Winslow Taylor (1856–1915). Taylor was neither the first nor the last to advocate these ideas. Analyzing the ideology behind the machine age, historian John M. Jordan has shown that Taylor's thought must be understood against the backdrop of a broad social trend called the efficiency movement, which aimed to advance the paradigm of the machine in all areas of US culture.[93] Advocates of the efficiency movement, most of whom were engineers or technocrats, followed in the steps of the social reformers who decades earlier had tried to radically redesign the city based on principles of hygiene. They saw the model of rationalized mechanical organization as an ideal instrument for creating a new society, with scientifically backed technological processes that would allow culture and business to flourish as never before.[94] Born out by the idea that impractical processes and redundant resources could be identified and eliminated, the guiding principle of efficiency supported technical, economic, and social objectives. Procedures promoting efficiency promised not only to optimize work, but to promote growth and prosperity as well.[95] Taylor has been acknowledged retrospectively as the most influential thinker of the movement, perhaps because his plans were so radical. He sought to give the demands of the system absolute priority over the needs of individual subjects. Furthermore, as Thomas Hughes emphasizes, Taylor pursued a model of socio-technical organization "in which the mechanical and human parts were virtually indistinguishable."[96]

Supporters of the efficiency movement were major agents in transferring the machine code to other areas of social organization, but they were not alone. Just three years after Taylor's groundbreaking study was published, industrial magnate Henry Ford (1863–1947) introduced the eight-hour workday and a daily wage of five dollars for all assembly-line workers at his automobile factory in Dearborn, Michigan. Although this moment represents the symbolic birth of Fordism, it took several decades to be established as a hegemonic regime of accumulation throughout Western industrial nations, as David Harvey has shown.[97] In 1922 Ford presented a well-received vision of rational social order based on mass production and mass consumption, along with new forms of organization, reproduction, and worker supervision.[98] Like Taylor, Ford's ideas followed from the belief that applying the principles of efficiency and standardization to work, everyday life, and even leisure activity would lead to overall prosperity and the improvement of society. As these ideas gained momentum, engineers joined the protagonists of social transformation for the machine age. In the United States, the engineer inherited some of the pioneer's iconic force, with the frontier shifted from the Wild West to the undiscovered potential of machines and nature.[99] Hughes has suggested that along with the engineer, the system builder became another twentieth-century American hero.[100]

The experts who brought their knowledge of socio-technical processes to bear on the subway system included engineer Bion J. Arnold, industrial designer John Vassos, architect and engineer Squire J. Vickers, and members of the City Club of New York. Employing the machine code, they succeeded in transforming masses of passengers into functionally equivalent units whose standardized distribution could be measured, quantified, and controlled. Engineering thus became a powerful tool of subjectivation in the machine age, along with logistics. These fields generated expertise in the organization of complex mobility chains, which proved relevant not only in the context of the subway, but also in administration offices, factories, retail stores, and private households.

In the course of the twentieth century, the discipline of logistics became such a powerful instrument of control for social processes that Paul Virilio has referred to this as an "epoch of logistics."[101] While the Babylonians and people of other ancient cultures already operated with certain logistical technologies, the history of logistics as we understand it today began with industrialization.[102] Eighteenth-century postal services established organizational principles such as consistent transportation chains and universally valid tariff systems, along with modalities of collection and distribution.[103] Nineteenth-century steamship and railroad operators worked to optimize processes related to timing, connection, and allocation. Transport logistics also gained valuable insights from military operations, specifically in terms of the replenishment of supplies, a topic explored in many eighteenth-century treatises.[104] Logistical knowledge was also crucial in the industrial production and distribution of food, with respect to the large-scale production of baked goods and organization of slaughterhouses, for example.[105] Around 1800 the implementation of methods to improve the efficiency of meat processing led to the "mechanization of death," as Giedeon has called it. The resulting knowledge was also relevant for the rational organization of production and distribution processes more generally, and eventually this same knowledge was applied to the control of passengers on the New York subway.

Early in the machine age, methods from logistics were applied in many areas of society without being labeled as such.[106] It was only in the 1960s that organization techniques converged under the banner of logistics.[107] As Joachim Radkau puts it, "many concepts exist before we have a term for them."[108] Today the discipline of logistics covers a number of systemized practices from engineering, economics, and technology.[109] They all rest on a specific notion of rationality, focused on organizing processes as efficiently as possible. While the meaning of efficiency depends on the specific context, it is always related to the reduction of costs, resources, or time, in addition to the optimization of circulation. The purpose of logistics is to organize mobility, to keep information, material, capital, energy, and even people moving. In the eyes of contemporaries, the New York subway was an obvious setting for the application of logistics, in part due to the highly artificial and malleable nature of underground territory. Logistics became a means of translating between the socio-economic logic of the system and its technology. Knowledge from the fields of logistics, engineering, and business formed a powerful dispositif that constituted passengers as "infrastructured subjects."

STANDARDIZATION AND NORMALIZATION

Passengers are human cargo.[110]

—*Paul Fawcett*

Although the original plan for the subway was to put an end to overcrowding in the city, it immediately became clear that the system could not manage the masses of passengers who were on hand from the opening moment. The first months of operation revealed the full extent of overloading, especially with respect to rush-hour passengers. Between eight and ten in the morning and four and six in the afternoon, the subway moved more than one-third of the entire volume of daily passengers.[111] Millions of New Yorkers endured the everyday tumult and chaos of long lines at the entrances, crowded platforms, and overloaded cars. The predicted maximum volume of 600,000 passengers per day was exceeded almost immediately, reaching more than 800,000 by 1908. Even with every available train on the tracks, subway operators could not get overcrowding under control.[112]

Gradually subway operators and passengers realized that they had no choice but to accept the situation and develop coping strategies. Logistics experts and engineers looked for ways to structure erratic crowd behavior, while passengers developed cultural techniques of civil inattention.[113] Taxed with a massively overburdened system, in 1907 the newly formed Public Service Commission gave Bion J. Arnold (1861–1942) the task of investigating inefficient procedures in the IRT and proposing solutions to improve operation. Arnold and his team presented a seven-part report, documenting efforts to impose a paradigm of efficiency that included passengers as part of the logistical infrastructure of the machine.[114]

A proponent of the efficiency movement, Arnold carefully analyzed the entire system, including power supply, tracks, signals, cars, and stations. He measured braking distances and studied train schedules, calculated the maximum number of trains that could run at any given time, and measured air circulation and temperature variations in stations and tunnels. The idea was to spot and eliminate weak points in even the most minute aspects of the system.[115] Arnold saw the subway as one gigantic, complex machine that would operate most efficiently when all of its component elements were integrated and coordinated as well as possible. This included the practices of passengers.

Before he could start integrating the masses, Arnold first had to quantify them by implementing technologies that could measure and regulate the position of individual passengers as well as their distribution and circulation in the system. With the help of his many coworkers, Arnold set up an intricate administrative system that made it possible to count passengers and record transit volumes for every station, platform, and train. They recorded loading times—getting on and off the subway—and train occupancy rates at every time of the day and year. Offering a kind of passenger set theory, they presented their results in complex diagrams (figure 2.7).

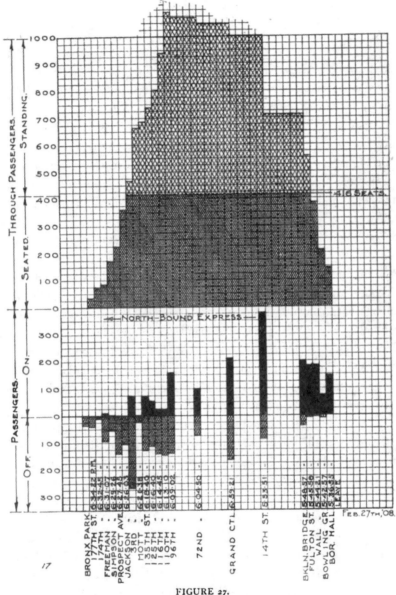

FIGURE 27.

WEST FARMS EXPRESS.

Leaving Borough Hall for Bronx Park at 5:36 P. M.

February 27th, 1908.

Total number of passengers	1869
Greatest number of passengers at any one time	1089
Length of time from Brooklyn Bridge to 96th Street	20 min. 5 sec.
Time lost between Brooklyn Bridge and 96th Street	4 min. 5 sec.
Average length of passenger travel	5.74 miles

   This diagram indicates that comparatively few passengers travel from Brooklyn to Manhattan during the evening rush hour and that a large portion of the load of a West Farms express is obtained at 14th Street. More passengers boarded this train at 96th Street than left it, but a large number were distributed to the four stations directly north of 96th Street and nearly twice as many passengers left the train at Third Avenue to transfer to the Elevated as left the train at any other station.

Figure 2.7 The occupancy rate of a train on the journey from Borough Hall to Bronx Park. Capacity overload is also shown by Arnold's expansion of the grid upwards, literally bursting the scale.

Arnold thus gained knowledge about passengers as a general population rather than as individual subjects, establishing a fundamental perspective for the technology of rule described by Michel Foucault as biopower.[116] This term designates a form of subjectivation and social order that emerged in the eighteenth century through quantitative census taking. Population here does not simply refer to the sum of all subjects to be governed. Quantitative methods of description bring new characteristics into view: "A population has a birth rate, a rate of mortality, a population has an age curve, a generation pyramid, a life-expectancy, a state of health, a population can perish or, on the contrary, grow."[117]

Simply put, biopolitical methods generate a specific form of knowledge by statistically determining averages and then presenting that information in patterns of classification.[118] In his work on the New York subway, Arnold drew on such methods to convert chaotic masses of passengers into a system of tables and averages, imposing standardizing, homogenizing features. As a result, the things that are known about passengers are averages: average number of rides per year, average duration of a ride, average number of rides per day, and so on. The statistical distribution of occurrences generates a normal range, or a tolerance zone around the calculated mean. What appears as the normal behavior of passengers is the result of methods that claim to be objectively rational and free of value judgment.

Literary scholar Jürgen Link has pointed out the key role of statistics in the development of technologies of rule in modern societies.[119] He identifies paradigms through a staccato of substantives: "homogeneity, continuity, one-dimensionality, scalability, quantifiability, computability, comparability with normal distribution, standardized prognostics, and controllability."[120]

Arnold's method of gathering statistics employed categories of social organization that can be described as technologies of normalization, to borrow from Foucault.[121] By transforming masses of passengers into charts for calculation, Arnold made them look less like an irrational mob. The ability to represent the flood of passengers in graphs and distribution curves made variations in passenger volumes visible; as a result, it was possible to generate detailed projections, even for specific months and days of the week.[122]

Arnold and his team were in a position to determine social normality by recording statistics and defining tolerance zones, thus marking the appearance of a new kind of norm. Neither a rule for governing social behavior nor a system of moral values, this was a quantifiable norm comprised of statistical probabilities and industrial standards. According to both Link and Foucault, the standardization of mass, weight, time, and industrial dimensions operates as a powerful tool of subjectivation. In the context of the New York subway, processes of industrial and technological standardization combined with processes of statistical normalization to form a complex instrument of control, dividing passenger behavior into compliant and deviant practices.

Arnold and his team contributed to the emergence of a new kind of norm in several respects. By representing a recurrent everyday moment in the lives of millions of passengers, they captured social normality. By making populations of passengers visible in charts, they also identified extremes and abnormalities, such as the massive overloading of the system during rush hour, irregular departures, and insufficient capacity. In turn, this data was used to increase the efficiency of circulation and make comprehensive adjustments to the system.

Arnold's methods contributed to a number of measures taken to optimize the circulation of trains and reduce overcrowding. The greatest obstacle to system efficiency was passenger behavior. Crowds were simply too big, and passengers moved too slowly and chaotically when entering and exiting subway cars. Arnold noted with frustration that "the delay at the station platform is caused largely by the passengers getting off slowly from crowded cars."[123] Pushing and shoving and an overall lack of passenger discipline seriously disrupted transit operations time and again.

In a section of his report titled "The Value of One Second," Arnold emphasized that increasing the efficiency of the system depended on coordinating and optimizing even the smallest detail.[124] Based on complex calculations and detailed lists of individual subway events, he argued that reducing the time a train stopped in the station by just one second would considerably increase passenger capacity across the entire system. Today Arnold's discovery of the statistical connection between individual passenger behavior and the system's overall performance may seem trivial, but it was a crucial step toward coding the deviant practices of individual passengers as a significant disturbance to the smooth operation of the entire system. Attributing relevant agency to each individual passenger, Arnold made it all the more urgent to find ways to implement standardization and discipline for passenger behavior.

Part of the solution was a method that Foucault calls "disciplining the body."[125] The aim is to produce compliant, disciplined, and productive individuals as subjects who can be integrated into the hegemonic social institutions of economics, politics, and culture. Like procedures related to liberal rule and governmental regimes of circulation, "disciplining the body" as a technology of rule predates the machine age by more than a century. The dispositifs that Foucault has famously described as central to discipline unfolded in eighteenth-century institutions including prisons, schools, barracks, and hospitals.[126] The implementation of a regime of discipline went hand in hand with the development of policing institutions allowing for direct access to the bodies of subjects. Arnold seems to have been aware of this when he made his suggestions regarding how to shorten loading times: "suitable police regulation upon the platforms of the station should be provided in order to properly control such individuals as may, through selfish motives, interfere with the prompt closing of the doors."[127]

Much to Arnold's dismay, it was the responsibility of uniformed station masters to get passengers into and out of subway cars, using loud instructions and occasionally violence. There was no independent transit police force until the formation of the New York City Transit Authority in 1953.[128] Until then, the IRT and BMT each had their own private security services. The IND, which was run by the city, employed its own unarmed officers recruited from police force applicants. In 1933 these officers were put under the control of the New York State Railway Police, and in the following years they acquired more and more police powers. Their main tasks were to oversee orderly entering and exiting from trains and to announce stops, but on IND lines they were also authorized to arrest passengers. These so-called special patrolmen had a reputation for brutal enforcement, and their often rough coordination of the masses warranted many written complaints.[129] They disciplined passengers with physical force and verbal orders. Incessant directives to "Watch your step!" and "Step lively" punctuated the passenger experience of New Yorkers.[130]

Paradigms of standardization and efficiency soon shaped interactions between guards and passengers. In 1917 regulations for the behavior of station officers codified the formulation of commands. Tellingly, the BRT published an "Efficiency Bulletin" specifying a "standard courtesy code" for all personnel.[131] This was essentially a long list of expressions to be memorized, with each instruction attached to "Please!," "Excuse me!," or "I am sorry!" The explicit aim of these simplified, standardized commands was to increase the rate of passenger obedience.

Along with these efforts to discipline passengers, two other strategies were pursued in order to promote homogeneous, compliant behavior among the masses. The first was to install a system of signs, including warnings, directions, and prohibitions. Proliferating in the territory of the subway during the decades after its opening, symbols and signs gave rise to a number of conflicts that we will discuss in chapter 3. The second strategy, no less successful, was to inscribe behavioral demands into the technical apparatuses of the subway itself.

PASSENGER PARCOURS

Subway operators, logistics experts, and engineers realized that in order to assure optimal functioning, it was imperative for passengers to move quickly and smoothly through the system. With masses pouring into the subway every day, it was necessary not only to constantly optimize separate segments of the system, but also to make them mesh. Engineers and logistics experts saw the arrangement of the system—its entrances, waiting areas, platforms, and cars—as a kind of parcours that passengers had to navigate with a steady rhythm. From this perspective, daily transit presented a complex sequence of movements in time and space, each with its own territory, norms, and requirements, constituting a choreography of multifaceted interactions between humans and machines. Only in the undisrupted execution of this choreography could the infrastructure reach its full potential.

We will consider the passenger parcours by reconstructing different areas of the system, exploring their technical and aesthetic transformations as well as their effectiveness in exerting control over movement. These areas were sequentially ordered: entrance gates, control areas, turnstiles, stations, and cars. All of these spaces produced specific information about passengers, and this knowledge was then inscribed, or built into the design of spaces and equipment. In the decades after the subway opened, its interior—furnishings, seating arrangements, lighting, and turnstiles—changed in ways that reflected the paradigms of the machine age; standardization, automation, and efficiency became crucial elements of subway design. Driven by the scripts of the subway's logistical and technological regime, these material changes were also decisive for the transformation of the passenger as a subject.

Along with the utopian idea that this new form of transit would promote hygienic urban circulation, the design of the subway was also shaped by the Victorian era. The system's first architects, Heins & LaFarge, implemented various modernist elements while pursuing an aesthetic strategy of functional grandeur. At the same time, the City Beautiful movement made its mark on subway design with tile mosaics, decorative ironwork, and ornamental entrances.[132] While these details were considered highly modern when the subway was first built, they emitted an anachronistic aura after only a few years. The shift away from ornamentation followed an increasingly functionalist understanding of architecture.[133]

The impact of this functional imperative became obvious when the first of the 133 grand entrance pavilions were dismantled just months after the subway commenced operation. Inspired by the design of the Budapest subway system and admired for their delicate iron and glass construction, these kiosks were touted as masterpieces of engineering worthy of a subway that was a monument to progress.[134] However, their sprawling structure took up space on already crowded sidewalks, blocking the view of traffic for pedestrians and drivers. Neither had people forgotten the congestion and rioting around these entrances on the first days of subway operation. For the sake of functionality, some of the original entrances were dismantled during the very first year of operation. This also reflected a shift in the coding of passengers as subjects. Rather than giving passengers an impression of sublimity as they entered, it had become more important to speed up the transition from the city above to the system below. Subway entrances soon became simple stairways signaled by comparatively spartan electric lighting.[135]

At the bottom of the steps, passengers entered a crowded area where a controller regulated entrance into the subway. Station managers controlled the flow of people at first, but as of the early 1920s, passengers were met with a new piece of equipment that in many ways embodied the paradigms of the machine age: the automatic turnstile.[136] As we will see, the turnstile separated people not only physically, but socially as well.

———

The turnstile became a powerful piece of equipment for counting and controlling subway passengers.

TURNSTILES

> To get in, as always, there was a price to be paid. The historical threshold
> of beatitude: history exists where there is a price to be paid.[137]

—*Michel de Certeau*

When the New York subway opened in 1904, entering the system was somewhat like entering a theater or cinema.[138] Passengers got in line, bought paper tickets at the counter, and then made their way to a station manager who tore the tickets in half and offered them admission. Yet with the system overloaded due to unanticipated floods of passengers, this process was constantly backed up. Police intervention was required to restore order as tumult broke out time and again.[139]

Untenable conditions and an enthusiasm for the possibilities of automation encouraged the system's engineers to begin experimenting with what was at first called a "passimeter."[140] The original apparatus already combined all of the basic elements that later went into the automatic turnstile. A simple frame with four bronze revolving bars, it could be released by an attendant pressing a foot pedal. During the night hours, it could be comfortably operated from the ticket counter by tugging a wire. The device's many advantages were highlighted in an advertisement by the Perey Company, which had applied for a patent and eventually installed their invention throughout the subway system. The ad promised the ability to control masses of passengers with just a tap of the foot: "One woman clerk can make change and pass 45 to 50 passengers per minute, all the while protected in a warm, well-lighted and safe booth. The attendant has absolute *mastery of the crowd*, through the foot pedal control on the turnstile."[141]

Despite the promise of crowd control, the device often failed to function properly, and at first it was rarely implemented.[142] Operators made another attempt in 1921, when innovations brought turnstiles driven by electricity. Dropping a coin into a box triggered contact that allowed the bars to turn.[143] But this technology also had its limitations. The first models in particular were disrupted by passengers whose bodies diverged from inscribed ideal dimensions: "it invited considerable congestion when fat passengers sought to make the passage."[144] Nonetheless, subway operators pursued further development of automated entry for economic reasons. In combination with the inflation that followed the First World War, the fact that they were bound by law to a fare of five cents drove subway enterprises to the verge of ruin.[145] Methods of logistical efficiency seemed to offer solutions to financial problems: optimize operational processes; adhere to the principles of standardization and cost-saving; and automate individual key elements of the system. At the same time, these methods allowed for reduction in the number of employees. Hundreds of people lost their jobs after an automatic system was introduced in 1921 for collecting fare and controlling entry. Instead of four attendants, only two were required for fare collection: one to supervise the turnstile, and another at a newly installed booth where other coins and bills could be changed for nickels.[146]

To the amazement of subway operators, within just a few months of installing the new turnstiles, passengers developed the habit of always having the right change ready at hand. People passed through the entrance areas more quickly, and crowding was significantly reduced. Subway operators were so impressed by the results that they broadcasted them in advertisements and press releases.[147] In addition to the public's fascination with automation, posters aimed at familiarizing people with the new technology contributed to the successful disciplining of passengers (see figure 2.8).

Inspired by increases in efficiency, soon the entire system was equipped with turnstiles, including the elevated trains.[148] A new model was introduced in 1931, the High Entrance/Exit Turnstile, also known as HEET. This tall, sturdy revolving gate inside a metal cage allowed passage in one direction only (figure 2.9). Invoking a torture instrument from the Middle Ages, passengers referred to it as the Iron Maiden. The machine soon became notorious for reducing the number of station attendants to zero. Passengers could not sneak past or climb over the stalwart Iron Maiden, an intimidating piece of equipment by many accounts. Passengers feared getting locked in or injured by the machinery, which did apparently happen sometimes.[149]

In response to passenger fears regarding the HEET, in the 1930s the Perey Company commissioned renowned industrial designer John Vassos (1898–1995) to devise a

Figure 2.8 Billboard from November 1921 instructing passengers in the correct use of turnstiles. Courtesy of New York Transit Museum.

Figure 2.9 The High Entrance/Exit Turnstile (HEET) or "Iron Maiden," ca. 1934. Exhibited in the New York Transit Museum. Photo by the author.

new type of turnstile that better suited the new aesthetic and functional paradigms of America's machine age (figure 2.10). Vassos streamlined the stocky units and covered them with smooth metal. He replaced the four-part horizontal stile with three vertical metal bars.[150]

The result was not only sturdier, it also made passage quicker. An icon of machine-age industrial design to this day, Vassos's apparatus embodied the ideals of movement, speed, and efficiency that subway operators and turnstile manufacturers were aiming to reach.[151]

The introduction of this new system of gates for entering and exiting the subway changed the spatial arrangement of its stations. Massive iron barriers were erected around the turnstiles to control the flow of passengers through the gates and then distribute them.[152] The measures proved so successful for regulating large numbers of passengers that they were soon copied by many other underground transit systems around the world.

For New Yorkers, passage through the system of gates became such a strong symbol of what it meant to be part of the masses that it found its way into American literature. For example, in Saul Bellow's 1964 novel *Herzog*, the eponymous protagonist remarks: "Innumerable millions of passengers had polished the wood of the turnstile with their hips. From this arose a feeling of communion—brotherhood in one of its cheapest forms. This was serious, thought Herzog as he passed through. The more individuals are destroyed [. . .] the worse their yearning for collectivity."[153] In the course of the machine age, filing past these turnstiles had become a collective daily rite for millions of subway riders. What were the scripts implemented along with this artefact, and what were the practices developed by passengers in their interactions with it? Some of the desired results had already been formulated in John Vassos's job description. In designing the new apparatus, his task was "to include new features that would eliminate unnecessary staff, increase entrance capacity, and control the flow of people as they went in and out of the turnstiles."[154]

Vassos also attempted to anticipate possible psychological properties of the passing subjects. He described how he tried to imagine the practices and sensations of future users, and then to incorporate this into his design. He wanted to avoid one anxiety disorder in particular: "Here my knowledge of the aichmophobic's reaction—fear of

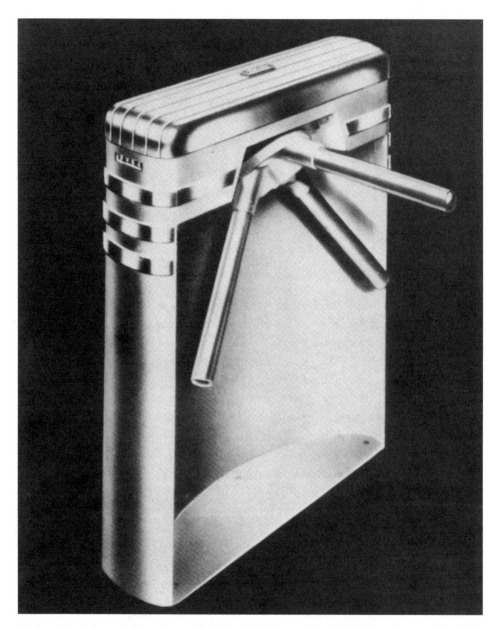

Figure 2.10 Design drawing of the turnstile by John Vassos on behalf of the Perey Company, ca. 1932. Courtesy of Perey Turnstiles.

pointed objects—guided me, and I produced a simple contrivance with gently curving surfaces, with any disturbing design around the feet of the user eliminated."[155] Along with the needs of future users, Vassos's inscriptions also anticipated possible deviant behavior on the part of station attendants, who were often suspected of pocketing collected coins.[156] In contrast, turnstiles were incorruptible, never tired, and less prone to mistakes.

Regulating subway access with turnstiles appeared simple: drop a coin into a slot and you were entitled to enter subway territory. But other scripts were also at work, particularly in relation to the bodies of passengers. One had to be capable of dropping a coin into a slot and passing through a physical barrier in order to enter the subway.[157] In addition to paradigms of efficient, economical circulation, a model of the ideal passenger was inscribed into the machinery. This passenger possessed a normalized body that was compatible with the turnstile, knowledge about how the technology worked, and correct change.

Another script built into the turnstile was that of individuation. German engineers dubbed it the *Vereinzelungsanlage*, which literally translates into "separating equipment." The term is telling.[158] Indeed, the apparatus separates a mass of passengers into sequenced units, making it possible to control and measure circulation into and out of the system.[159] With the installation of counting machines in the turnstiles, it became possible to register the exact number of passengers, supplying engineers and logistics experts with a wealth of data. While Arnold and his team had toiled to collect such biopolitical information, quantification mechanisms in the turnstiles generated an almost endless stream of detailed data regarding the distribution and circulation of bodies. The transformation of the chaotic flow of passengers into a measurable population was a dream come true in terms of governmentality. Bernhard Siegert has articulated the new biopolitical function of doors around 1900, noting that with the advent of turnstiles and automatic sliding doors, an entrance no longer addresses a person "as a persona but treats, forms, and monitors him or her as 'bare life.'"[160]

Forerunners to the turnstile offer important insight into the development of separating equipment as a powerful instrument of control, embodying the spirit of mechanization as described by Sigfried Giedion.[161] Giedion sees the origins of such equipment in nineteenth-century slaughterhouses before it appeared in many areas of everyday

---

human life in the early twentieth century. The mechanical butchering of animals constituted a kind of inverted precursor of the production lines in Ford's factories. Turnstile systems were also already at work in this context, allowing for an unprecedented increase in efficiency through the serial separation, slaughter, and butchering of pigs, sheep, and cows.

The Perey Company itself indicated that its technology originated in the keeping of livestock, while locating the mythical beginnings of the turnstile in the British Isles during the first century of the Common Era.[162] Cattle farmers there had placed wooden crosses on posts to create passages between pasture enclosures that were too narrow for animals, but wide enough for people (figure 2.11). The company emphasized that Vassos's turnstile could be traced back to this setting, with the crossbars of the apparatus also resembling a three-legged milking stool turned on its side.[163]

Precedents for passenger turnstiles can also be found in the train stations that had already been present in Western metropolises for more than fifty years before underground transit became a reality. Although not automatic, there was equipment that separated the territory of the city from that of transit.[164] In the subway there were similar arrangements in miniature, with the primary function of regulating circulation between the mechanical elements of the system below and urban space above. Automation sped up circulation, made it more efficient, and eliminated the separation of passengers into different classes. At the same time, the equipment retained the original property of biopolitical control. Defining societal and architectural differences between inside and outside, passage marked a transition into another territory, coded with its own norms, rules, and behavioral expectations.

The automatic turnstiles developed for subways also call to mind other forms of separating equipment that proliferated in the course of the twentieth century, for instance in laboratories and hospitals. They share the specific hygienic function of denying access to certain impure elements. Such biopolitical machines illustrate what Siegert calls "a conception of architecture as a thermodynamic machine and a shift from the nomological function of the door to a control function."[165] This shift of agency from subject to device represents an attempt to preserve public order by maintaining control, and to ensure that the bodies of passengers remain uninjured.[166] The first function can be designated as security, the second as safety.[167] In the separating equipment of

Figure 2.11 A turnstile in the cattle pastures on Block Island, Rhode Island. Photo by Ed Hendrickson. Courtesy of Perey Turnstiles.

the subway, these two functions together constitute the powerful dispositif of crowd control.[168]

Although automatic separating equipment was designed to give newfound efficiency and structure to the masses, passengers thwarted the plans of engineers again and again. Turnstile manufacturers explicitly claimed: "The intending passenger cannot pass without paying the Company," but passengers found subversive ways to do just that.[169] Techniques of deinscription involved jumping over barriers, entering the subway via emergency exits, or "piggy backing" by squeezing two people at once through the gate to save on one ticket.[170] Passengers also quickly discovered that turnstile mechanisms could often be turned back before latching, enabling the next in line to slip through without paying.[171]

During the first decades of operation, such deviant behavior was relatively rare, not least due to the extremely low fare price of just five cents. When making the case for automatic turnstiles, engineers focused on how the equipment structured and accelerated passenger movement while reducing personnel costs. They hardly addressed the need to prevent people from dodging fares.[172] It was only when the price of a ticket was raised in 1948 that "fare beating" became more and more of a problem.[173] Until then, turnstiles reliably collected coins from passengers. And once these passengers passed into the station, the imperative of maximally efficient circulation remained in force as they navigated the increasingly complex system.

FUNCTIONAL SPACES

The first interior design of subway stations still bore the signature of the Victorian era, with decorative mosaics and other ornamentation, but the spirit of modernist rationality soon prevailed. Heins & LaFarge, the architects of the first segments of the subway system, had barely fulfilled the terms of their contract in 1906 when responsibility for design was handed over to in-house architect and engineer Squire J. Vickers (1872–1947). An advocate of the efficiency movement and believer in the aesthetic tenets of the machine age, over the next thirty-six years Vickers changed the aesthetic and functional design of the subway to incorporate the paradigms of sober form and rational structure.[174] Vickers laid out principles of easy maintenance, hygiene, clarity, and durability in a speech he delivered in December 1917 for the Society of Municipal Engineers of New York City, declaring: "Bearing in mind the utilitarian nature of the subway, this severity of design seems to us the most appropriate treatment."[175]

One of Vickers's first steps was to reorganize stations in accordance with Arnold's suggestions. He made areas subject to high volumes of passengers larger and widened narrow corridors. He reduced the number of sharp corners and turns in order to accelerate the flow of people and reduce the risk of physical injury even when crowded.[176] Some of these design interventions required a lot of effort, but they promised to significantly increase capacity. Vickers designed new stations with mezzanines that offered space for entrance control, waiting areas, and restrooms. These areas were furnished with simple, sturdy wooden benches, which often had arm rests to prevent people from lying down.[177] By the 1930s, vending machines for beverages and candy were also ubiquitous.[178] Personal scales were another very popular addition, and passengers could weigh themselves for just a penny.[179] In retrospect it seems fitting that this apparatus of quantification was widespread throughout territories of transit, not only in subway stations but also in the stations of the elevated train and other railways. Coin-operated telephones were also installed beginning in 1921. Intensively used, they soon stood in almost every station.[180]

Mezzanines were designed not only as waiting areas but also to facilitate access to platforms and tracks. This proved to be a very economical form of construction that also improved the distribution of passengers between cars. As Vickers was proud of pointing out, the rearrangement of space substantially increased circulation in the subway.[181] The spaciousness and ornamentation of City Hall station, the undisputed jewel of the system, had amazed early passengers. In contrast, Vickers's plans for new stations were spartan and functional, resembling "utilitarian boxes" more than cathedrals of progress.[182] Stations were never absolutely identical, but increasingly similar due to the principles of standardization.[183] All of these changes were made in an effort to discipline passengers' individual behavior and increase the efficiency of the system as a whole.

In the carefully thought-out arrangement of turnstiles, barriers, and corridors, we can see the technology of discipline that Foucault describes as the "art of distributions."[184] Strategies for ordering individuals were already at work in the disciplinary spaces of eighteenth- and nineteenth-century schools, prisons, and factories. The purpose of such strategies, according to Foucault, was to "eliminate the effects of imprecise distributions, the uncontrolled disappearance of individuals, their diffuse circulation, their unusable and dangerous coagulation; it was a tactic of anti-desertion, anti-vagabondage, anti-concentration."[185] It is clear why similar strategies were deemed appropriate to control crowds of subway passengers. Yet in contrast to schools, prisons, and factories, the focus in transit was not on keeping subjects in a stable space, but on controlling distribution in order to regulate circulation.[186] This was particularly important when it came to the direction and speed of movement between entrance gates and subway cars. Vickers's layouts were in line with machine-age architectural ideology that upheld the priority of circulation for more than just transportation infrastructures. Analogous functions and designs went into modern apartment buildings, skyscrapers, factories, and offices.[187] Siegert also underscores the fundamental redefinition of architecture during this period: "Building no longer belongs in dwelling but in the passage. Existence is designed from the point of view of transit."[188]

By the time Vickers began designing stations for the newly founded IND in 1925, mechanisms of efficiency and functionalism had been implemented throughout the system. All entrances and exits were close to the tracks, which considerably reduced crowding on platforms. Every station was coded with a different color in order to facilitate passenger orientation. Distances between stops were carefully calculated.[189] A complex yet rational path had been forged, enabling passengers to move from the entrance gate to the cars as quickly and smoothly as possible.

Following Vickers's interventions, the spatial structure of subway stations remained largely the same for decades, with changes limited to furnishings and visual elements.[190] Subway cars, in contrast, underwent a number of transformations. Looking for ways to maximize capacity and optimize interiors, focus shifted to passengers once again.

CONTAINER BODIES

We are soft, and construct softening boxes.[191]

—*Michel Serres*

During the subway's planning phase, it quickly became clear that the desired speed of transit would necessitate an entirely new type of car.[192] While the underground fleets in London and Berlin were made up of refitted elevated trains, the New York subway needed a new kind of car, one that was faster, sturdier, and safer than any model built before.[193] For logistics experts, the greatest potential for optimization lay in accelerating the process of loading and unloading. Arnold had already lamented the inefficient design features that contributed to most delays: "The crowded condition of the car entrances and station platforms results in passengers leaving the cars in single file and with considerable difficulty and discomfort."[194]

In addition to the disciplining of passengers by station police, Arnold advocated optimizing the material components of the cars. He made a case for cars to be rebuilt with four doors on each side, with signs marking them alternately as entrances or exits, in order to separate the flow of passengers into those getting on and those getting off. He also suggested that seats be rearranged in a way that would significantly increase load capacity (figure 2.12).

Impressed by Arnold's models, IRT directors executed many of his suggestions in the following years. Signal equipment was improved and more cars were added per train. However, the idea for separate entrances and exits proved to be impractical; a test run with modified cars in February 1909 showed that passengers simply ignored the signs.[195] In the end, all future cars were equipped with an additional door in the middle, which made circulation significantly more efficient.[196]

Based on the results of Arnold's analysis, more and more elements of the subway were subjected to the logistical processes of technical optimization, calculability, and standardization. But unlike technical equipment, passengers often resisted efforts to fit them into the subway's operational logic. To address this problem, technical

FIGURE 16.

CAR WITH DOUBLE DOORS NEAR ENDS. COMBINING CROSS AND LONGITUDINAL SEATS. (48 SEATS.) MORE COMPACT SEATS FOR FUTURE CARS IN PRESENT SUBWAY.

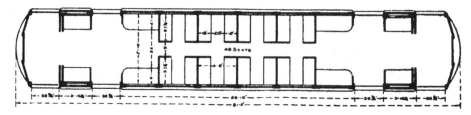

FIGURE 18.

SHOWING ARRANGEMENT OF GUIDING RAILS ON STATION PLATFORMS, TO BE USED IN CONNECTION WITH THE RECOMMENDED TYPE OF CAR.

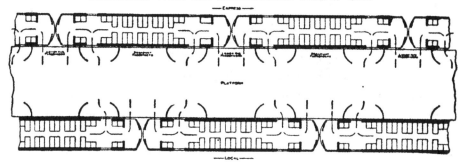

Figure 2.12 According to Arnold, the best and most efficient arrangement of seats in subway cars as well as trains in subway stations. Taken from Arnold, *Report No. 1–7 on the Subway of the Interborough Rapid Transit Company of New York City.*

equipment was implemented to amplify announcements by subway staff and broadcast them throughout the entire system. Around 1900, inspired by the success of radio transmission as a new medium, there were experiments on elevated trains with loudspeaker systems that transmitted orders given by staff through all of the cars. The noise produced by megaphones dangling from the ceiling was so incomprehensible, however, that the equipment was soon removed.[197] Additionally, there were frequent complaints about the unbearable heat and lack of sufficient lighting in subway cars. Subway operators often lacked the financial resources to equip cars beyond the bare necessities.[198] Nonetheless, the principles of automation were also applied to subway cars. In 1915 doors were installed that opened and shut either automatically or by remote control. This reduced the number of necessary train attendants to just one, and like the introduction of turnstiles, it meant that many subway employees lost their jobs.[199]

None of these innovations stopped crowding and congestion from getting worse by the day. Implementing Arnold's suggestions allowed 33 trains to run per hour instead of 29, and yet demand continued to grow. The subway broke one capacity record after another.[200] Engineers sought frantically for ways to improve efficiency, applying the logistical imperative of mechanization to design car interiors with a standardized maximum capacity. The task was to achieve an ideal distribution of passengers, and this meant taking a closer look at their bodies (figure 2.13).

Figure 2.13 shows how the standardization of load capacity for subway cars was linked to defining norms for passenger bodies. When calculating maximum capacity, Arnold had defined the space required per passenger as 12 × 18 inches.[201] In 1914 the Public Service Commission defined the standard space per standing passenger as four square feet.[202] The establishment of such norms was central to the machine coding of passengers as subjects. With bodies perceived solely in terms of measurability, passengers' individual features disappeared behind their logistical status as standardized cargo.

Subject codes were instrumentalized and inscribed into the design of many technical elements of the system, from seating arrangements and in-between spaces to entrance gates and sign systems. Together these elements formed a powerful body dispositif that defined the range of what is normal around the calculation of ideal passenger measurements. Beyond a certain tolerance value, deviation from the norm was subject to sanctions. By distinguishing between normal and abnormal bodies, the system excluded all bodies beyond the normal range. Such discrimination, which often affected people with disabilities and elderly people, was not addressed in contemporary discourse, indicating the effectiveness of the standardization process.

Once passenger bodies had been described, standardized, and quantified in terms of "load," the resulting knowledge blocked out passengers' individual features and black-boxed their inner lives, to employ a term from science and technology studies. A central concept of cybernetics in the second half of the twentieth century, black-boxing operates as a method to reduce complexity: the basic premise is that a system is easier to manipulate when observed solely in terms of input and output, with its internal functions obscured.[203]

The black-boxing of subway passengers operated on several levels. With the subway itself as a collection of containers moving rapidly through a vast network of dark tunnels below the city, passengers also came to be seen as "container subjects," or a mass of circulating black boxes.[204] Similarly to other prominent descriptions from the time, such as Wilhelm Reich's "character armor" or Max Weber's "iron cage," black-boxing describes the solidification of the boundary between subjects' inner lives and the largely engineered surrounding environment. These ideas are also related to psychological models from behaviorism, which coded the human psyche as an incomprehensible apparatus, limiting analysis to behavioral input and output. A necessary step toward standardizing passengers, black-boxing also made it possible to homogenize them. This involved a specific form of abstraction. Separated from the attribution of race, class, and gender, passengers appeared as standardized material entities to be logistically integrated into the workings of the transit machine.[205] Yet while logistical operations consistently neglected the emotional states and subjective experiences of passengers, these elements became a central focus for contemporary arts and human sciences, as we will see in chapter 4.

While the tenets of logistics thus cast passengers as standardized container subjects guided by an external authority, entities to be controlled in terms of input and output, it would be wrong to interpret this as desubjectivation. On the contrary, comprehending individuals as reified cargo is a technique that assigned clear subject positions to passengers, equipping them with standardized features and demand profiles.[206]

Turning passengers into standardized functional units within the overall subway system also facilitated the application of another tool from logistics, namely, modularization. Simply put, modularization divides complex systems into smaller units in order to optimize the performance and control of the overall apparatus. This involves standardizing individual elements and coding them as functionally equivalent. These procedures are essential for dynamic organization, grouping, and potential substitution. Developments related to modularization played a major role in mechanizing American society in the first half of the twentieth century. As cultural theorist John C. Blair notes: "The American emphasis shifted from whole to part, or more precisely, from a predictable whole sanctioned by tradition to an assemblage of parts. [. . .] Assuming that compatibility of components could be assured, the major conceptual leap was to a whole as an aggregate of individual subcomponents."[207] According to Blair, modularization as an organizing principle underlies a range of machine-age cultural phenomena, from the organization of courses at elite universities and the layout of factories to the structure of skyscrapers, poetry, the blues, and jazz. The immense power of modular structures in all cultural spheres went hand in hand with the development of machine logic.

Modularization in transit meant that masses of passengers came to be understood as compositions of functionally equivalent units. With passenger bodies standardized, engineers could focus on optimal arrangement. On the one hand, there was a desire to make transit comfortable for as many passengers as possible. On the other hand, fewer seats meant greater load capacity for each car, which was especially urgent during rush hours. The earliest seating arrangements for elevated trains had followed the so-called European model, with a pair of two-person benches facing one another.[208] This arrangement raised logistical problems, wasting valuable space and slowing people down as they entered and exited the train. There were similar issues with the American model, which had rows of seats facing in the direction of travel that could be folded to face in the opposite direction when the train changed its orientation.[209]

The imperative to efficiently increase capacity led to a number of experiments with seating arrangements. The IRT began substituting transverse seating with longitudinal seating along the outside walls of the cars. In an effort to combine comfort with functionality, most BRT cars followed the Manhattan model, with a row of seats running along the outer walls at the ends of the car and double seats cross-cutting the car in the middle.[210] A quick success, the seating models used by the IRT and the BRT were used in all subsequent layouts for New York subway cars. While engineers and builders chose these arrangements for logistical reasons, the seating options also influenced interaction among passengers.

Historian Wolfgang Schivelbusch has drawn attention to the social power of seating arrangements in public transportation.[211] He shows that European seating arrangements began disappearing from railway cabins and coaches around 1850. These first arrangements encouraged the development of travel conversation, a form of communication that was gradually lost in the course of the nineteenth century: "The face-to-face arrangement that had once institutionalized an existing need for communication now became unbearable because there no longer was a reason for such communication."[212] Sitting directly across from one another, passengers felt increasingly uneasy and embarrassed. With the Manhattan model, long rows of seats along the outer walls of the car put distance between passengers and limited interaction across the aisle; one could converse with the person in the next seat, at most. However, any relief from social pressure that this configuration might have provided was counteracted by the fact that cars were usually massively overcrowded, so that passengers were all squeezed together.[213] New Yorkers complained that it would be a serious legal offense to transport animals in this way, but subway passengers were subjected to it day in and day out.[214] Perhaps people suspected the origin of the models being used to govern their movement.

Subway operators continued to seek ways to increase passenger capacity.[215] Technical specifications for the system's subsystems differed considerably during the first years of subway operation, but by the time the three networks were combined under the Transit Authority in 1953, most BMT, IRT, and IND car models had become very similar.[216] Through processes of standardization and homogenization, the subway's technical norms were applied to subjects as well as equipment. The next step was to implement a kind of container ethics.

Despite repeated efforts on the part of subway engineers, massive overcrowding remained a problem for decades. The slogan of "A seat for each passenger!" was dropped shortly after opening day. A group from the City Club of New York formed to tackle the problem, including technocrats, politicians, engineers, and planners. Like Arnold and Vickers before them, they subscribed to the tenets of the efficiency movement. Their report from 1930 carefully documented the problem of subway overcrowding and presented possible solutions.[217] Quantification, rationalization, and standardization were marshaled as magic bullets, with the group insisting: "it is not too late now to set up standards."[218]

One urgent priority of the City Club was to introduce a binding standard for maximum car loads that would not only reflect technical requirements, but also take into account the health, comfort, and safety of passengers. They did so with recourse to logistical tools from Arnold's 1908 study. The City Club went a step further than Arnold's diagnosis of inefficiency, describing conditions in the system as "inhuman" and "evil."[219] Expressing concern for the fragile bodies of passengers, the report lamented the injuries caused by ruthless subway attendants as well as fellow travelers: "The actual physical injuries incurred from violent methods used by stronger passengers in boarding and leaving trains and the strong-arm methods of guards who pack the passengers in, are numerous and serious."[220]

Accounting for passengers in terms of standardized masses while taking up the perspective of individual subjects, the City Club in some ways carried on the legacy of late nineteenth-century social reformers. However, passengers were no longer anticipated as moral heroes, appearing instead as exhausted subjects who deserved a minimal ethics given the strain of daily transit. The City Club proposed a set of minimum standards for utilitarian passenger rights:

1. Every passenger should have some handle or support that he can grasp to steady himself.

2.   Every standing passenger should have enough room to move his arms freely, for instance, to reach to his pockets for a handkerchief if necessary.

3.   Every standing passenger should be far enough from his neighbor so that his face shall not be directly before that of another person.

4.   There should be room enough for a passenger to move freely towards an exit as the train approaches its destination.[221]

In the eyes of the City Club, these principles were not based primarily on rational, logistical calculation, but on a "standard of human decency."[222] Significantly, however, they still saw this standard as something that could be precisely calculated. Taking the four principles into account, one could simulate complex scenarios for the distribution of passengers across subway cars, experiment with possible variations in interior furnishings, or analyze passenger action while entering and exiting the train. Employing careful measurements of subway car dimensions and calculations of average load, the report suggested that each passenger should be allowed a volume of 8.5 cubic feet.[223] Diagrams served to illustrate this ethics of distribution, showing passengers only as standardized, abstract bodies to be arranged in various formations (figure 2.14).

These calculated distributions of human bodies represent a remarkable rationalization of appropriate proximity. They spell out a kind of container ethics aimed at defining a minimum of passenger dignity in terms of quantifiable "territories of the self."[224] The logistical, technocratic ethics behind the City Club's proposals for standardization generated a norm based on the natural and universal dignity of passenger-subjects. This norm also relied on a link between spatial distance and moral decency, combining the technical norm for maximum car loads with a social norm for physical proximity. This is a powerful example of the way that machine paradigms structure spheres of human togetherness. By aligning technical standardization and logistics with norms for social behavior, the dispositifs of the subway as a machine produced infrastructured bodies that could be made calculable, compliant, and productive.

Comparing the interiors of subway cars depicted in figures 2.13 and 2.14, we can see that it took decades for the process of standardizing abstraction to evolve. While figure 2.13 still represents passengers as individually discernable bodies, figure 2.14

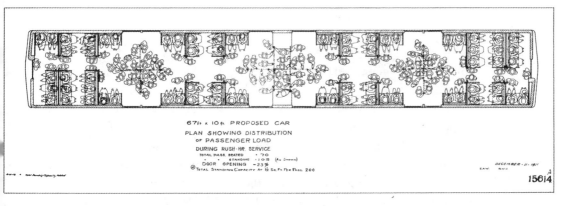

Figure 2.13 Proposed BRT Subway Car Passenger Load During Rush-Hour Service (1913). This drawing shows the desired passenger capacity in the standard BRT cars introduced in 1915. Courtesy of New York Transit Museum.

PLATE II

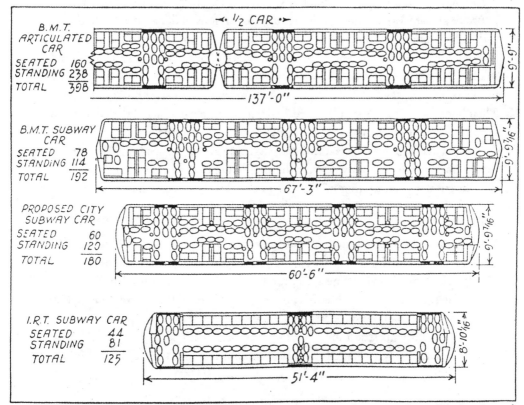

Diagram of 4 types of subway cars showing arrangement of seats and possible location of standing passengers according to the principles determined by City Club for standard load.

Figure 2.14 The distribution of passengers according to the principles of container ethics articulated by the City Club of New York (1930).

shows only identical container-like entities arranged in rank and file. The interconnection of social and technical norms makes it possible to draw a direct line between individual passengers and the biopolitical totality of the subterranean population. Foucault ascribes a special function to norms, as a hinge between the disciplining of bodies and the biopolitical regulation of populations: "The norm is something that can be applied to both a body one wishes to discipline and a population one wishes to regularize. [. . .] The normalizing society is a society in which the norm of discipline and the norm of regulation intersect along an orthogonal articulation."[225]

The City Club's proposal to allow individual subjects to exercise specific bodily practices, such as grasping a rail or reaching into one's pocket, reveals a technology of rule that disciplined passengers in a very specific way. Such procedures subordinated the bodies and practices of subjects to a rigid socio-technical regime, while also aiming to activate and engage individual strengths. The technologies of discipline that Foucault saw at work in eighteenth-century institutions underwent a decisive shift. In earlier disciplinary regimes, the body constituted an organism whose natural strengths and organic capabilities were to be made productive. But whereas eighteenth-century soldiers, patients, and delinquents were primarily addressed in terms of their own vital productivity, subway passengers were coded as elements in a technological apparatus. Their bodies were cast as standardized, interchangeable parts of an infrastructure with technical specifications they had to meet in order to be transportable.

Viewed in this light, it is clear that the City Club's central motivation for improving the system was not primarily care for the well-being of passengers, but rather concern that the strain of commuting might lessen their ability to work. Under the title "The Great Economic Loss to the Riding Public," the group wrote:

> Who can estimate the reduction of work capacity of an employee who has to travel for an hour in the morning with the struggling mass that is forced into our inadequate transit facilities? If it could be measured in money values, it would undoubtedly run into millions of dollars—lost—wasted. How much happiness is destroyed, how much leisure time is spoiled by the terrific nervous strain involved in getting home after a hard day's work! To translate these things into charts is impossible, but their economic and

social importance cannot be overestimated—and they are on the loss side of the ledger.[226]

According to these telling lines, it is regretfully impossible to put a price on the loss of happiness (although it must be millions of dollars), and misery is primarily an economic issue. Of course, this might have been a strategic argument to give the club's demands more weight. In the end, the group's suggestions for limiting subway car capacity had no real impact, as subway operators needed every paying customer and were not interested in limiting the number of passengers.

Even though some of these measures were not implemented, the passenger as a subject was shaped in lasting ways by processes of quantification, standardization, black-boxing, and rationalization. While daily transit quickly became an everyday experience for millions of New Yorkers, there were complex historical dynamics at work in the adjustments people made to fit the system. As we have seen, technological and logistical paradigms of the early twentieth century played a decisive role. Like factory workers, whose smallest steps were analyzed according to the principles of scientific management, the practices of subway passengers were coded and optimized.

Things grew even more complicated in the second half of the twentieth century, with the advent of scientific methods also allowing for the individualized study of heterogeneous groups of passengers. Additionally, beginning in the 1970s, more attention was given to passenger opinions that were gathered to inform management decisions.[227] Nonetheless, logistical methods of quantification remain dominant to this day, in large part due to their claim to scientific objectivity and political neutrality.

In the discourses and procedures examined thus far, we have considered passengers primarily as logistical subjects. Yet the subway also represented novel social and perceptual experiences. In overcrowded cars, passengers developed containerization practices that surprised even the members of the City Club: "That such conditions prevail, might suggest that the citizens of the city are apathetic to a horrible experience occurring daily and inevitable in the lives of a vast number of them. It is impossible to believe that such apathy exists."[228]

Indeed, apathy was perhaps an appropriate response to a logistically organized and thoroughly rationalized environment. Passengers practiced containerization in order

131

to reinforce the boundary separating their inner lives from harsh surroundings. Closely considered, what at first appears to be apathetic behavior turns out to be complex set of cultural techniques devised for coping with a particular experience. The territories of the subway offered an unfamiliar sensory landscape, structuring the perceptions of passengers in a specific way. Chapter 3 reconstructs the perceptual techniques of transit, demonstrating how the logistical regime of the subway shaped the modalities of user experience. This was particularly true of visual signals and eye contact.

Techniques of the Senses

Not expression—but signals; not substance—but motion![1]

—*Helmut Lethen*

Like all subjective experiences, the meaning of visual, acoustic, and olfactory stimuli is subject to historical change. Karl Marx articulated this dynamic already in his time with the proclamation: "The cultivation of the five senses is the work of all previous history."[2] Transformations in the history of sensory perception are notoriously difficult to track, however, especially when it comes to subjective interpretation.[3] Nonetheless, this chapter aims to reconstruct the sights and sounds surrounding subway passengers in the first half of the twentieth century. We initially explore the impressions that passengers were exposed to in underground urban territories, and the tactics they developed in order to cope with these impressions. We then turn to the subway's visual regime to discuss the inscriptions and historical dynamics of signs, maps, and advertisements. Analyzing such artefacts allows us to draw conclusions about the practices and perceptions of passengers, while also revealing different aspects of subject constitution that took effect in these practices and perceptions.

Calling on signals over expression and movement over substance, cultural theorist Helmut Lethen's points to a paradigm shift in social order through the establishment of a subject code of New Objectivity. This call finds particular resonance in the context of the New York subway. Chapters 1 and 2 of this book explored some of the implications of placing top priority on the idea of circulation. We now turn to a perceptual shift favoring the idea of the signal. Subway passengers attempted to subdue their emotions when interacting with one another, cultivating an air of dispassion and disinterest in their gestures and facial expressions. At the same time, a new organ of social control emerged through the visual signals and sign systems used to address subjects in the subway system.

Subjectivation of passengers according to the machine code meant standardizing and quantifying them under the subway's logistical and technical regimes. This process shaped forms of perception and social interaction, and confronted passengers with a new kind of rationality, associated with protection against stimuli, industrial consciousness, and New Objectivity. We can further identify the phenomenon of containerization among subway passengers. The cultural techniques of encapsulation mobilized by people in the subway appear as forms of emotional containment.[4] Practices of isolation and subduing emotion form a complex of subject codes that Helmut Lethen has aptly described as the "conduct code of the cool persona."[5] When faced with the impertinences of modern society, "staying cool" provided individuals with some orientation, shaping their responses among the masses. The technologization of social life reproduced the model of the container subject. The subway's logistics and control mechanisms reduced the passenger to a standardized, black-boxed entity, reinforcing the barrier between inner life and subject exterior. Passengers also developed practices to uphold this barrier themselves. Reconstructing this process shows how the subway evolved into a powerful tool of subjectivation, a kind of infrastructural dispositif of the senses.

Before exploring these phenomena, however, some remarks are in order regarding the considerable difficulties involved in reconstructing any history of subjective sensory perceptions. It is complicated to map passenger techniques of cultural perception, and there are several pitfalls to avoid, as emphasized by historian of the senses Alain Corbin.[6]

In the first place, we must be wary of positivist accounts of the evolution of sensory perception. We should also be wary of the idea that history has refined and civilized sensory experience, as Norbert Elias has argued. We must also reject the notion that sensory perception has evolved continually through the increasingly complex disciplining and training of subjects. There are problems with identifying historical dynamics or genealogies in the evolution of the senses. Sociologists and ethnographers have a number of proven methods at their disposal, but exploration of the history of the senses must rely largely on what has been passed down in writing. The danger lies in taking such textual sources at face value, assuming that the discourse of the times was identical to the structure of passenger sensations and perceptions.[7] It is tempting to draw direct conclusions about how people experienced their sensations based on text, overlooking the frequent difficulty of translating sensory experience into writing. This obstacle is even more pronounced when the sensory experiences in question were novel and unfamiliar. In addition, the description of sensations deemed trivial or those associated with shame is often relegated to the margins of discourse, at best. We will encounter this problem again in chapter 5, when we explore how subway management processed written complaints.

Keeping these complicating factors in mind, the current attempt to describe passenger sensations in historical terms is aimed at gaining insight into passenger subjectivity. For this purpose, it is also necessary to lay out the norms, values, and imaginaries associated with these sensations. Anthropologist Mary Douglas has shown, for instance, that many cultures associate certain olfactory impressions with the locally dominant idea of cleanliness.[8] The emergence of specific dispositifs of perception in historical subject cultures decisively depends on the constitution of the material environment. Corbin has shown this to be the case for the sense of smell, reconstructing the massive efforts of Western metropolises in the eighteenth and nineteenth centuries to regulate exposure to odors or ban them entirely from the urban sensory landscape.[9] Controlling the urban sensory dispositif—or deodorizing the city—can be seen as a remarkably successful governmental technique, changing the structure of perception for residents of the city, as well as the material environment. Similar moments stand out in the history of early New York subway passengers, whose sensory impressions were forged to a great extent by the technology that surrounded them.

As passengers became exhausted by what they saw, heard, and smelled, they began modulating their five senses in order to cope. The subway environment was primarily structured by what one could see, advancing a general shift toward the dominance of sight in modern metropolises.

The focus on vision for the purposes of this study is also based on the availability of source material. The New York Transit Museum Archives contain predominantly texts and visual material, such as documents and photographs. Very few audio recordings and traces of olfactory or tactile impressions have been preserved, and this also contributes to the dominance of the visual in historical accounts. We nonetheless take a brief look at the sensations of hearing and smelling in the subway before moving on to discuss strategies of seeing. The second half of the chapter takes a closer look at signs, advertisements, warnings, and so on, investigating how they were employed for the control and moral education of passengers. Last but not least, we will consider the long-running poster series of the famous Miss Subways contest, and the effects it had on passengers' self-perception.

ECONOMIES OF PERCEPTION

In order to reconstruct passenger experience during the first decades of subway operation, this study draws on groundbreaking studies in the cultural formation of human perception by Alain Corbin, Michel Serres, and Jonathan Crary.[10] As these authors and others have demonstrated, forms of anthropological perception are anything but naturally given. Like human nature in general, perception is a complex cultural construct. Perceptions change as the environment and context of life change. Subjects are also actively involved in the establishment of situational and historical practices of perception. We are dealing with the questions of which dispositifs of perception were most powerful at a particular time, and how passenger-subjects obeyed, adapted, or undermined the regimes of perception and knowledge imposed upon them. Cultural techniques of perception, adaptation, sensitization, and isolation are central features of the subject constitution of individuals. As cultural theorist Susan Stewart puts it, "the opening and modulation of the senses takes part in a dynamic that is at the core of subjectivity."[11]

In the following pages, we will explore the interpretation and standardization of perception from the passenger side, as well as the dispositifs of perception and the sensory regimes in the subway system. Although the system coded its ridership as a normalized, controlled crowd, it would be incorrect to characterize passengers as merely fatalistic, passive individuals. As we have already seen in the previous chapters, subway infrastructure facilitated new and sometimes transgressive forms of perception and experience. These developments must be viewed against the background of massive upheaval in the subject cultures and sensory perceptions in Western metropolises around the turn of the twentieth century.[12]

Georg Simmel and Walter Benjamin were two of the earliest chroniclers of sensory regimes in modern metropolitan culture. While their work focuses on modern urbanization in European contexts, their observations and reflections nonetheless provide relevant insights into dynamics playing out simultaneously in other metropolises, such as Buenos Aires or New York. Moreover, many aspects of the sensory adaptations to modern urban life that Simmel and Benjamin observed, particularly with respect to technologization and containerization, were amplified by the transit environment of underground tunnels and cars.

The impact of the transit environment on passenger perception was preceded by a numbing of the senses as described by Simmel around 1900. He presents this as the result of the strong yet fleeting sensory impressions belonging to life in the big city, which also altered the significance of the individual senses.[13] "Herein necessarily lies a significant factor for the sociology of the metropolis. Going about in it, compared with the small city, manifests an immeasurable predominance of seeing over the hearing of others; [. . .] above all through the means of public transportation."[14]

The need to direct one's attention to specific impressions led to a modulation of the senses. Historian Jonathan Crary shows that the cultural practice of attentiveness became a vital resource in the early twentieth century.[15] Sociologist Herbert A. Simon speaks of a general "economy" of attention and perception.[16] In an environment full of excessive sensory impressions, this refers to the circumstances of subjects forced to concentrate their perceptions based on rational calculations.[17] According to Simon, subjects faced with an abundance of impressions and information are compelled to make selections and form preferences, but they must also be able to ignore stimuli, block out information, or postpone their responses. "Hence, a wealth of information creates a poverty of attention and a need to allocate that attention efficiently among the overabundance of information sources that might consume it."[18]

Informed by these perspectives, and in keeping with Simmel's account, the transitory perception of passengers can be described as fleeting, blasé, and numb. For Simmel, this disposition of "metropolitan individuality" arises from an "intensification of emotional life due to the swift and continuous shift of external and internal stimuli."[19] We will see how passenger subjectivity was increasingly characterized by attempts to preserve autonomy and keep a distance from the surrounding environment. This resulted in the development of a mode of subjectivity that Wolfgang Schivelbusch has called "industrialized consciousness."[20] This refers to the formation of complex procedures of protection against stimuli, on the one hand. On the other hand, as Simmel describes, subjects distinguish themselves through "heightened sensitivity" and aesthetic reactions to select sensory impressions: "It is of significance, still insufficiently noticed, for the social culture that the actual perceptive acuity of all senses clearly declines but, in contrast, the emphasis on sense pleasure or lack of sense pleasure increases with the refinement of civilization."[21]

Numbing and indifference go hand in hand with a heightened sensitivity that generates aversion, promotes alienation, and marginalizes social intimacy. Simmel sees the modern subject's perceptual economy as shaping personal social behavior, especially in the context of urban transit.

For New York subway passengers, visual and olfactory perceptions were used to assess social rank, defining how much physical and emotional closeness was acceptable. Such perceptions were also effective as instruments of sanction and discipline, and complex sensory challenges often led to practices of isolation and avoidance. For example, reading quickly became a favorite barrier to social interaction on the subway. Passengers attempted to take control of the transit experience by implementing fragile private territories for themselves.

The unfamiliar artificial territory below the city placed many new demands on the senses. For the first subway riders in particular, high-speed travel in loud, poorly ventilated containers generated sensory overload, amplified by the experience of being confined with many others in a crowded space. It was not just a matter of feeling lost in the crowd, or the loss of orientation or a sense of place in transit. The subway was hot and dark, and there was no escape from the smells and sounds of machines. For many passengers, it represented an industrial Hades.

In the Hades of Names

Walter Benjamin casts the urban underground as a modern incarnation of the ancient underworld, describing the Paris Métro "as another system of galleries [that] runs underground through Paris, where at dusk glowing red lights point the way to the underworld [Hades] of names."[22] The metro's system of underground signs was complicated and confusing. The "Hades of names" was a place of sensory overload, disorientation, and isolation. Despite close physical contact, passengers were lonely and apathetic. In Benjamin's drastic words, "Here, each one dwells alone; hell is its demesne."[23]

This was true not only in Paris. In the New York subway as well, the experience of sensory exhaustion, spatial fragmentation, and alienating machinery elicited scores of associations with the inferno of the outcast and the damned. To this day, on both sides of the Atlantic, the image of hell remains an extraordinarily persistent topos in urban mythology.[24] For machine-age passengers, the association was warranted in part by the darkness, stench, and heat that they endured day in and day out. According to reports from the time, the underground regions of the subway were unbelievably warm in the summer, especially.[25] The system that had been heavily overloaded from the start soon literally overheated. The body heat of millions of people mingled with the waste heat of huge electric motors generated by countless starts and stops. To make matters worse, the first tunnels and stations were lined with watertight material to keep electrical equipment dry, but this lining also retained heat.[26]

In 1905 the *New York World* published a cartoon captioned "Dante tries the subway; finds the air hotter than he found it in Hades" (figure 3.1).[27] The cartoon suggests that passengers had more than one reason for comparing the subway with Hades. Besides the heat, there was the feeling of being at the mercy of a hellish machine. Dante asks an instructor: "What are these poor tortured people?" He is told in reply that they are "condemned [. . .] to hang for a half hour, morning and evening."

Subordinated to the unfamiliar and abstract organizational principles of machines, subjects evidently experienced anxiety and fear of a loss of control, condensed in the

Figure 3.1 V. P. Whitney: "Dante tries the subway; finds the air hotter than he found it in Hades." *New York World*, June 26, 1905.

image of subway hell. This was compounded by the fact that the subway was also a complex social setting. Jean-Paul Sartre's famous remark that hell is other people certainly had resonance for many subway passengers. In addition to verbal threats and the danger of assault, the olfactory landscape in cars and stations contributed to associations with the underworld. The stale air was permeated by bodily perspiration, machine exhaust, and much more. Above all, the body odors of other passengers gave rise to frequent complaints. Being densely packed among countless other human bodies was an olfactory challenge, and the interpretation of odors carried strong social connotations. Simmel observes that smelling other people "forms one of the sensory bases for the social reserve of the modern individual."[28] In modern, big-city life, the sense of smell became a "dissociating sense" that unveiled the antipathy among classes and races.[29] Along similar lines, George Orwell writes that odor can determine social hierarchy: "That was what we were taught—*the lower classes smell*. And here, obviously, you are at an impassable barrier. For no feeling of like or dislike is quite so fundamental as a *physical* feeling. Race-hatred, religious hatred, differences of education, of temperament, of intellect, even differences of moral code, can be got over; but physical repulsion can-not. You can have an affection for a murderer or a sodomite, but you cannot have an affection for a man whose breath stinks—habitually stinks, I mean."[30]

As an elementary factor of social hierarchy, olfaction worked not only to degrade the working class and the socially disadvantaged, but to disparage people of different races as well. In the United States, the racist belief that African Americans had some scent of their own appears to have been so widespread that even Simmel knew about it far off in Germany.[31] Indeed, in the 1960s as well, many passenger complaints pertained to the alleged body odors of African Americans and Latinxs.[32] These complaints provide evidence for Simmel's assertion that the social question "is not only an ethical one, but also a nasal question."[33] According to Simmel, the sensory impressions of metropolitan inhabitants are predestined to evoke aversion and disgust, and this seems to have been the predicament of subway passengers in particular. Despite all efforts to put up barriers, revulsion at the sweaty bodies of others was coupled with embarrassment and self-stigmatization regarding personal odor.[34] The battle against body odor was part of the deodorization of urban space. It was no coincidence that advertisements and vending machines in the subway offered mostly hygiene products, including soap, deodorant, mouthwash, and breath-freshening chewing gum.[35]

Along with olfactory stimuli, acoustic impressions required a high degree of affect control on the part of passengers. Historical documents pertaining to subway noise frequently emphasize that operations were unspeakably loud. Motors, brakes, and rail friction produced a deafening soundscape previously familiar only to factory workers in heavy industry.[36] Above the machine thunder, loudspeakers added screeching announcements from station masters and conductors to the din. The *New York Times* published the first noise complaint just one day after the subway opened: "Much is being said in jest about the new ailment which will result from travel in the Subway. There is no joke, however, in saying that horrible or terrible confined noises in the Subway will injure the hearing of thousands. Wise people will plug their ears with cotton until some inventive genius produces a guard that will do the business. It will be impossible for passengers to engage in conversation, because they cannot be heard."[37]

Beyond the grating acoustics, the complainant also attributes the demise of traveler conversation to the subway's industrial soundscape. As mentioned in the previous chapter, Schivelbusch dates the end of travel conversation back to the railway.[38] However, this only applied to upper-class riders in expensive compartments; there seems to have been no lack of lively communication in the cheaper open cars of third and fourth class on trains. Even the industrial sensory landscape of the subway could not entirely squelch transit conversations. While passenger perception may have been primarily fleeting, blasé, and distanced, the isolation and individuality of urban subjects also facilitated new forms of social interaction.

For Simmel, travel conversation is a unique form of social interaction that occurs only in transit. It can be extraordinarily intimate and candid when suddenly one finds oneself speaking with great familiarity to a total stranger.[39] Simmel lists three conditions that may encourage such exchange: being outside of one's usual milieu, sharing momentary impressions and experiences with others, and knowing that parting is imminent: "The traveling acquaintance—from the feeling of being obligated to nothing, and of being really anonymous in relation to a person from whom one will be separated for ever in a few hours—often entices one to quite remarkable confidences, giving in unreservedly to the impulse to speak what only experience has taught, through their consequences, to control."[40]

The temporality of the encounter underscores the here and now. Travel conversation is not characterized by anonymity and distance, but rather determined by these factors. Taking place in the encapsulated, deterritorialized subway car, conversation can become intense precisely because it is transient, inconsequential, and may be cut off at any moment.

For some passengers, involuntarily overhearing other people's conversations resembles the "tyranny of intimacy" that Richard Sennett has diagnosed as characteristic of the deterioration of twentieth-century public conduct.[41] The ear picks up signals from its surroundings more or less unfiltered. Furthermore, as Simmel points out, "hearing is supra-individualistic in its nature: all those who are in a room must hear what transpires in it, and the fact that one picks it up does not take it away from another."[42] With no escape from the auditory environment, one is at the mercy of the noise of the city and its traffic.[43] Involuntary participation in the conversations of others—the aversion it often triggers— is a constitutive element of the subject form of the passenger. As a result, one often prefers to move silently through the territories of transit.

One sound, however, was absent from the subway's often unspeakably loud soundscapes during the first years of operation, namely, music. Anthropologist Susie J. Tanenbaum has shown that street music was a common part of New York's public life going back to the colonial era.[44] Music flourished in the streets, parks, and squares of New York during the years following waves of immigration around the turn of the century, but it was explicitly prohibited on the elevated trains. And although music was a big part of the opening ceremony for the subway, it was banned from the system the very next day.[45] In the 1930s, despite his otherwise liberal reputation, Mayor La Guardia banned live music and street performances throughout the city. The ban was rigorously enforced for more than thirty years. Tanenbaum writes that music was banned from subway stations and cars primarily in order to keep passengers circulating smoothly through the system.[46] With auditory impressions fading to the background, visual stimuli came to the fore, generating new techniques of perception. Passengers

143

mobilized specific vision strategies. Gazing, looking the other way, and staring became instruments of social control.

STRATEGIES OF THE GAZE

The First Commandment: Thou Shalt Not Stare.[47]

—*Jim Dwyer*

As was the case for auditory and olfactory impressions, complex dispositifs of visibility and legibility emerged in the territories of the subway as well. These were also subject to technical and situational dynamics and changed considerably over time. As cultural theorist Regula Burri has noted: "From this perspective, the act of looking becomes a historical, cultural phenomenon because every epoch has had a unique relationship to the sense of vision and has developed its own regime of looking, that in part is shaped by developments in technology."[48] If this is true, we should find that specific sensory regimes and economies of attention emerged among subway passengers of the machine age. There is indeed abundant evidence of new visual techniques, which can be split into two categories. One includes learning how to interpret material semiotic systems, such as informational signs, directional posts, and advertisements. The other includes passenger practices of seeing and being seen. We will return to the first category below, first focusing on complex strategies of the gaze and associated social interactions among passengers.

As many scholars of urbanism have emphasized, vision has a special status as a mode of perception among the anonymous forms of interaction in modern metropolises.[49] Eyes are highly social sensory organs; they fulfill a unique social function by allowing us to look at one another. The eye is not only challenged by the visual impressions of transit, but also capable of instigating social interplay among fellow travelers. The impression of a mutual glance is personal and intimate. In Simmel's words, "The eye unveils to the other the soul that seeks to unveil the other."[50]

This intimacy is largely due to the fact that one's glance at another person is itself an expression that can only be recognized in the reaction of the other. Looking at someone else becomes a reflection and perception of oneself. Thus, it is an intimate act, to be avoided in fleeting contact with a large number of people for the purpose of keeping one's distance. For Simmel, the "ostrich-like attitude" of sticking one's head in the sand by looking away and avoiding eye contact can be understood as an expression of self-protection and a desire for anonymity on the part of passengers.[51]

In the subway, especially, staring at others was increasingly interpreted as aggressive and rude. The cloak of isolation and anonymity that passengers drew around themselves proved delicate and thin, requiring techniques that would allow them to give their faces the expression of unmoved distance. Only by controlling one's gaze was it possible to observe social distance from others while tolerating spatial proximity. Looking at one another is tolerable with mute acquiescence, but eye contact and conversation are to be avoided.[52] In historical terms, Simmel sees this as a new kind of behavior that was encouraged by the development of modern transit: "Before the development of buses, trains, and streetcars in the nineteenth century, people were not at all in a position to be able or to have to view one other for minutes or hours at a time without speaking to one another. Modern traffic, which involves by far the overwhelming portion of all perceptible relations between person and person, leaves people to an ever greater extent with the mere perception of the face and must thereby leave universal sociological feelings to fully altered presuppositions."[53]

Sociologist Erving Goffman shares Simmel's assessment of the mute gaze, but he goes a step further in identifying the stare as an impolite and transgressive act of public communication. According to Goffman, staring at a fellow passenger violates social norms, but only when that person notices: "The implication is that whereas direct staring is to be avoided, one is free to be exposed in one's staring before those whom one is not staring at. Now a standard defense against being caught staring is to enact a scan that gives the appearance of happening to fall upon the victim the moment he happens to look at the scanner."[54]

From Goffman's detailed account of the rules governing visual interaction, it appears that staring directly at other passengers is considered much more improper and embarrassing than it had been around the turn of the century, according to Simmel's description written more than fifty years earlier. Silently staring has been replaced by cultural techniques of evasion and avoidance of eye contact. This shift suggests that the practice of staring at someone has come to be seen increasingly as an aggressive and invasive act. Evoking shame in its object, the practice of staring can be employed as a means of exercising social control.[55] Goffman offers a powerful example from the biography of a person of short stature describing the offensive stares of fellow subway passengers:

> There were the thick-skinned ones, who stared like hill people come down to see a traveling show. There were the paper-peekers, the furtive kind who would withdraw blushing if you caught them at it. There were the pitying ones, whose tongue clickings could almost be heard after they had passed you. But even worse, there were the chatterers, whose every remark might as well have been 'How do you do, poor boy?' They said it with their eyes and their manners and their tone of voice.
>
> I had a standard defense—a cold stare. Thus anesthetized against my fellow man, I could contend with a basic problem—getting in and out of the subway alive.[56]

Testimonies such as this one vividly demonstrate why the first subway passengers were forced to develop techniques to restructure the economy of their senses. In an effort to minimize or eliminate the duration and strain of transit, they cultivated practices of contemplation and turning inward. More than anything else, reading proved to be the best way to isolate oneself from social and visual assault. This required a certain composure on the part of the passenger, however. Psychologist Susan Saegert emphasizes that the "experienced subway rider, for example, may learn to read on a crowded subway, but not, I suspect, without learning to increase control of his attention."[57]

As Schivelbusch has shown, the practice of reading in transit is at least as old as the railway itself.[58] Half a century before the subway came into existence, reading in the train was already a response to the demise of travel conversation, especially in first-class compartments. Additionally, in contrast to railroads and elevated trains, the subway in its tunnels did not offer panoramic views of the outside world, and reading became even more widespread as a result. The *New York Times* reported that people started reading on the train the very day it went into operation: "'Mark my words,' said the observant citizen, 'the Subway is going to boom the newspaper business. When you get in, there's nothing to look at except the people, and that's soon a tiresome job.'"[59]

This prediction proved correct. In 1919, just a few months after the subway broke the record of one billion passengers per year, the *New York Daily News* was launched as the city's first small-format tabloid. In contrast to the large-format *New York Times*, the compact *Daily News* was easier to handle during transit and very popular among commuters.[60] Making reading part of daily life became a common response to the social and material conditions in subway stations and cars. Within a short time, passengers were directing their attention to newspapers, magazines, and books, which allowed them to shut out the strain of the transit environment by turning inward. Reading material provided a kind of physical protection against stimuli that helped passengers to avoid undesired contact: "The Newspaper is the chador of the subway car, the perfect psychic veil, armor against unwanted intimacy. No one is fully dressed in the subway without something to read."[61] A series of photographs taken in 1946 by the young director Stanley Kubrick foregrounds newspapers as an indispensable daily companion for many passengers (figure 3.2).

Thus, the sensory techniques of industrial awareness already established among early railroad passengers took effect in the territories of the New York subway in heightened form. At the same time, new techniques of sensory modulation and containment developed into central strategies for passenger-subjects to cope with the particular challenges of machine transit, in terms of both perception and social interaction. Observing these new patterns of behavior in the metropolises of the East Coast, Goffman applies the term "civil inattention."[62] Civil inattention consists of communicating acknowledgment of the presence of people in order to subsequently ignore them completely. Common in waiting rooms and elevators as well as in the subway, the practice is usually nonverbal: glancing briefly at one's fellow passengers, sidestepping them and navigating through the crowd without physical contact, standing up to offer a seat to someone who might need it, and so on. On the overcrowded New York subway, turbulent motion made it difficult to avoid bodily contact despite one's best attempts. This was generally not considered impolite, but accepted as part of daily urban life.[63] Such acceptance required modulation of one's sense of touch as well.[64]

Figure 3.2 Life and Love on the New York City Subway. Passengers reading in a subway car. 1946. Stanley Kubrick for LOOK magazine. Museum of the City of New York. X2011.4.10292.30D ©Museum of the City of New York and SK Film Archives.

The remarkable indifference developed by passengers in response to the various sensory challenges of transit serves as evidence of containment, which cultural theorist Hannes Böhringer has identified as an inherent function of a container: "Wherever it lands, a container is a magnet for indifference. Indifference stains its contents and its surroundings. Things inside and things around it disconnect from one another. They seem isolated and strange and themselves become indifferent containers with interchangeable meanings, inner functions, and outer disguises."[65]

The subject code of containerization includes a kind of distanced objectivity, which Lethen has described as a reaction to a urban new culture of mobility: "traffic transforms morality into objectivity, compelling behavior appropriate to function."[66] This is true not only for the impressions received from other passengers, and the consequent reactions of isolation and encapsulation. It also applies to passenger responses to the range of signs and signals spreading all over the subway as soon as it opened.

VISUAL REGIMES

The subway's very first passengers were already confronted with an abundance of visual signals as soon as they entered the system: directions, warnings, advertisements, and many more. Providing passengers with the information they needed to navigate the system, signs also called upon them to behave in a certain way. As we will see, they were also a means of subjectifying passengers. Combining all of these functions, such sign systems make up a visual regime. As cultural theorist Doris Bachmann-Medick points out, this term brings the subjectifying qualities of perception into focus: "It highlights the overall conditional connection of visualizations in culturally specific techniques and practices, and in the societal power relations embodied in a glance, for example."[67]

The various visual regimes installed throughout the system over the course of the machine age, closely related to historical semiotic and material dynamics, shaped the perception and practices of passengers in multiple ways. Signs connected information and knowledge to the real movements and interactions of passengers, constituting subway spaces as hybrid settings linking architecture and signs, texts and territory, into complex interconnected structures. Sociologists Jérôme Denis and David Pointille highlight this as well: "The signboards or the street nameplates can thus be seen as utter components of the performance of modern public space. Urban signs both order physical spaces and configure the action of their dwellers."[68]

This is achieved by providing specific scripts that define social and semiotic order in transit, regulating passenger morals and conduct. Madeleine Akrich has shown that scripts configure both users and the space in which they move.[69] This also applies to the visual regime of the subway. With its directions, maps, and advertisements, the subway's regime of signs coordinated a host of heterogeneous inscriptions that addressed passengers in very different ways, running the gamut from bans on spitting and advice for subway etiquette to patriotic appeals and consumer cues. The latter in particular often aroused angry protest.

To do justice to this situation, it is necessary to classify the sign systems used in the subway.[70] The first category of signs primarily served orientation purposes, such as transit maps, direction signs, and signs identifying stations and lines. Especially in the first decades of subway operation, these often-confusing signs were a frequent topic of complaint. Signs were not well coordinated among the subway's three different operating companies, and sometimes even contradicted one another, much to the frustration of passengers. Secondly, there were warnings and signs communicating prohibitions, informing passengers of forbidden or dangerous behavior, schedule changes, delays, and other disturbances. The Subway Sun poster series played a special role in these attempts. Issued by the hundreds for many decades, these posters not only conveyed news from subway operators, they also tried to educate passengers by promoting the values of health, patriotism, and courtesy. Finally, there were advertisements all over the place, promoting everything from beverages to modeling agencies. Despite strong initial resistance from passengers, advertisements aimed to integrate the sphere of transit into the consumer culture of the expanding city.

The primary function of subway signs is *discursive*: their purpose is to communicate with passengers through prohibitions, warnings, or directions. From this perspective, signs operate first and foremost as text, introducing discursive order into the passenger environment.[71] With the spatial order of the city blending into the textual order of transit, schedules, signs, and maps became indispensable instruments for the navigation—and imagination—of the urban underground.

In addition to their discursivity, the scripts built into the subway's system of signs were also *performative*. Not only did they provide information, but they also aimed to shape passenger behavior in order to ensure smooth circulation throughout the system. This idea was based on an understanding of passengers as reactive subjects who could be controlled through such a regime. The behavioral scripts built into the subway's sign systems were also meant to structure interactions with the material environment, other passengers, and subway staff.

Finally, the subway's visual regimes were *normative* in nature: as powerful instances of subjectivation, they communicated prohibitions, recommendations, and behavioral norms. They presented an image of the ideal passenger that went far beyond merely dealing appropriately with subway machinery. The Subway Sun and commercial advertisements addressed people in the subway not only as system users, but also as taxpayers, consumers, and patriots, propagating behavior in accordance with general societal norms and values. Additionally, the ad campaign of the Miss Subways contest presented female passengers from the working class as both ideal passengers and objects of the male gaze.

While the scripts of the subway's visual regimes were intended to structure the discursive perceptions, performative practices, and normative subject positions of passengers, their validity and effectiveness was by no means assured. In order for the imperatives inscribed into signs to be acknowledged and followed, they had to be backed by authority and legitimacy. On the one hand, this resulted from passengers' strong need for safety and orientation. On the other hand, given the visible authorship of subway operators, signs were endowed with imperative force that could be enforced by the police and courts if necessary. This was particularly true of ubiquitous instructions and prohibitions displayed in stations and cars. Only by acknowledging the authority of such signs and following their scripts could passengers integrate themselves into the social and semiotic order of the subway. The fact that they did so voluntarily was due in large part to the experience of foreignness and disorientation, as Gillian Fuller has emphasized: "Graphical signage cools down the anxiety of unfamiliar terrains and replaces it with a familiar authority—the sovereign structures of transit systems."[72]

151

In order for sign systems to provide orientation and self-assurance, subjects have to be able to decode their discursive, performative, and normative functions. A number of preconditions must be fulfilled before one can decipher and interpret signs, maps, and instructions. One must have a command of the language as well as an understanding of pictograms, which further depends on a basic idea of how modern urban space is organized. At the same time, one must have a sense of the technical functioning of subterranean mass transit in order to effectively relate these signs to the city above. The visual regime can only fulfill its performative and normative functions when all of these preconditions are met. The first subway passengers struggled to master these cultural techniques. At the same time, it was also no simple task to establish an efficient, standardized system of orientation. This was especially true of the New York subway, but reports from the early phases of subway operation in Paris, London, and Berlin also contained many descriptions of passenger confusion and disorientation.[73] In addition to hell, the labyrinth became a dominant metaphor for the experiences of the urban underground.

SEMIOTIC LABYRINTHS

In the notes for his *Arcades Project*, Walter Benjamin writes: "The city is the realization of that ancient dream of humanity, the labyrinth."[74] If this was true of nineteenth-century Paris, humanity came even closer to the realization of the dream with the construction of underground railways in Western metropolises over the following decades. With its multiple operating companies, dozens of lines, hundreds of stations, and elaborate network of local and express trains, the New York subway system in particular generated confusion and disorientation among passengers. At first it was the newfound speed and unfamiliar terrain that caused passengers to lose their sense of where they were—and where they were going—as reported for example by the *New York Times*: "'The funniest thing to me,' said an employee of the Subway at Grand Central, 'is the questions the people ask. Summarized, they are: Where am I at? They want to know whether they are on the east or west side of the city, which is up town and which is down.'"[75]

A wide variety in the signs addressing passengers was one factor in the confusing experience of deterritorialization. Remarkably, the imperatives of efficiency and standardization that governed many aspects of subway expansion generally did not extend to the design of maps, directions, and signs. While the paradigms of the machine age were implemented in the design of subway cars and stations, signs and maps were anachronistic artefacts in a system alleged to have the world's most advanced technology. Right after the system went into operation, passengers began complaining that signs were confusing and chaotic, and these complaints only increased in the decades to come.[76] In the early years, a range of additional sign systems were installed, including notices, warnings, and advertisements. Often placed directly alongside the signs that were meant to provide orientation, this multiplicity only added to the system's labyrinthine effect.

To facilitate orientation, the system's first architects, Heins & LaFarge, sought to give each station its own look. As a common design principle, they used large-format ornamental ceramic mosaics with inlays for the station name and direction of travel.[77] Although this design struck people as impractical and anachronistic from the outset, the next architect, Squire J. Vickers, was reluctant to abandon it entirely. In hope of reducing confusion, he did replace the often playful lettering and ornamentation with clear, unembellished fonts. Meanwhile, the thicket of signs continued to grow with the expansion of the system and the addition of the BRT network. When it came time to design the IND, Vickers dropped the ornamentation entirely, creating a complex visual code that assigned a specific color to each station. The reasoning behind the color scheme remained a mystery to most passengers, however.

A 1907 report from the City Club of New York shows how confusing and overwhelming the subway's visual schemes were for passengers. According to the report, the signs in cars and stations were too few and poorly placed, with practical consequences: "The insufficiency of these is shown by the number of questions that are asked of the guards, resulting in delays to the trains."[78] The City Club's logistics experts saw communication between subway passengers and staff as an indicator of how poorly the sign system fulfilled its purpose. They had in mind an ideal passenger who would be so competent in decoding and following signs that social interaction would be entirely unnecessary. In reality, however, passengers needed to actively exchange information in order to navigate the underground labyrinth; signs not only differed significantly in design, but they also contradicted one another (figure 3.3). Directions, station names,

Figure 3.3 Sign systems in the New York City subway in the 1940s. Numerous overlapping displays contributed to the labyrinthine character of the system, as did constantly multiplying information. Courtesy of New York Transit Museum.

and abbreviations were frequently ambiguous, and signs were often half-heartedly pasted over or repainted.[79]

The subway companies tried many times to devise guidelines for the design and placement of signs, but these were often ignored by station masters and maintenance workers. Instead, subway staff often improvised signs or posted illegible handwritten notices. These drove not only order-loving architects and system engineers to desperation, but millions of passengers as well.

Even transit maps were of little use in this labyrinth. They were unavailable in most stations, and more than ten years passed before subway companies began distributing pocket-size maps to passengers. Hotels, department stores, and travel agents began printing and distributing subway maps instead.[80] Years later, when the IRT and BRT finally began printing their own maps, the layouts differed considerably. In 1919 the BRT distributed a highly abstract transit map with no landmarks, streets, or scale details for orientation, showing only an oval outline of Manhattan with BRT lines and stations (figure 3.4).[81] In contrast, the IRT's version showed only a simple street map to which their subway lines had been added (figure 3.5).[82]

These early transit guides reveal a central dilemma of all map design: how to combine accuracy with abstraction.[83] To make matters worse, neither of these maps showed Manhattan in its proper geographical position, but lying on its side, with the west at the top. This did not change until 1924, when the BMT issued a clearer map that showed Manhattan vertically from north to south. Adding another layer of complication, the different subway companies focused on representing their own networks, with the lines of the other operators sketched in a rudimentary fashion or ignored all together.

Meanwhile, Europe's underground transportation systems had made significant progress in the graphic design of sign systems and transit guides. The Paris Métro and the London Underground came up with new ways to depict the abstract information that passengers needed, making great strides toward modernization and standardization in visual regimes.[84] In 1932 electrical engineer Harry Beck devised a map for the London Underground that marked a milestone in the history of information design (figure 3.6).

Figure 3.4 The first transit map of the BRT from 1919.

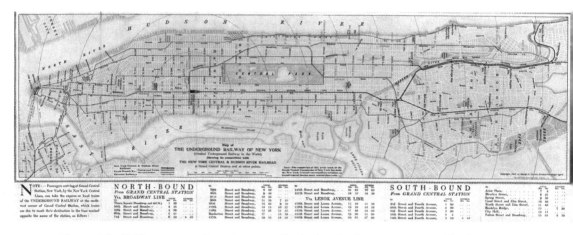

Figure 3.5 IRT transit map from 1905. Initially displayed only on trains, maps like this were not distributed to passengers until the 1920s.

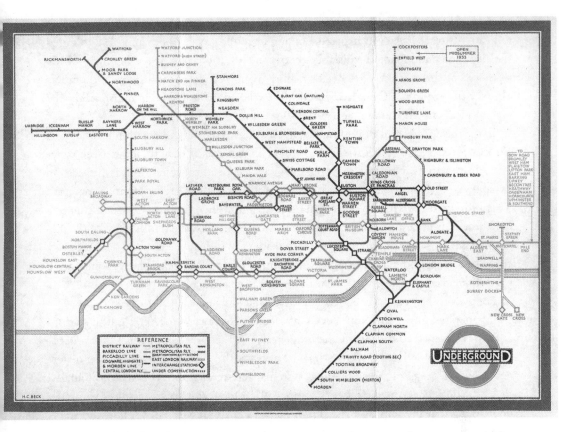

Figure 3.6 Harry Beck's iconic 1933 transit map, presenting London as a complex circuit system. This map was an immediate success with the passengers of the London Underground. ©TfL from the London Transport Museum collection.

Beck came up with a revolutionary layout using an austere geometrical grid that radically distorted or even negated London's ground-level topography. He enlarged the city center and made the distances between all stations consistent. Seminal for the aesthetics of the machine age, Beck's map remains a model for information designers all over the world: "although it was not the first, nor the last, it is indisputably the archetypical Underground map."[85]

In the meantime, almost nothing changed in the design of information for the New York subway. New York's street grid might have made it easy to adopt Beck's layout, but the subway operating companies saw no need to do so. Yet as the system expanded, particularly in the 1930s, passengers began asking for more map material. This prompted the IRT to rework its transit maps, incorporating Beck's principles half-heartedly at best. Nonetheless, from this time onward, the transit maps of the different companies became an indispensable element in the subway's visual regime. They could soon be found in every station and subway car, and the portable version became part of the everyday inventory of millions of passengers.

Following the economic crisis of the 1930s, and even after all three systems were conglomerated under the New York City Transit Authority in 1953, signs in stations and cars remained confusing and heterogeneous. Although the principles of standardization and rationalization were otherwise at work throughout the system, the different design practices of the component systems were retained for cost reasons. Finally, in 1957, a designer named George Salomon (1920–1981) composed an angry pamphlet demanding that the subway's visual regime be radically reworked.[86] A burning advocate of the spirit of standardization and increased efficiency, Salomon sharply criticized the look of signs and maps, listing their deficiencies in detail, from the muddle of different lettering, codes, and symbols to contradictions and the use of the same name for more than one station or line (figure 3.7).[87]

According to Salomon, the nuisances culminated in a visual cacophony that was not only aesthetically catastrophic, but also incomprehensible to anyone but experts.[88] For people from out of town in particular, subway maps were "the most bewildering thing in a bewildering city."[89] Even for passengers with years of experience, the confusion of the sign system was a constant source of frustration. Salomon compiled a list of the inconveniences that the semiotic chaos caused subway passengers: "Loss of time and

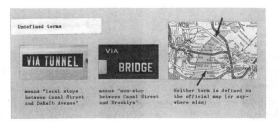

Undefined terms

VIA TUNNEL

means "local stops between Canal Street and DeKalb Avenue"

VIA BRIDGE

means "non-stop between Canal Street and Brooklyn"

Neither term is defined on the official map (or anywhere else)

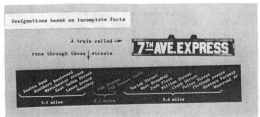

Designations based on incomplete facts

A train called → 7TH AVE. EXPRESS

runs through these streets

Boston Road
Southern Boulevard
Westchester Avenue
East 149th Street
Lenox Avenue
Broadway

9.8 miles

7th Avenue
7th Avenue South

5.2 miles

Yorick Street
West Broadway
Park Place
William Street
Clark Street
Fulton Street
Flatbush Avenue
Eastern Parkway
Nostrand Av

9.4 miles

Needed information lacking

Agents' makeshift signs result from lack of official channels for determining the public's information needs

DOWN TOWN

DOWN TOWN

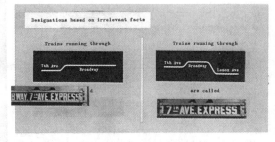

Designations based on irrelevant facts

Trains running through

7th Ave
Broadway

B'WAY, 7 AVE. EXPRESS

Trains running through

7th Ave
Broadway
Lenox Ave

are called

7 AVE. EXPRESS

One name for several things ...

86TH ST
86th St. & Central Park West

86TH ST
86th St. & Broadway

86TH STREET
86th St. & Lexington Ave.

(and there are two more stations called 86th Street in Brooklyn)

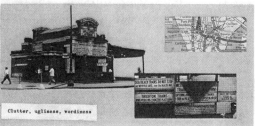

Clutter, ugliness, wordiness

SEA BEACH TRAINS DO NOT STOP

BRIGHTON TRAINS

... and several names for one thing

BROADWAY JUNCTION = BROADWAY EAST NEW YORK

EASTERN PARKWAY = FULTON-EAST N.Y.

Use of inappropriate styles and materials

BMT LINES
UPTOWN TO 59TH ST QUEENS PLAZA ASTORIA & FLUSHING
DOWNTOWN TO WHITEHALL ST S'FERRY BROOKLYN & CONEY ISLAND
VIA SEA BEACH WEST END BRIGHTON BEACH & 4TH AVE. LINES.

Long lines of text are hard to read, especially in narrow capital letters

CANAL STREET BROADWAY

Mosaic lettering cannot be added to without disfigurement

Figure 3.7 Montage of some of the elements of visual chaos identified by Salomon. In his view, if the labyrinthine character of subway was to be reduced, the main task was to standardize these elements. George Salomon: "Out of the Labyrinth: A Plea and a Plan for Improved Passenger Information in the New York Subways," ca. 1957. Courtesy of New York Transit Museum.

energy in retracing steps, cost of rides taken in error, inconvenience of asking the way, fear of getting lost, inconvenience of being blocked by persons not sure of their way, nervous strain, sometimes, danger as a result of all these."[90]

Salomon imagined passengers so overwhelmed and frightened by inconsistent signs that they were ultimately a danger to themselves and others. The only solution was to thoroughly redesign the entire visual regime of the subway, radically standardizing the terminology used for stations and routes as well as the design, material, and placement of signs and maps. Rigorous implementation of the aesthetic principles of efficiency, clarity, and consistency were required to turn the "bewildering puzzle" into an "intelligible, easily remembered system."[91] Yet despite his urgent plea and alarmist tone, Salomon did not make much of an impression on the Transit Authority. They did, however, allow him to design a new version of the transit map, published in 1958 (figure 3.8).

Salomon's map was a revelation. Not only was it the first to show all of the lines now subsumed under one system, but it also offered a radical interpretation of the city as a machine system.[92] Influenced by Beck's principles, Salomon distorted New York's topography in favor of a better representation of subway routes, and consistently eliminated all ground-level landmarks like parks and streets. As a kind of imaging device for machine aesthetics, the map presented the city as an abstract and complex set of circuits. While it required sophisticated prior knowledge of modern urbanity and machine logic on the part of passengers, Salomon's map was so successful that it was reprinted several times until 1972.

PROHIBITIONS AND WARNINGS

In addition to sign systems devised to facilitate orientation, passengers were confronted with a growing number of signs prohibiting certain kinds of behavior in the subway. These signs aimed to influence user practices and establish standards of behavior that would promote efficient mass circulation. The exact number and precise distribution of different prohibition signs cannot be reconstructed; aside from a few references and some photographs in the Transit Museum Archive, few traces of them remain. Nonetheless, it is possible to identify two major categories of warning signs. Some were

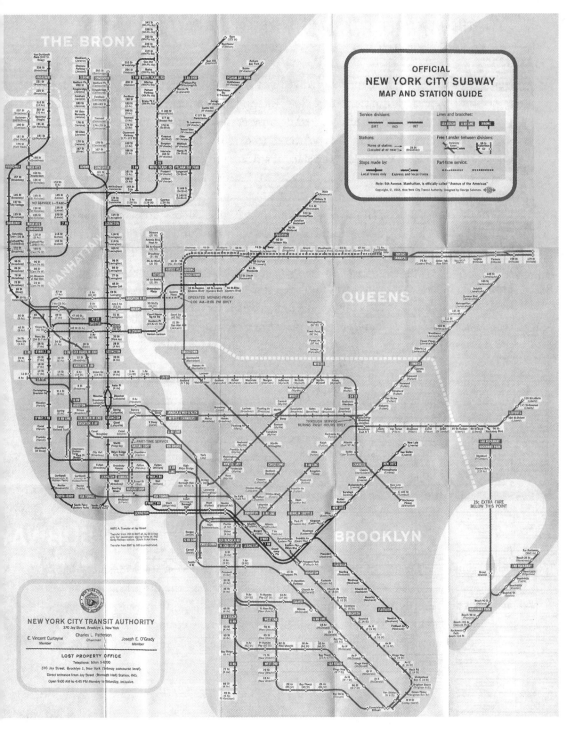

Figure 3.8 George Salomon's transit map for the New York subway from 1958. Courtesy of New York Transit Museum.

signs that warned of the dangers involved in using the system carelessly. These pertained for the most part to stepping onto tracks and into tunnels. Especially in the first years, a number of deaths had occurred because passengers either underestimated the speed of the trains, or because they ventured to explore the new underground terrain for themselves. Hood writes that the alarming death toll eventually led to the installation of large signs warning in clear letters: "All Persons Are Forbidden to Enter upon or Cross the Tracks."[93]

The other group of prohibitive signs from the first decades of system operation dealt with passenger habits that were incompatible with the subway's hygienic regimes. These signs directed passengers not to litter, smoke, or spit (figure 3.9). Smoking and spitting were apparently a serious problem in the new subway. Smoking had been prohibited in elevated trains and omnibuses since the late nineteenth century, but the regulation was not generally observed. In the years after the subway opened, smoking remained an everyday practice in all of New York's transit systems.[94] Complicating the situation, the per-capita consumption of cigarettes in the United States increased more than sixteen-fold between 1910 and 1930.[95] Over the decades, the smoking ban appears to have been enforced to some extent in subway cars and underground stations, with the threat of draconian punishment by subway staff.

However, for subway management, the habit of spitting caused greater alarm than smoking. In order to stop spitting—a behavior exhibited for the most part by male passengers—a large number of variously worded signs were placed throughout the subway well into the 1940s.

Don't Spit!

Just a few weeks after the subway opened, the *New York Times* published an article titled "One Real Danger of the Subway," expressing indignation at a widespread habit among passengers: "We refer to the nasty habit of spitting, the disgusting evidences of which are becoming more manifest as travel increases [. . .] If there is any place in the city where our severe ordinance against promiscuous spitting needs to be rigidly enforced, it is in the closed Subway and I hope the police and Magistrates will show no mercy to the unforgivable offenders."[96]

CHAPTER 3

Figure 3.9 Prohibition sign in the subway, ca. 1937. Courtesy of New York Transit Museum.

This strong emotional reaction to spitting in public, and the coding of the habit as disgusting, present a symptomatic challenge for the historical reconstruction of passenger practices. If prohibition signs and the threat of punishment came in response to a practice that was so widespread at the time, how did this practice come to be seen as a problem? Until the late nineteenth century, spitting was not considered unhygienic or uncivilized in Western cultures.[97] In many societies, spitting in public is still a common habit, especially among men. In order to understand why the prohibition against spitting was inscribed into the subway, it is necessary to take a closer look at the history of the battle against tuberculosis in North America.

After Robert Koch scientifically demonstrated the existence of the tuberculosis bacterium in 1882, disproving the previous belief that this was a hereditary illness, it became clear that the epidemic spread through germs. Saliva was quickly identified as the primary carrier of infection, and American health authorities mobilized massive resources to curb the spread of TB. Despite awareness campaigns and appeals to the public to change habits of personal hygiene, tuberculosis ranked as the most frequent cause of death among the American population around 1900. As medical historian Jeanne E. Abrams has shown, all across the country panicked authorities began issuing ordinances against spitting in public.[98] New York, America's largest and most densely populated city, represented the epicenter of infection. New York had issued an ordinance against spitting as early as 1896. Less than twenty years later, almost two hundred American cities followed suit with bans on spitting, threatening people with high fines and prison sentences.

These ordinances, however, often caused tension between the belief in individual freedom and the need to protect the population. For most health experts, the severity of the disease justified subordinating individual claims to freedom to the primacy of the common good.[99] Spitting constituted an act of ignorance and negligence, and soon it was further linked to vulgarity, repugnance, and primitiveness. Charged with moral imperatives, the crusade to ban spitting went beyond a purely pragmatic approach for the protection of the population. Geographer Matthew Gandy has pointed out that these campaigns also perpetuated prejudice against poor people and the working class: "The control of TB amplified middle-class antipathy toward the 'lower classes' and heightened anxieties over immigration and racial mixing."[100] This was particularly true for New York City, where in 1911 half of those infected with the disease came from lower classes, often becoming poorer still as a consequence. Yet although spitting in public was seen as a typical habit of the working class, whose members were already stigmatized as vulgar and parasitic, in reality the practice of spitting was widespread in all ranks of society.[101]

While around 1900 spitting was so widely accepted in New York that even the police barely pursued violations, things changed after the subway opened. The territories of industrial mass transportation were seen as sites of potential infection. In order to check the spread of TB as quickly as possible, it was urgently necessary to reeducate passengers.[102] When the subway opened in 1904, Dr. Thomas Darlington from the New York health department warned: "In the subway the danger of disease from expectoration is fully as great, if not greater, than on surface roads. When the sputum dries the germs it contains are free and become dangerous. We cannot have them blowing about the tunnel. A very close watch will be kept to see that this habit does not get a foothold on the underground system."[103]

The implementation of warnings and prohibition signs in the subway supported a biopolitical regime that addressed passenger-subjects in multiple ways. It once again reflected the idea that the behavior of passenger masses could be altered, in this case by implementing signals. It also suggested that each passenger could be a potential hazard for public health. This notion gained visibility in the debates over hygiene that arose again immediately after the subway opened. Chemistry professor Charles F. Chandler had already proved that the quality and flow of oxygen through the system was beyond reproach.[104] Nonetheless, new objections sprung up that the subway was an ideal environment for infection by germs of all kinds.[105]

In response to these concerns, authorities commissioned a more comprehensive study. Renowned sanitation engineer Dr. George A. Soper (1870–1948) led a large team working with the newest laboratory methods to investigate the system for all kinds of germs. Four years later, Soper's more than 300-page study offered an all-clear signal. His analyses proved beyond a doubt that there were only half as many microbes per cubic meter in the subway as in the urban areas above ground.[106] And yet, while Soper had once again demonstrated that the subway's hygiene regime was excellent in itself, he warned that passengers posed the greatest threat of contamination to the system. They brought germs into the otherwise hygienic underground territory, where germs and diseases found a perfect environment to spread.

The combination of an overloaded system and the unhygienic practices of passengers made the subway into a potential focus of infection: "It is practically certain when great crowds are packed together, as they often were in some stations and most cars, that dangerous bacteria are, at least occasionally, transmitted from person to person. An obvious feature of this danger lies in the fact that people talk, cough, and sneeze into one another's faces at extremely short range under such circumstances."[107] The motif of threatening, infectious passenger masses reemerged in this context, calling for discipline through procedures of appeal. In the eyes of the subway companies, there was hardly anything to be done about the overloading of the system, and so it was necessary to try to control the practices of passenger by means of prohibition signs and serious penalties (figure 3.10).

As a result of the health department's campaign, by the end of the 1920s passengers had integrated "Don't spit!" into their system of norms. Medical historians still do not agree on whether the sharp decline in the rate of tuberculosis that began in the 1930s was related to successful implementation of the ban on spitting, or to an overall rise in the standard of living.[108] And while the original justification for introducing the ban has been largely forgotten, the conviction persists that spitting in public—including in the subway—is disgusting and obscene. As the signs prohibiting spitting gradually disappeared even before the machine age came to an end, warnings about smoking, other forms of pollution, and other dangers remained an everyday sight for passengers as part of the system's visual regime.

THE SUBWAY SUN

Besides notifications and signs, there was another visual element that played an important role in the early twentieth-century subjectivation of passengers: the Subway Sun, a newspaper-style series of posters that offered information and promoted the positions of the IRT. In 1918 the company began hanging these posters in every car, and from then on they were a permanent fixture in the system's visual regime. Journalist Ivy L. Lee (1877–1934), who was in charge of public relations for the IRT—at the

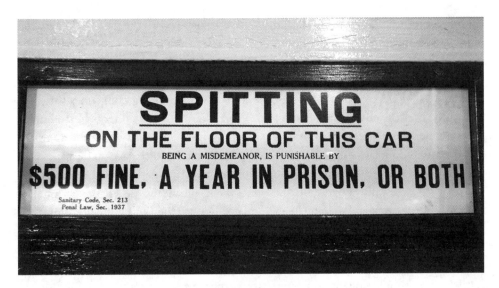

Figure 3.10 One of the last instances of the ban on spitting in a subway car from 1939. Courtesy of New York Transit Museum. Author's photo.

time a relatively unfamiliar responsibility—had come up with the idea.[109] Lee saw the Subway Sun and another series of poster-series called the Elevated Express as the ideal instrument to present the IRT as a likeable, down-to-earth company while paving the way for fare increases (see figure 3.11 and figure 3.12).[110]

Despite the serious financial losses that all subway operating companies suffered, especially after the end of the First World War, it was politically impossible to implement an increase in fares. Any politician arguing in support of higher subway prices could expect public protest and lost votes. Discouraged, Lee also gave up the campaign after a few years, but Subway Sun posters remained part of the visual regime until 1932. In 1947 the Subway Sun returned once more, providing passengers with information on the system and shaping behavioral norms until the 1960s.[111] In hindsight, the Subway Sun's strategy for communicating with users appears visionary. These posters made it possible to address passengers directly, educating them regarding details of the system, the immense cost of operation, and much more. Lee's preference for facts and statistics over simple slogans also demonstrates the primacy of logistical knowledge at the time.

The images and texts of subway posters confronted passengers with a heterogeneous range of effective appeals, demands, and codes of conduct.[112] Furthermore, they portrayed an idealized version of the typical passenger whose behavior rose above simply conforming to the rules of the subway. This was already obvious in the Subway Sun's third poster, titled "The Call to War," calling all male passengers between the ages of 18 and 45 to register for military duty (figure 3.13). This was the first in a series of posters that addressed passengers as patriots (figure 3.14).

The call to enlist was only one appeal among many made by the almost 400 posters. Passengers were encouraged to visit the zoo, go to the beach, and exercise regularly. They were reminded of their civic duties, such as the timely submission of tax returns (figure 3.15). Many posters also highlighted the advantages of the subway over the increasingly popular automobile (figure 3.16). Additionally, posters appealed to decency and courtesy, asking passengers to be friendly and polite (figure 3.17), to keep aisles clear, and not to hold doors open (figure 3.18).

Through recommendations and imperatives, subway posters spread the notion of an ideal passenger who conformed to more than just the technical demands of the

Crowds, Taxes and Rents

New York was never so congested as now. No wonder rents are going up!

Two million strangers, the papers say, were in town to see the Twenty-Seventh parade. Half a million commuters and other outsiders are here every day.

Every one of them riding on the Subways and Elevated is carried at a loss, and New York City must help pay the bill.

The City has $250,000,000 invested in Subways and the 5-cent fare does not pay the cost of the service these strangers receive.

Unless the fare is increased to give the City—out of Subway earnings—interest, etc. on its investment (about $13,000,000 a year,) it must be raised by taxation.

INTERBOROUGH RAPID TRANSIT COMPANY

Theodore P. Shonts
President

Figure 3.11 *Subway Sun*, volume 2, no. 8, May 1919. Courtesy of New York Transit Museum.

Figure 3.12 *Subway Sun*, volume 23, no. 19, May 1920. Courtesy of New York Transit Museum.

Figure 3.13 *Subway Sun*, volume 1, no. 3, September 7, 1918. Courtesy of New York Transit Museum.

Figure 3.14 *Subway Sun*, no. 34A, ca. 1924. Courtesy of New York Transit Museum.

Figure 3.15 *Elevated Express*, volume 6, no. 9, April 1923. Courtesy of New York Transit Museum.

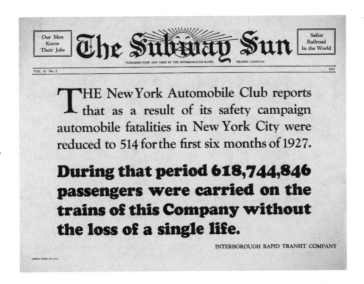

Figure 3.16 *Subway Sun*, volume 10, no. 3, ca. September 1927. Courtesy of New York Transit Museum.

Figure 3.17 *Subway Sun*, no. 52, ca. 1924. Courtesy of New York Transit Museum.

Figure 3.18 *Elevated Express*, volume 11, no. 32, 1928. Courtesy of New York Transit Museum.

system. The subway educated passengers in matters of patriotism, civil duties, and proper conduct. Creating a link to the rules and norms of the city above, these appeals also contributed to the coding of the subway as public space. Among its many appeals, the Subway Sun did not appeal to passengers as consumers, but it was not long before passengers became the target of subjectivation through advertisement.

CALLS TO CONSUME

On opening day, newly minted passengers were puzzled by the many holes they detected in the beautiful mosaics that adorned the walls of the subway stations. The next day, they saw advertisements for products of all kinds.[113] Remarkably, passengers as well as the urban elites were outraged at the sight.[114] High-ranking public officials found the marketing "cheap and nasty."[115] Within days subway operators were inundated with letters of protest and threats of boycott.[116] It soon became clear that the IRT had not only concealed its intent to display advertising on subway walls from the public, it had not even informed the architects of the system. When this news spread, local art associations called upon passengers to destroy the advertisements, claiming that they had all the right in the world to defend themselves against such barbarism—with force, if necessary.[117]

In addition to advertising, passengers were infuriated by all forms of consumer offerings in the subway, from kiosks with newspapers, flowers, and other goods, to vending machines with chewing gum and soft drinks. The public outcry finally led Mayor McClellan to order the IRT to remove all of these elements from the system within forty-eight hours. If his ultimatum were not met, the mayor threatened to send an "axe brigade" to chop the stands and signs to pieces.[118]

Banned from subway space temporarily, eventually a court ruling decided in favor of the IRT, and vendors returned. By contract, city officials had bound subway management not to set up any objects that would obstruct the visibility of station signs, but this regulation did not eliminate advertising.[119] The IRT proceeded to place advertisements everywhere except directly over existing subway signs. In terms of jurisdiction, this practice was acceptable, and advertising subsequently found its way not only into stations, but also into car interiors, entrance areas, and transit maps as well.[120]

Retrospectively, public outcry against advertising in the subway might seem strange, especially as it was an everyday sight in the aboveground city by 1900.[121] This response demonstrates the effective coding of this infrastructure as a civil monument at the beginning of the machine age. The territories of the subway represented an almost sacred space, and the profanation of this space with advertisements was seen as act of shameless barbarism. At first glance, it might appear that this conflict primarily involves the social significance and aesthetics of the subway, but it can also be seen as a symptom of the profound change taking place in the idea of the passenger. Not long before, the passenger had been anticipated as the moral hero of a golden age to come; now the passenger was merely a potential consumer of everything from soft drinks to laxatives.[122]

The installation of advertising in the subway and the almost hysterical reaction against it reveal a rupture between bourgeois and post-bourgeois subject culture. By the end of the First World War at the latest, "consumer culture" was a dominant topic of sociological analysis, and this tendency had already been visible since the turn of the century.[123] Many intellectuals of the times were wary of the spreading practices of mass consumption, in part because these practices emphasized the symbolic character of goods over their utility. Thorstein Veblen's 1899 study, *The Theory of the Leisure Class*, remains one of the most powerful critiques of this development.[124] Employing the concept of "conspicuous consumption," Veblen explores and attacks consumer practices aimed solely at demonstrating social status.[125] His work provided the blueprint for those who upheld more traditional values to diagnose modern consumer culture around 1900 as decadent, artificial, unproductive, and hedonist. Reckwitz provides a concise summary: "From a bourgeois perspective, the consumer must appear as a parasite."[126]

Despite resistance, advertising in the subway quickly became a banal, everyday sight, no longer provoking emotional outbursts among passengers. The photo collections in the New York Transit Museum Archives offer powerful testimony of the rampant spread of advertisements in the subway system in its early decades.[127] From billboards in entrances and stations to the "car cards" in subway cars, passengers were surrounded by advertisements promoting an ever-growing assortment of products and events: from sports events, theater productions, and movies to personal hygiene products, candy, cigarettes, beer, and other beverages. These advertisements offer a powerful impression of the fast-growing world of commodities for American consumer society in the first half of the twentieth century. They also show how passengers had transformed into a valuable target group for a cornucopia of products within just a few years. Passengers now appeared as subjects whose needs could be shaped by advertisement. The semantics of the posters—*Live the simple life! Enjoy! A positive cure for constipation!*—spoke directly to passengers, addressing their presumed needs, desires, and aspirations.

Advertising messages thus added another semiotic level to the territories of the subway by recoding the passenger as the subject of consumer appeals. They illustrate how quickly the world underground had adjusted its visual regimes to match those of the city above. To the extent that the subway was aesthetically assimilated as an integral element of New York, it lost its allure as a civilizing achievement and became a trivial part of everyday urban culture. Practices of flanerie also shifted into the underground structures of the subway, marked in part precisely by the eager study of advertisements.[128]

The stations and cars of the subway—as well as the public spheres of the city above—became evermore enriched with complex surface stimuli. As the machine age progressed, it became an everyday practice of urban subjectivity to decipher them in passing. The worlds of perception emerging for the inhabitants of Western metropolises were for the most part oriented around visual impressions: illuminated signs, fashion, motion pictures, shop windows, and department store displays.[129] As Reckwitz emphasizes, aesthetic experiences were woven into the aggregate of consumer activity, as seeing was often followed by selecting and then consuming the advertised commodities.[130] In a way, this applied not only to the consumption of goods advertised in the subway, but also to the use of the subway as one means of transportation within a range of mobility infrastructures.

The constitution of the passenger-subject as the addressee of a Fordist mass culture of consumption was astonishingly compatible with the coding of the same subject though knowledge defined by logisticians and technocrats (see chapter 2). Both modes of subjectivation shared certain assumptions. In the eyes of logistics experts and advertisers, passengers and consumers were primarily constituted by their capacity to be directed from the outside. The idea of the masses was active here; passengers were seen first and foremost as potential consumers of mass-produced goods. The strategies that marketing experts used to determine the characteristics of consumers also overlapped with the logistical strategies of engineers. Market research, which began emerging in the first half of the twentieth century, was keen on using survey methods from logistics. With some modification, such quantifying methods put market researchers in a position to define standardized target groups, calculate distribution, and predict market dynamics. The individual consumer could thus be addressed as a functionally equivalent constituent of mass consumer culture. Historian Jackson Lears aptly describes this dovetailing of biopolitical governmental technologies with the accumulation regimes of Fordist mass society: "National corporations employed advertising agencies to represent factory-produced goods to a mass market: the fables they fashioned merged personal and social health, individual and nation, creating narratives of adjustment to a single efficient system."[131]

The strategies used by the New York Subways Advertising Company clearly demonstrate the extent to which the subject coding of consumers and passengers followed the same principles of standardization and quantification. Founded in the course of the conglomeration of systems in 1940, the firm was commissioned with coordinating the posting of about 35,000 advertisements in stations and on the countless surfaces provided for ads in subway cars throughout the system as a whole. In 1941 the advertising company hired statistician and pioneer of public opinion research Elmo Roper (1900–1971) to study the effectiveness of advertising in the subway.[132] To quantify passengers for his study, Roper employed the same logistical procedures that Arnold had elaborated three decades earlier.

Among other details, Roper discovered that the average passenger volume per month was 5,638,000, every New Yorker rode the subway on average twenty-six times a month, and a typical ride lasted exactly 23.26 minutes.[133] He also found that 80 percent of New York City's female population and 98.9 percent of its male population used the subway. An additional 375,000 people from out of town also used the subway every month. This time, the purpose of these findings was not to increase efficiency in the circulation of passengers, but to produce arguments to persuade potential advertising clients of the subway's potential. Persistent overloading was not presented as a problem, but as proof that the subway was an ideal place to reach consumers. To put this claim to the test, the New York Subways Advertising Company decided to experiment with a special kind of ad campaign that would become iconic for both the subway and the city in the decades to come, aimed at promoting particular passenger ideals.

"Our Loveliest Subway Rider"

In the early 1940s, alongside warning signs and by-then ubiquitous advertisements, the first in an exciting new series of posters appeared in the subway as well as busses. Under the headline "Meet Miss Subways," it included a headshot of 14-year-old Mona Freedman, some short biographical notes, and the following lines: "Mr. Powers selects Miss Subways from among those who use the greatest transportation system in the world. Look around this car. Next month's selection may be riding with you."[134]

Initiated as a campaign by the John Robert Powers Model Agency to promote its business, and supported by the New York Subways Advertising Company, the posters for the Miss Subways pageant became an instant eye-catcher, and a consistent element of the subway's visual regime for decades to come.[135] For the advertising company, the pageant was envisioned as an instrument "to increase eye traffic for the adjoining ads,"[136] which promoted a variety of goods and services, from chewing gum to typing classes. The first posters were very well received by the public and became a common topic of conversation among passengers. Capitalizing on this success, the company soon announced the competition and its winners on 14,000 posters plastered in nearly every bus and subway car, where they became an everyday sight for millions of passengers.

After Mona Freeman was announced as the first winner in May 1941, a new Miss Subways was chosen out of several hundred submissions—usually sent in by friends, colleagues, or family members—every month or two for the next 35 years. In the first years, the head of the modeling agency, John Robert Powers, personally selected the winners. In 1963 the process became more democratic. Now passengers could cast their votes by postcard, and later by telephone, choosing among six contestants who were introduced in large posters under the headline, "Who will be 'Miss Subways'???

Our Loveliest Subway Rider." These posters featured the women's head shots, names, and employers.

Winners were awarded bracelets featuring gold- or silver-plated subway tokens, and professional portraits were distributed widely throughout the system alongside a short text featuring the winner's interests and occupation. Sometimes the women's marital status or number of children was featured as well, but the focus of the descriptions was clearly on their careers and ambitions.

According to the organizers of the campaign, this was a deliberate decision. They aimed to set their competition apart from other beauty pageants, such as the prominent Miss America contest. The chosen women were supposed to look like "the girl next door," or in the next subway seat, rather than a model or celebrity. According to the New York Subways Advertising Company, they also wanted Miss Subways to "reflect the girl who works—what New York is all about." They also emphasized, "Prettiness per se is passé. It's personality and interest pursuit that counts."[137] Winners were portrayed as hard-working, independent New Yorkers employed as clerks, teachers, nurses, and secretaries, or striving for academic degrees.

For New Yorker novelist and journalist Melanie Bush, who had already been deeply impressed by the posters as a child, the focus on the female passenger's ambitions made the campaign "a rare rose to find clamped in the teeth of mass advertisement."[138] Going through the posters for a *New York Times* article in 2004, she notes:

> I was jolted back to the thrilling rose that Miss Subways offered to a young, identity-seeking passenger. [. . .] What I waited for each new month was: What did she do? What were her goals? The Miss Subways l wanted to be was the airplane pilot. Or how about "travel writer"? "Scientist"? "Surgeon"? Right there on the subway! For me. Any mention of their appearance flew back into oblivion like dead planets. Maybe next month she'd plan to be an astronaut. Or president!
>
> What was actually going on here, I saw, was women, real New York women, talking to each other about their intentions and transmitting these messages through the medium of some men's advertising campaign. And where were they doing it? Underground![139]

The emphasis on the independence and career ambitions of female passengers was in part a reaction to shifting gender roles during the Second World War, when women entered the work force—and also the subway—in far greater numbers than ever before. Posters from the 1950s and 60s were often less radical. Responding to the increasing prevalence of housewives, they featured references to hobbies and consumer desires such as holidays in Bermuda or new skis.

Officially, the only criteria for submissions were that contestants had to be female New Yorkers and frequent subway riders. Passengers soon noted, however, that the selected contestants and winners were exclusively white. In 1942 the African American social club All-Ears started a campaign by submitting profiles of African American subway passengers to Powers' agency, along with letters urging him to allow these women into the competition and onto poster displays. These petitions were initially met with a rebuff. People from civil rights initiatives continued to write streams of letters to the mayor and subway operators, pushing for a selection that more closely resembled actual passenger demographics.[140]

Finally, in April 1948, the first African American passenger was crowned as Miss Subways: Thelma Porter, a young student and president of the Brooklyn Youth Council of the NAACP. According to sociologist Maxine Leeds Craig, this was seen by many people at the time as a crucial breakthrough toward recognition and civil rights. Porter's achievement of "beauty" status transcended discriminatory barriers, raising her to a level of fame and recognition previously reserved for white women. Her portrait not only featured widely on posters in subway cars and busses but also on the cover of the NAACP journal *Crisis*, followed by reports on discrimination in education and labor, and calls for civil rights.

While it would take another 36 years for an African American woman to be crowned as Miss America, in the following years, more and more contests around the country opened up their formerly all-white line-up of candidates and winners. In November 1949, Helen Lee became the first Asian American Miss Subways, and Yolanda Revson was the first Latina Miss Subways in January 1951. In the decades to come, while still strongly biased toward white passengers, Miss Subways posters showed the diverse faces of New York's female commuters: Irish, Italian, Jewish, Catholic, Latina, African American, and Asian. In a way, the advocacy of civil rights groups to include passengers of color in the competition can be seen as a proxy for claiming their right to be recognized as a legitimate part of the subway crowd. While African American and Latina women were never officially excluded from riding the subway, they were disproportionally targeted by insults and harassment from other passengers as well as personnel, and also excluded from the subway's workforce until the late 1950s.[141] In addition, popular contemporary representations of passengers, for example in the Subway Sun, depicted exclusively white and predominantly male riders. Along with an almost all-male, all-white subway staff, this helped to code the territories below the city as a dominantly white male space.

In its thirty-five-year run, around two hundred women held the title of Miss Subways. The program was discontinued in 1976 due to a proclaimed "lack of relevance," as well as changing gender politics; through the growing influence of the women's rights movements, the campaign increasingly came to be seen as "socially unacceptable."[142] Yet especially in its heyday during the thriving pin-up culture of the 1940s and 1950s, the Miss Subways campaign was not just a ubiquitous element of the New York subway, but iconic of the city itself. As a result, it found its way into countless products of popular culture, for example serving as a central plot device in Leonard Bernstein's 1944 musical *On the Town*. Here, a sailor on shore leave falls in love with a "Miss Turnstiles" after seeing her poster in the subway. He sets out to find her, only to discover that she is not a model or celebrity but a struggling yet ambitious young woman trying to make her way in the city.

As a central element of the visual regime in the subway for decades, accompanying the daily commute of millions of people—and perhaps their daydreams as well—the Miss Subways contest impacted passenger culture in multiple ways. First of all, it reinforced the coding of the subway as potentially romantic territory, similar to the motifs of infrastructural romance and erotics that had already appeared in subways songs, movies, and dances in the early years after the system's opening.

The female passengers displayed on the posters were framed as objects of desire for a male gaze, in several respects. Presented alongside advertisements for consumable goods and services, these women were displayed in a similar fashion, with their assets and features listed in detail. They were shown in the best light, ready to be selected from a range of options, just like the adjoining ads for soda or suits. Yet while the campaign codified women as similarly consumable objects, the posters also often presented powerful role models for many female passengers, as Melanie Bush's testimony vividly demonstrates. Independent and ambitious, they represented a specific ideal of the modern female passenger that went beyond traditional gender roles, reflecting changing attitudes toward women in the workplace. Furthermore, the Miss Subways contest broke the color line long before national beauty pageants; at least in its intentions, it also advanced a different, more down-to-earth beauty standard. In presenting working-class women of various demographics as ideal passengers, the campaign also enforced the notion that the subway was a space where women were welcome, emphasizing that they were an integral—and desired—part of the subway crowd.

The Miss Subways campaign can be regarded as progressive in the sense that it acknowledged woman's career goals, aimed for inclusivity, and offered a relatively representative portrayal of the diverse demographics of subway passengers. However, it did not radically confront the moral codes and gender norms of the times. Positioned among calls to consume products and services of all kinds, the Miss Subways campaign primarily served as a—successful—strategy to promote the subway as a space for advertising. Enriching the surfaces of subway cars and stations with exciting visual stimuli, the posters also provided a welcoming target for the passenger's gaze, especially given that staring at others was considered inappropriate or even rude. With the Miss Subways posters printed extra-large and accompanied by biographical information, riders could study the portrait of a fellow subway passenger in detail, be it with aspiration, longing, curiosity, or just out of sheer boredom. For advertisers and operators, the popularity of the Miss Subways campaign clearly demonstrated that passengers were not only eager for such visual messages, but also highly receptive to them. At this point, all that was needed was clients who would buy up the vastly available ad space in subway cars and stations to promote their goods and services. And where better to find them than in the subway itself?

EYE TRAFFIC

In 1946, keeping in mind Roper's study, the New York Subways Advertising Company released an elaborate system of posters in the subway designed to persuade companies to advertise in these highlighted spots. Today these colorful and intricate "car cards" are regarded by historians of design as significant examples of modernist aesthetics.[143] They also reveal a specific vision of the ideal passenger and passenger receptivity to visual messages (see figure 3.19 and figure 3.20).

Despite its rather cartoon-like style, the images used in the campaign often have a totalitarian undertone. The staccato of slogans like "Selling . . . Selling . . . Selling . . . Sold" (figure 3.19) and "Repetition means Remembrance" (figure 3.22) demonstrate repetition as a successful strategy for appealing to consumers. Potential advertising clients of the day received the image of a passenger who would surrender without resistance to the promises of consumerism based solely on the persistence and redundancy of visual signals. This reflects a behavioristic understanding of the passenger as a black-boxed container subject, whose consumer choices are the necessary consequence of daily confrontation with the stimuli presented by advertising messages in the subway. The subway itself is presented as a panoptical territory: slogans such as, "No hiding place down there" (figure 3.21) or, ". . . in sight . . . in light . . . all day . . . all night" (figure 3.23) told potential advertisers and passengers that it was ultimately impossible to escape the appeal of visual regimes.

The iconography of these posters demonstrates the extreme transformation that the idea of the passenger had undergone since the onset of the machine age. When the subway first opened, passengers were often seen as an undisciplined mob. Forty years later, they had been standardized, qualified, and black-boxed as predictable and controllable subjects. Their interactions had been increasingly objectified. Their behavior could be predicted based on statistics, and driven by visual and auditory signals. The subject culture of passengers was an integral part of an overall societal process of technologization so strong that it overshadowed events in Europe. Visiting New York in 1950, French artist Francis Picabia remarked: "I have been profoundly impressed by the vast mechanical development in America. The machine has become more than a mere adjunct of human life. It is really a part of human life—perhaps the very soul."[144]

Figure 3.19   Subway Car Card by Paul Rand, 1947. Courtesy of New York Transit Museum.

Figure 3.20   Subway Car Card by E. McKnight Kauffer, 1947. Courtesy of New York Transit Museum.

Figure 3.21   Subway Car Card by Erik Nitsche, 1947. Courtesy of New York Transit Museum.

Figure 3.22   Subway Car Card by Paul Rand, 1947. Courtesy of New York Transit Museum.

Figure 3.23 Subway Car Card by Sascha Maurer, 1947. Courtesy of New York Transit Museum.

As we will see in the next chapter, many people in the United States agreed with Picabia's diagnosis regarding the omnipresence of machines and their dominance in public and private life, but they did all not share his enthusiasm. In the 1940s, more and more people were critical of machine-age paradigms. An increasing number of American artists, intellectuals, and scholars were alarmed by the incorporation of machines into every aspect of life, which they saw as an instrument of alienation and control. Remarkably often, the codified, isolated passengers of the New York subway became emblematic of such critique.

4

LONELY ROBOTS

The big ride to freedom led straight to the world of passengers and commuters.[1]

—*Wolfgang Sachs*

In the decades following the jubilant opening of the subway, euphoria quickly abated and gradually devolved into quite the opposite. Liberal ideals of subjectivity had raised expectations of great personal freedom, but as the machine age wore on, those models were eyed with more and more suspicion. In this chapter, we will see how criticism of mass society often crystallized around the passengers of the New York subway and a subject culture devoid of emotion and ruled by technology. The subway itself, once a herald of liberation and prosperity, increasingly resembled an all-powerful apparatus requiring cultural techniques of self-protection and containment. While the new mobility offered by the subway certainly had emancipatory aspects, opening up the expanse of the metropolis as a space of pleasure and adventure, these aspects soon faded from view. What remained was the daily grind of commuting and a growing awareness of the toll it took on passengers, especially commuters.

While logistics experts and technocrats addressed the subway masses primarily as a logistical and organizational problem, disregarding their individual dispositions, writers and visual artists began exploring the inner world and experiences of subway passengers more and more. As we will see, they often depicted passengers as lonely, exhausted introverts, reading or staring apathetically into space. The paintings and drawings of young New York artists in particular gave dramatic expression to the passions and fears of passengers. This was especially true for the artistic movements forming around the paradigm of social realism, which focused on the miseries and ordeals of commuter masses down in the tunnels below the city. By the 1940s at the latest, disillusionment and cynicism regarding the machine age in general and the subway in particular were widely manifest in the visual arts. But what was behind this critical perspective?

In their monumental study *The New Spirit of Capitalism*, social philosophers Luc Boltanski and Ève Chiapello explore various critiques of modern capitalism, many of which capitalism later incorporated for its own purposes.[2] Boltanski and Chiapello identify two veins of critical reflection, arguing that their respective themes have been fairly consistent since the nineteenth century. The first is artistic critique, focused on the loss of authenticity and suppression of self-determination and liberal ways of life, along with alienation and disenchantment. The second is social critique, which addresses the authoritarian and exploitative structures produced by capitalism, along with growing social inequality and poverty.[3]

In this chapter, we will see how artistic portrayals of the impositions of subway travel can be understood as instances of artistic critique inasmuch as they primarily depict the experiences of alienation and isolation that had become a visible element of passenger culture in the first half of the twentieth century. Literary texts by Christopher Morley and E. B. White, as well as artistic works by Fortunato Depero, Walker Evans, and George Tooker, represent passengers as subjects who become isolated elements of the standardized masses. Working with aesthetic strategies, this form of criticism is more likely to evoke emphatic reactions than argumentative engagement.[4] Boltanski and Chiapello see this emotional dimension as an indispensable precondition for critical reflection and analysis: "The formulation of a critique presupposes a bad experience prompting protest, whether it is personally endured by critics or they are roused by the fate of others. This is what we call the source of *indignation*. Without this prior emotional—almost sentimental—reaction, no critique can take off."[5] During the first decades after the subway opened, critique of its passenger culture surely did take off. Critical descriptions of individual and collective everyday experiences of the machine age were not limited to literature and the visual arts. Various US filmmakers, such as Gore Vidal and Charlie Chaplin, also brought the subjectifying impositions of the culture of Fordist mass production and mass consumption to the silver screen.[6] At the same time, social and cultural theorists in the United States strove to articulate the specific forms of socialization that these cultural models produced. As of the late 1940s, the question of the disposition of the subject in capitalist mass culture was the key issue for contemporary critical social analysis.[7]

This form of critique became a central focus for academic sociology, especially in the United States. Within a few years of each other, popular books like *The Lonely Crowd* (Riesman, 1950), *White Collar* (Mills, 1951), and *The Organization Man* (Whyte, 1956) offered detailed descriptions of urban subject culture in the United States. Around the same time, historians of technology such as Sigfried Giedion and Lewis Mumford published groundbreaking works on the genesis and historical dynamics of modern, high-tech societies. Striking a tone of skepticism toward technology, they often turned to the New York subway to illustrate the human suffering caused by mass culture and the spread of machines.

These critical portrayals of New York's passenger culture are worthy of closer inspection. We will begin by examining a selection of artistic positions, and then proceed to a reconstruction of various analyses from contemporary sociology and cultural theory. With their emphasis on alienation in the masses and fear of technological overreach, these standpoints correspond in many ways with artistic critique as defined by Boltanski and Chiapello. At the end of the chapter, we will look at the socio-economic crisis in New York at the end of the machine age, and the beginning of an exodus of passengers from the subway. First, we will take a closer look at one type of subject that stood out more and more distinctly from the spectrum of passengers, namely, the commuter. As commuter culture gained momentum, the daily chore of getting to and from factories and offices became a central motif of modern urban experience, specifically with respect to the impositions of Fordist mass society.

COMMUTER PATHOLOGIES

The commuter is the queerest bird of all.[8]

—*E. B. White*

While commuting for work first became an everyday practice for many inhabitants of New York in conjunction with subway expansion, the genesis of the modern commuter subject can be traced back to more than half a century before the subway opened. Commuter culture began to form when railway infrastructure reached the expanding settlements in the outer boroughs. Likewise, the term "commuting" dates back to the mid-nineteenth century, when workers from these new suburbs began taking the train to and from work in inner-city factories.[9]

By the 1920s, the experience of commuting had become so common that artists and writers began devoting more attention to it, generally with sympathy for the passengers who subjected themselves to these mobility regimes day in and day out. Author Christopher Morley, for example, wrote of masses of commuters pouring out of the subway every morning: "But as the crowd pours from the cars, and shrugs off the burden of the journey, you may see them looking upward to console themselves with perpendicular loveliness leaping into the clear sky. Ah, they are well trained. All are oppressed and shackled by things greater than themselves; yet within their own orbits of free movement they are masters of the event. They are patient and friendly, and endlessly brave."[10]

Contemporary accounts of the subway highlight the discrepancy between the discipline and control of passengers on the one hand, and the dignity with which commuters shouldered the daily inconvenience of subway transit on the other.[11] In his iconic text *Here Is New York* (1948), essayist E. B. White (1899–1985) depicted the subway commuter in such terms. For White, New York had to be understood not as a homogeneous unity, but as a composition of a variety of urban worlds of experience and ways of life, each with its own unique subject culture. Besides the city of native New Yorkers and the New York of migrants and adventurers, there was the New York of commuters, who had their own unique experience of the city. Commuters knew New

York almost exclusively from transit experience and office views. In White's eyes, they were pitiful creatures indeed, who experienced the city only in functional relation to their work, with little conception of the adventure and pleasure that New York could afford. In machine-like movement between home and work, the commuter constituted the antithesis of the flaneur. In White's words, "The commuter dies with a tremendous mileage to his credit, but he is no rover."[12]

The flocks of commuters who doubled Manhattan's daytime population also gave the city a certain rhythm. White's metaphor for their temporality was not flattering: commuters transformed New York into a city "that is devoured by locusts each day and spat out each night."[13] Descriptions from the time also typically gendered the commuter as male, as in White's concise 1929 rhyme:

> Commuter—one who spends his life
> In riding to and from his wife;
> A man who shaves and takes a train
> And then rides back to shave again.[14]

White's caricature of the passenger as patriarchal breadwinner once again recalls the promise of salvation borne by the passenger as a hero of pastoral, bourgeois morality. By the end of the First World War, however, women constituted a vital part of the workforce, for example as secretaries, saleswomen, and administrative employees who relied on the subway to get to work every day. While White and Morley took little notice of them, the visual arts of the day were somewhat more attentive to the experiences of female passengers.

Passenger Images

Created during his two-year stay in New York, the painting *Subway, Crowds to the Underground Trains* (figure 4.1) by Italian painter, writer, and designer Fortunato Depero (1892–1960) offers a particularly drastic depiction of the infrastructural territories beneath the city.[15] The uniform bodies of the passenger masses are squeezed into the smallest of spaces, trapped in the labyrinthine halls of the subway, surrounded by

Figure 4.1 Fortunato Depero: *Subway, Crowds to the Underground Trains*, 1930. Courtesy of Archivio Depero, Rovereto.

gaping stairwells and tunnels. Although Depero considered himself indebted to Italian futurism, the painting does not express any technology-induced enthusiasm. Instead, the cold faces of passengers display disinterest, while their bodily movements and postures appear uniform and monotonous.[16]

In *Many Are Called*, a series of photographs taken between 1938 and 1941, photographer Walker Evans (1903–1975) captured the stoic demeanor and emotionless faces of subway passengers on film. Eighty-nine portraits show the heterogeneous composition of the subway ridership, with passengers exercising techniques of transient contemplation and self-containment (figure 4.2–4.4).

To capture these intimate moments, Evans took these pictures secretly, hiding his camera under his coat with the lens positioned behind a buttonhole.[17] He worked mostly in IRT cars because the seating arrangement in rows along the outer walls created the perfect distance for taking pictures without people being aware of it.[18]

Some commentators of Evans's work have drawn attention to the remarkable ambivalence in these passenger portraits.[19] These hardworking individuals have an air of composure and dignity about them, despite the ordeals of life in New York during the Depression era. But at the same time, in these supposedly unobserved moments, their faces admit to daily ordeals. The title of the series, *Many Are Called*, is from a Bible verse (Matthew 22:14—"For many are called, but few are chosen"), applying the motif of the damned and outcast to subway passengers. Less than a decade later, George Tooker (1920–2011) painted *The Subway* (1950), which represents the moment of threat and a sense of being at the mercy of the system (figure 4.5).

The central figure in Tooker's painting is a middle-aged white woman surrounded by a maze of barriers, gates, and the menacing gazes of several male figures. Her face suggests an experience of panic, depression, and pain. Tooker portrays the narrow halls of the subway station as a terrifying place of no escape—a space of fear for female passengers, in particular.[20] The architecture as well as the presence of other passengers create an intimidating, brutal effect. The experience of being a passenger comes across as alienating, claustrophobic, and potentially dangerous. Social interaction is to be avoided at all costs; isolated subjects encounter each other with caution and suspicion.

Figure 4.2   Walker Evans, *Subway Passengers*, New York City, 1938. Gelatin silver print. 12.2 × 15.0 cm (4 13/16 × 5 15/16 in.). The Metropolitan Museum of Art. Gift of Arnold H. Crane, 1971 (1971.646.18). © Walker Evans Archive, The Metropolitan Museum of Art. Digital image © Whitney Museum of American Art / Licensed by Scala / Art Resource, NY.

Figure 4.3 Walker Evans, *Subway Portrait*, 1938–1941. Gelatin silver print, 12.2 × 17.6 cm. The J. Paul Getty Museum, Los Angeles. © Walker Evans Archive, The Metropolitan Museum of Art. Image source: Art Resource, NY.

Figure 4.4   Walker Evans, *Subway Portrait*, January 13–21, 1941. The Metropolitan Museum of Art. Gift of Arnold H. Crane, 1971 (1971.646.19). © Walker Evans Archive, The Metropolitan Museum of Art. Image source: Art Resource, NY.

Figure 4.5   George Tooker, *The Subway*, 1950. © The Estate of George Tooker. Courtesy of DC Moore Gallery, New York. Courtesy of Art Resource.

Tooker's painting demonstrates many aspects of the critique of passenger culture in New York at the time. In this work, the ambivalences of the subway—with its experiences of liberation and new spatial possibilities alongside real threats and uncertainty—clearly tip in the direction of a negative assessment of passenger existence. The themes in Tooker's painting were typical of many contemporary works of art, from Reginald March's series *Subway Sunbeams* (1922–1927) to Ida Abelman's *Wonders of Our Time* (1937).[21] While these works differ greatly, their portrayals of passengers have something in common, interpreting the subway as a place of social alienation in the middle of a densely packed crowd. A recurrent theme in artworks related to the New York subway, the lonely crowd was also a leitmotif in contemporary sociological critique of modern mass culture.

Lonely Crowds and Cheerful Robots

Few sociological studies from the mid-twentieth century were as influential as David Riesman's *The Lonely Crowd*.[22] Coauthored by Nathan Glazer and Reuel Denney and published in 1950, the study became an instant bestseller. The book's clear and simple claim is that as population growth started to consolidate in the twentieth century, cultural dynamics emerged that would significantly change the characteristics of the American people. Riesman distinguishes three types of characters that more or less followed one another historically: the tradition-directed type, the inner-directed type, and the other-directed type.[23]

For Riesman, prior to industrialization, an individual's structures of meaning and personal rites were for the most part determined by tradition. Population growth in the eighteenth and nineteenth centuries, however, encouraged the development of a new kind of character directed by an inner compass formed by strong morals and learned norms and values. Once parents and other figures or institutions of authority had firmly implanted morals in a child, those morals directed a person's behavior throughout life, even in times of drastic social transformation.[24] The inner-directed character type that resulted from the loss of traditional models, increasing division of labor, and social differentiation came up against a further crisis at the beginning of the twentieth century.

The intertwined transformations of capitalist acquisition structures and modern socialization ultimately led away from tradition and inner direction to the other-directed character type, according to Riesman. Primarily associated with Fordist mass culture, this model first emerged in big US cities like New York and Boston, especially among people of upper income levels.[25] But by the 1940s, this other-directed character type had also become a dominant model for blue-collar and white-collar workers.

With the concept of the other-directed type, Riesman emphasizes how subjects are controlled primarily by their immediate surroundings. This may mean a peer group, but it may also include instruments of mass communication, such as advertising and political propaganda. While inner-directed types still had a deeply internalized gyro-compass to help them find their bearings and stay the course, other-directed types exchanged their inner compass for a radar system, using external signals to determine their own position.[26] Developing this technological metaphor, Riesman describes the new "radar type" as follows: "The goals toward which the other-directed person strives shift with that guidance: it is only the process of striving itself and the process of paying close attention to the signals from others that remain unaltered through life. This mode of keeping in touch with others permits a close behavioral conformity, not through drill in behavior itself, as in the tradition-directed character, but rather through an exceptional sensitivity to the actions and wishes of others."[27]

In contrast to the nineteenth-century bourgeois subject culture of personality, these new types of characters are not merely ready but eager to accept direction from others. Their conduct is no longer controlled by shame or conscience, but by "diffuse anxiety."[28] They are in a constant state of alert, prepared to receive signals at any time, and to adapt their behavior accordingly.

Riesman's narrative of the change in character from inner-directedness to other-directedness also raises the question of personal autonomy. A readiness to receive and follow directions does not necessarily imply a loss of autonomy and self-determination.[29] The outer-directed subject can gain emancipation from the traditional authority of religion, family, and state. At the same time, the demands and signals that proliferated as the machine age progressed were so diverse that elements of choice opened up for urban subjects.[30]

Post-bourgeois forms of subjectivity—with their new freedoms and choices—come at the price of a lack of social orientation and personal development.[31] Simmel had already recognized that a plurality of urban lifestyles and liberation from rigid role models also enables new forms of autonomy.[32] Riesman borrows from Simmel in his assumption that other-directed characters evolve in order "to cope with the social demands of modern urban culture."[33]

Riesman's analysis clearly struck a nerve with contemporaries. The book's huge instant success indicates that Riesman offered readers interpretive models that helped them to understand who they were—and where they were. The book's title quickly became a prominent symbol of modern urban work culture and related social pathologies.

The image of the lonely crowd seemed nowhere more fitting than in the subway. To this day, Riesman's term is mobilized to describe social interactions observed in transit situations, especially with respect to the isolation and control of subjects.[34] We still view passengers as filled with diffuse anxiety, overwhelmed by competing offers of orientation and potential social contacts—as powerfully captured by George Tooker. We can also see other-directedness at work in the control of passengers inscribed into the subway's visual regimes in the early twentieth century. The emergence of the ideal of a black-boxed, standardized passenger-subject whose practices can be shaped by visual signals from warnings to advertisements reflects Riesman's account of a withdrawal from moral subjectivity in favor of behaviorist objectivity.

Just one year after Riesman's influential depiction of mass culture in machine-age cities hit the bookshelves, sociologist C. Wright Mills published another powerful work of analysis and critique: *White Collar: The American Middle Classes*.[35] Like his friend and colleague Riesman, Mills found new forms of subjective culture especially pronounced in large American cities, especially in New York.[36] But even more than Riesman, Mills saw the masses of white-collar employees driven by fear and intimidation.[37] According to Mills, the culture of the modern metropolis produces a form of social life so alienated that social interaction becomes irrevocably pathological. Mills's diagnosis for the modern subject could hardly be more pessimistic: "Intimacy and the personal touch, no longer intrinsic to his way of life, are often contrived devices of impersonal manipulation. Rather than cohesion there is uniformity, rather than descent or tradition, interests. Physically close, but socially distant, human relations become at once intense and impersonal—and in every detail, pecuniary."[38]

The picture that Mills sketches of American urban work culture is that of a uniform mass devoid of all individual thought. People are nothing but "cheerful robots," standardized, automated, and yet highly motivated.[39] Once again the metaphor chosen to describe individuals in the machine age is technological. To explain modern mass society in the United States, contemporary American sociologists such as Riesman and Mills often cited the entanglement of bureaucratic regimes and technological innovation.[40] While criticizing how Fordist modes of production and consumption alienated and exploited subjects, cultural theorists and historians of the day were primarily concerned about the disproportionate power of technology.

The First World War had already made many scholars acutely aware of the ambivalences of machine culture; it was entirely clear that the mechanization of war was inseparably tied to the rise of the machine code as a model for society. The first total war under the paradigm of mechanization and technological innovation had left horrific numbers of victims, with new military machinery such as tanks, lethal gas, and automatic weapons maiming and destroying human bodies in an unprecedented manner. As philosopher of technology Thomas P. Hughes has shown, many technological innovations that found their way into private homes and public use as signs of modern prosperity were originally developed by the defense industry for military purposes.[41]

After the Second World War, the voices of those lamenting the catastrophic effects of the spread of machines across all areas of life had become legion. Intellectuals such as Günther Anders, Martin Heidegger, Sigfried Giedeon, Herbert Marcuse, and Lewis Mumford blamed modern technology for creating a new form of control and domination. According to their arguments, the extreme power intrinsic to this technology not only exploits subjects and forces them into dependency, but also threatens the very existence of humanity. These thinkers and others radically rejected the belief in technology-based progress that had been so characteristic of the first decades of the machine age.

In retrospect, the work of New York cultural theorist and historian of technology Lewis Mumford (1895–1990) can be seen as paradigmatic for the widespread skepticism of technology that marked the final years of the machine age. While Mumford arrived at radical cultural pessimism in his late works, his attitude toward the technologization of civilization was deeply ambivalent for many years. His early writings in particular were marked by utopian hope in the liberating potential attributed to new technologies of electricity, mobility, and mass communication.[42] Gradually shifting from euphoria to disillusionment to radical renunciation of the phenomena of the machine age, Mumford's views on the subway and subway passengers also became increasingly bleak in the course of his cultural diagnoses.

Mumford's considerations on technology culminated in the concept of the "megamachine," laid out in the two volumes of his almost eight-hundred-page magnum opus, *The Myth of the Machine* (1967/1970).[43] According to Mumford, human history has seen several instances when highly complex and resource-intensive organizations combined hierarchical administration and authoritarian "monotechnics" to carry out enormous projects. He locates the first such megamachine five thousand years ago in ancient Egyptian pyramid construction.[44] On a tour de force through thousands of years of global history, Mumford identifies megamachines at work in organizing the early Christian church, recruiting the armies of the Middle Ages, and supporting the campaigns and colonization efforts of absolutist regimes. A megamachine can be understood as a dispositif of organizational power that serves the ends of technical rationality only secondarily.[45] The function of such a machine consists in coordinating the accumulation and concentration of resources and labor, as well as implementing bureaucratic procedures and complex large-scale technology.[46]

In the modern megamachines of the nineteenth and twentieth centuries, Mumford sees a new quality related to the exploitation of people as masses. Industrialized wars, modern forms of settlement, and gigantic manufacturing systems were in a better position than ever before to accumulate resources and subject people to the centralizing principle of the machine. They succeeded in large part by applying the principles of modern administration, standardization and automation.[47] In Mumford's view, these dynamics peaked in the development of nuclear power plants ("air conditioned pyramids," as he called them),[48] and the monstrous form of urban accumulation found in the Western hemisphere: the "megalopolis."[49]

For Mumford, living through the radical metamorphosis of the city through the subway and skyscrapers, New York in particular was the embodiment of all of the unfortunate trends and mistakes of modern capitalist society and its alienated mass culture. As early as 1934, he remarked that New York City had become nothing less than "the center of furious decay, which [is] called growth, enterprise, and greatness."[50] Although the gigantic machinery of the subway had once been seen as a solution to the problems of the megalopolis, in Mumford's eyes it had instead become an instrument of standardizing power and universal control.[51]

In a text from 1919, Mumford explored whether or not the subway system played a positive role in integrating urban inhabitants.[52] He found that subway expansion primarily served the interests of capitalist real estate speculators by turning rural pastures along the outskirts of the city into housing projects. The blessings of the new circulation machinery thus fell upon the economic elites above all, while the inhabitants of the subway suburbs were left to shoulder the daily burden of commuting.[53]

In his later writings, Mumford continued to examine the enormous social transformations brought about by the subway. Not only did it promote economic exploitation and control, but its infrastructural environment also altered aesthetic experiences and habits in a deeply ambivalent manner.[54] The later Mumford wrote about the subway, the more he characterized it as an inhuman megamachine: passengers had to endure the "pulping mill of the subway," as well as brutal "Swedish massages"[55] by guards.[56] In 1952, just a few months before the three underground systems were united once and for all under the Transit Authority, Mumford compared the subway to the deadly technology of the weapons of mass destruction used in Hiroshima and Nagasaki. In a collection of essays titled *Art and Technics*, he laments the extreme power built into technical artefacts and elements of infrastructure, noting with regret "the overdevelopment of subways and multiple express highways and atom bombs [. . .] in our civilization today."[57]

Problematic from today's perspective, Mumford's position appeared justified against the backdrop of disillusionment with the euphoria of progress and the ubiquitous sense of threat posed by technology. The unspeakable horror and suffering of the Second World War had revealed the destructive power of modern technology and military logistics. Along with Mumford, many philosophers and intellectuals of the day were deeply skeptical of a culture in which they recognized new forms of social control and forces of destruction.[58]

With arguments echoing those of Mumford, engineer and philosopher of technology Sigfried Giedion (1888–1968) took a radical stance on the machine age and its unredeemed promises of a better life. His monumental work *Mechanization Takes Command* from 1948 traces the development of mechanization from its ancient beginnings to the point of "full mechanization" through the spread of Taylorist management and Fordist work organization.[59] Like Mumford, he finds the outcome of this development highly dubious, with the individual becoming increasingly dependent on production and overwhelmed by technology. "Never has mankind possessed so many instruments for abolishing slavery. But the promises of a better life have not been kept. All we have to show so far is a rather disquieting inability to organize the world, or even to organize ourselves. Future generations will perhaps designate this period as one of mechanized barbarism, the most repulsive barbarism of all."[60]

In conclusion, Giedion warned of the dangers of excessive technological power and mechanized barbarism out of concern for the subjects ensnared and exploited by these processes. Yet despite their shared cultural pessimism, Mumford and Giedion were not fundamentally hostile to technology. Their criticism was focused on large-scale technologies and the capitalist logic of exploitation realized through these technologies. They were far from alone in this negative assessment of the mass culture of the machine age. The discourse of the times was replete with pessimistic diagnoses of the modern subject—from Wilhelm Reich's "character armor" and "body armor,"[61] to Erich Fromm's *Man for Himself*,[62] to the entirely alienated subjects of Adorno and Horkheimer's *Dialectic of Enlightenment*.[63]

Despite their differences, these works all centered their critique on the consequences of modern capitalist mass production and consumption. Boltanski and Chiapello also emphasize the reciprocal relationship between the standardization of products and the massification of subjects: "The standardization of objects and operations effectively entails a similar standardization of uses and, consequently, of users, whose practices are therewith massified, without them necessarily wanting it or being aware of it. With the development of marketing and advertising at the end of the interwar period, and especially after the Second World War, this massification of human beings as users via consumption extends to what seems to be among the most particular, the most intimate, of the dimensions of persons, rooted in their interior being: desire itself, whose massification is denounced in turn."[64]

The artistic critique of Riesman, Mumford, and Giedion had two primary components. On the one hand, they addressed the alienation, inhumanity, and incapacitating technologization of machine-age subject culture in general and the culture of employees and passengers in particular. On the other hand, they criticized the oppression promoted by Fordist mass culture, especially in terms of the suppression of human freedom, creativity, individuality, and autonomy.[65] This critique was fueled by the diagnosis of cultural conformity and the technological obliteration of distinctions between human beings and machines. In the eyes of early twentieth-century contemporaries, these dynamics were remarkably apparent with respect to the subway and its masses of passenger-subjects, standardized by the regime of logistics.

Solidifying into a counterculture by the early 1960s, this critique of a thoroughly technologized and rationalized society fed into a wider critical current. Works such as Herbert Marcuse's *One-Dimensional Man* (1964) and Guy Debord's *The Society of the Spectacle* (1967) offered a drastic assessment of life in so-called late capitalism as inauthentic and alienated. These works were enthusiastically received by those involved in budding protest movements in the United States and Europe.[66] For Marcuse, who developed his main ideas on the subjectivity of one-dimensional man while living in New York between the late 1940s and early 1950s, subway passengers served as a prime example of what technologized society did to people. "The subway during evening rush hour. What I see of the people are tired faces and limbs, hatred and anger. I feel someone might at any moment draw a knife—just so. They read, or rather they are soaked in their newspaper or magazine or paperback. And yet, a couple of hours later, the same people, deodorized, washed, dressed-up or down, may be happy and tender, really smile, and forget (or remember). But most of them will probably have some awful togetherness or aloneness at home."[67]

Especially as of the 1950s, portrayals of the New York subway passenger as an emblem of alienation and isolation in capitalist mass society fit into a wider intellectual reckoning with the tenets of the machine age, colored by realization that the end of this era was coming into sight. It was no accident that the peak in subway standardization and the creation of a unified New York transit system occurred just when the paradigms of social massification, standardization, and mechanization were also reaching an apex. The 1950s marked the beginning of a period of serious crisis in New York passenger culture and gradual decay of the system. Society had discovered a new ideal that left the subway looking outdated. The new machine for the new times was the automobile.

FAREWELL TO THE MACHINE AGE

New technological developments gave people living in Western industrial nations of the 1950s the sense that they were witnessing a second industrial revolution. Heralds of this era included novel chemical methods for producing synthetic materials, and highly efficient combustion motors, along with technologies of nuclear energy, space exploration, transistors, and the first microchips. The dominant model of knowledge for the machine age, at its core an idea of the machine code, was gradually replaced by the idea of complex cybernetic systems.[68] The machine age was slipping into the past, and it was time to start writing its history.

Alongside Mumford and Giedeon, architectural theorist Reyner Banham (1922–1988) saw the middle of the twentieth century as such a significant turning point that he endeavored to place the previous decades into historical perspective from the distance of only a few years. In his book *Theory and Design in the First Machine Age* (1960), Banham declared the beginning of a new epoch, claiming that the megastructures and large-scale technical systems of the vanishing machine age would soon be complemented by a host of miniaturized gadgets that would open up a new technological dimension of everyday life.[69] This process of miniaturization would make all kinds of new conveniences possible, from radios and telephones to electric shavers and pacemakers. Banham located the greatest shift in private households. The "machine for living in" that Le Corbusier had envisioned in the 1920s for life in the future was steadily becoming reality for a growing portion of the population. Private households in the United States, especially, were filling up with technical novelties, from televisions and telephones to kitchen appliances, vacuum cleaners, and washing machines. As Banham dryly noted, the number of machines in the home now put more horsepower at the disposal of every housewife than an industrial worker at the turn of the century.[70]

Banham rightly acknowledged that many of these conveniences had been invented decades earlier, and some of them had already been enjoyed by a small portion of the population for some time already. But they were now spreading throughout society with such force that it was impossible to remain indifferent in the face of the transformation they carried with them. The shift away from large-scale technology to small machines and household gadgets also brought about a return to the human being as the measure of all things. In a certain sense, this also met Mumford and Giedion's demands.

With all of these developments unfolding as signs of the emergence of a new era, scholars began looking for appropriate terms of description. Banham wrote of a "second machine age," others suggested "late modernity" or "ultra-modernity" or even "nuclear age."[71] More critical voices from the Frankfurt School and some Marxist intellectuals in the United States preferred the term "late capitalism."[72] Whatever one calls it, the period from the 1950s until the late 1970s was marked by contradictory trends and developments, especially in the United States. Following the Second World War, the United States experienced an unimagined economic boom promoted by the New Deal and Keynesianism. As the country came close to full employment and the middle class grew and grew, suburbanization also expanded massively, with fatal consequences for urban centers.

Thus, the end of the machine age also marks the preliminary end of the prosperity and cultural hegemony of Western metropolises. Suburbanization and the erosion of mass urban culture set cities like Paris, London, and New York on the road to a long phase of crisis marked by economic depression, continual exodus of the middle class, and increasing impoverishment of the population left behind in decaying inner cities. This wave hit New York particularly hard, especially as a city that had not been greatly damaged by the effects of the Second World War. In 1950 New York was still considered the most powerful and influential metropolis in the world, but that soon changed.

With the end of the Great Depression in the late 1930s, Mayor La Guardia and his newly appointed city planner Robert Moses (1888–1981) had initiated a phase of public investment that would dramatically alter the fabric of the city. Public establishments such as parks, swimming pools, highways, and bridges emblematized the slogan of the 1939 New York World's Fair: "Building the World of Tomorrow!" A subway line built especially for the fair brought more than forty million visitors to experience the technological utopia of a coming New York, where the automobile was celebrated as the means of transportation that would shape the future of life in the city.[73]

Robert Moses, who saw himself as a visionary of car culture (although he never got his driver's license), began radically rebuilding New York to accommodate automobiles.[74] The subway held little interest for him. While the city invested gigantic sums in infrastructure for automobile traffic, the system most dependent on public subsidies went largely without. Moses was not alone in this attitude: the city's political elites, technocrats, and entrepreneurs all began dropping their support for the glitchy, cost-sucking subway. More and more, passengers themselves also judged the loud, grueling, overcrowded subway—and the life of the train commuter—to be an outmoded way of life. The promises of a better life now lay elsewhere, in the single-family homes of highway-laced suburbs. Factory owners and investors also began moving production out of the cities and into the outskirts and backcountry.[75] In the drastic words of Henry Ford: "The modern city is the most artificial and unlovely sight this planet affords. [. . .] The ultimate solution is to abandon it. We shall solve the city problem only by leaving the city."[76]

Although they had little else in common with Henry Ford, many intellectuals of the times shared this opinion. In his narration for the 1939 film *The City*, Lewis Mumford declared that the joys of life were to be found far away from the slums of industrial metropolises.[77] Inspired by the concept of the Garden City, he projected the future of Western culture onto pastoral suburbs held together by the imaginary idyll of communal life.[78] Modern suburbs also promised residents a kind of technological comfort that the old infrastructure systems of the big city could not match. Instead of cinemas, subways, and skyscrapers for the masses, people wanted their own individual televisions, automobiles, and high-tech suburban homes.

At the top of the list of emerging desires was the automobile that Ford's now fully developed production methods had made affordable for middle-class households.[79] This machine quickly became such an iconic possession that in 1951, Marshall McLuhan called it the "mechanical bride" of the American people.[80] After a century of exponential growth, the affordable automobile caused a previously unimaginable exodus from New York City. In the 1950s, a wave of migration fed the growth of enormous suburbs all around the city, extending much further than the subway suburbs of the past. No longer reachable by rail, highways were all that connected these new suburbs to the urban center.

Due to suburban migration, as of 1950 New York's population growth began stagnating. Prior to that point, the number of inhabitants had more than doubled since the subway's opening to reach almost eight million. This number even declined over the next decades.[81] Especially after the 1952 opening of Levittown, the first grand suburb, less than a ninety-minute drive from Manhattan, an increasing number of white middle-class residents began turning their backs on the city.[82] While Western industrialized nations experienced extraordinary prosperity in the late 1950s and 1960s, with fully developed Fordist regimes of accumulation producing stable economic growth and a steady rise in the average standard of living, parts of America's inner cities began to slowly sink into desolation and poverty.[83] The exodus from New York also had consequences for the subway, as the number of passengers began to fall for the first time in its history.

PASSENGER EXODUS

For many years, it had seemed unimaginable that the massive overcrowding in the subway would ever be reduced. After decades of growth, in 1946 and 1947 passenger numbers had reached absolute peak capacity, and the system even became briefly profitable as a result. The highest record, unbroken to the present, was reached on December 23, 1946. 8,872,244 people rode the subway on this day, many of whom were certainly doing their last-minute Christmas shopping.[84] From this point on, however, ridership dropped continually. The numbers of passengers recorded by turnstile counters and reflected in statistics compiled according to Arnold's methods showed a dramatic decline in volume year after year (figure 4.6).

The decline in subway ridership came to an end in the late 1980s and numbers have continually rebounded ever since, but the system never again attained the load capacity of the late machine age. The peak in system capacity was not the only indication of the end of an era. The unification of the subway's subsystems to become the New York City Transit Authority (NYCTA) in June 1953 also marked a new phases of subway management. Created explicitly in order to improve efficiency and cover costs, the Transit Authority strove to reach these goals by dismantling the elevated trains connections, which were seen as redundant and antiquated. The Transit Authority also halted ambitious plans for expansion and limited its activity to consolidating the existing network.[85]

Historians of subway technology also view this moment as the beginning of a downturn that lasted for decades.[86] While already considered deficient in terms of maintenance and cleanliness, until 1953 the system was still seen as modern and safe. By the 1970s, as the result of various crisis dynamics in the intervening years, the New York subway would come to be seen as the most dangerous and shabby underground railway in the world. Several drastic hikes in fares could not balance the serious debt that had mounted between the two world wars and accumulated even more due to precarious economic conditions. Each additional hike in fares accelerated the exodus of passengers, while the system's finances plunged ever deeper into the red. The emphasis on cost reduction led to reduced maintenance and repair, leaving the system even more prone to technical failures. By the late 1950s, the desolate state of the subway was visible to all who entered: paint peeled and plaster crumbled from station walls, while broken lamps and seats were replaced with long delay or not at all. Cleaning took place so infrequently that walls and seats got dirtier and dirtier, and the floors of subway cars and stations were often covered with a thick layer of trash.[87] Despite all the warnings, fines, and patrols that prohibited spitting and smoking on the subway, clouds of smoke and filthy surfaces had become a daily sight.

The exodus of the middle class and the increasing neglect of the infrastructure changed the demographic make-up of subway passengers, as well as their subject forms.

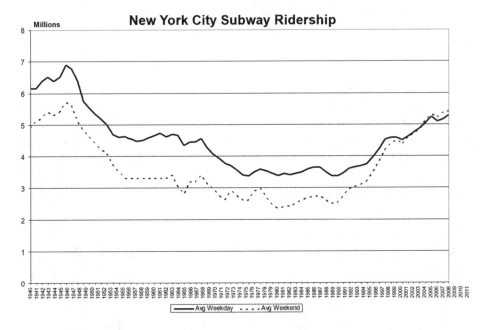

Figure 4.6 The development of passenger numbers on the New York subway between 1940 and 2011. Courtesy of New York Transit Museum.

The upper class had already largely abandoned the subway in favor of the automobile in the 1920s, and in the 1950s the white middle class followed in becoming drivers instead of passengers. They preferred to commute via new highways instead of riding the dirty and unreliable subway. The automobile was perceived as a vehicle of escape from standardized mass society; it came with the promise of individuality and freedom. In many respects, however, the same fate that had met the subway lay in wait for car culture as well. Initially welcomed as a liberating convenience that would create a better society, the automobile could not outrun disillusionment and criticism. At first, the automobile promised autonomy and self-determination. The freedom represented by the subject form of the motorist stood in sharp contrast to the everyday experiences and contortions of the passenger. In the eyes of contemporaries, the automobile offered privileged escape from the industrial Hades of the subway, on the highway to individuality and independence.

But the many residents of New York who could not afford automobiles had no choice but to continue to accept the humiliation of subway transit. Walter Benjamin provides a particularly powerful description of the fatalistic acceptance of being at the mercy of underground transit machinery, figuring the return of ancient rituals of sacrifice: "This labyrinth conceals in its innards not just one, but dozens of blind, rushing bulls, into whose jaws not once a year one Theban Virgin, but every morning thousands of anemic young cleaning women and still sleepy salesmen are forced to hurl themselves."[88]

As the subway began to decay, the daily ritual of getting to work became such an ordeal that it provoked massive protest from the remaining passengers. In the decades following consolidation under the New York City Transit Authority in 1953, witness accounts of the permanent crisis and mass exodus began to pile up in an unusual source: thousands of complaint letters were written by New York's subway ridership between 1954 and 1968. These letters spoke to the increasing strains, conflicts, and dangers associated with the subway, and they also served as an instrument of recognition and self-assertion for passengers during times of crisis.

CRISIS AND COMPLAINT

I used to feel so proud of our subways . . .[1]

—*letter from Cecile G. Cutler to the Transit Authority, November 29, 1961*

The City has become a place to avoid whenever possible.[2]

—*anonymous letter to Mayor Wagner, July 24, 1964*

February 1965 was just an ordinary month for the complaints department of the New York City Transit Authority. On February 10, the *New York Times* had once again reported an exorbitant increase in crime in the subway: in 1964 the rate of crime in the subway had risen by fifty-two percent, compared to only nine percent in the rest of the city.[3] But since the increase was only slightly higher than it had been in previous years and the system was considered very dangerous anyway, the report drew little public response. Not until seventeen-year-old passenger Andrew A. Mormile was brutally murdered by a youth gang in the subway on March 12, 1965, unleashing a flood of angry letters to authorities, did the city decide to react with a comprehensive program to save the subway.[4]

As of February, however, the department was still routinely working its way through complaint letters from disgruntled passengers describing the anxiety, harassment, and delinquency they experienced in the subway, and calling for more police protection. Among the almost one hundred complaints filed for that month alone were denunciations of fare dodging among children and adolescents, and expressions of indignation at the resurfacing spitting habit.[5] One woman also reported that she felt intimidated by the growing number of "colored people and Puerto Ricans" dozing or sleeping in the subway: "This is greatly adding to the fear of security. Your continuous, rigorous, handling of this problem will be greatly appreciated."[6]

In a letter dated February 17, another woman reported a traumatic event that she and her husband had experienced the evening before in the subway. They were attacked by several drunk adolescents and pulled the emergency brake in their panic. While the offenders ran off through the tunnels, the police officer who arrived on the scene responded to the couple with scorn, refusing to even document the incident. Despite her anger and frustration, the woman felt compelled to write to the authorities and report the episode in order to demand more security. She closed her letter with the lines "we must get rid of 'fear' by showing the lawless element that we have law and order."[7]

The next day, the Transit Authority received a memo from the Transit Police reporting the successful arrest of a group of young criminals who had been attacking and robbing passengers on the subway for months.[8] The day after that, a woman named Marlene Connor reported a subway incident involving an exhibitionist. In a long letter, she suggested stationing a police officer in every train, and installing an electric warning system in every car.[9]

A few days later, a passenger named Carl Heck sent an angry complaint in the name of all senior citizens from Brooklyn, Queens, and the Bronx, expressing his fear of using the subway at all. He wrote that he and his friends would gladly frequent the department stores and restaurants in Manhattan, but they "would not dare to set foot in that jungle called the subway."[10]

Also among the documents received in February was a letter from a passenger who identified himself as Paul Stone, expressing outrage at the fact that the restrooms at Christopher Street station served as meeting points for homosexuals.[11] When the Transit Police tried to contact Mr. Stone for more information later that same day, however, there was no resident of that name at the address given. They asked employees at the hotel next door, but there was no employee with that name, nor had there been a Paul Stone registered as a guest in the last five years.[12] The officials decided to check out the matter anyway, but when inspecting the restrooms they found no indication of moral offenses. They did make note of plans for increased surveillance and the installation of signs prohibiting "loitering."[13]

In mid-February the Transit Police received a detailed letter from an employee at Columbia University, reporting a violent assault by adolescents of color that almost cost him his sight. The police, he complained, were hardly interested in following up on the incident, remarking that it was just a trivial occurrence not worth their investigative effort. Frustrated, the man declared: "And is there anyone under these dreadful circumstances who would disagree with a citizen's right and intention to carry with him a weapon of defense?"[14] In the 1960s, numerous passengers actually did begin to arm themselves for subway transit, as demonstrated by a document that arrived two days prior to the university employee's letter, confirming the successful return of a revolver lost on the subway.[15]

Today we may find these incidents disturbing to read about, but in February 1965, none of these complaints represented anything particularly unusual in the eyes of authorities. They handled complaints routinely, replied with standardized formulations, and filed the letters away. However, one document filed in February 1965 did stand out among the others, and it was quickly circulated among the authorities: this was a report on the discovery of a wallet in the subway containing the driver's license of a man named Herman L. Reynolds, his discharge papers from the US Army, several receipts from pawn brokers, and several slips of paper on which was written:

> I have a gun. I want $2,000 in large bills.
> If you don't, I will give it to you in the head.
> I am a drug user. Wait 15 minutes to sound alarm.[16]

Encouraged by the Transit Authority, the police immediately started an investigation, going so far as to consult a graphologist for his expertise. Yet as in the other cases mentioned above, this encounter between a passenger and the administrative power structure ended without a further trace—at least in terms of written records.

Taken alone, these few letters from the holdings of the New York Transit Museum Archives already vividly testify to the erosion of passenger culture after the end of the machine age. In total, eight thousand pages of documents from the years between 1954 and 1968 are preserved in these archives.[17] Explored here for the first time, this body of complaints offers a unique view of passenger culture from the period. In these letters, passengers expressed matters in their own words, while also submitting their experiences to the procedures of administrative framing and evaluation. In a way, they were revenants of the "infamous men" of pre-revolutionary France who stirred the interest of Michel Foucault.[18]

This paper flood of indignation, accusation, and denunciation, along with the bureaucratic responses that washed back in return, provides insight into the subjective experience of passengers within the framework of their conflicts with the powerful dispositifs of the subway. The complaint letters operated as a recording mechanism, discursively registering offenses ranging from trivial daily inconveniences to extreme experiences of fear and violence. All together, they constitute an extensive catalogue of every imaginable evil, humiliation, and offense found in New York City's underground. To different extents, they pertain to passengers of nearly every race, class, gender, age, and level of education. For example, the corpus contains a letter from prominent media philosopher Marshall McLuhan, who used official stationary from Fordham University to propose a campaign for eliminating the "litter bugs" in New York (figure 5.1).

For our examination of the historical dynamics implicated in the development of the subject forms and subject positions of New York subway passengers, these complaints and the bureaucratic responses to them are revealing in many respects. When passengers describe themselves as increasingly overtaxed and fragile subjects, this relates primarily to their experiences of the material dimension of transit. Numerous complaints lament brutal machine noise, incomprehensible announcements, dark trains and stations, and pervasive stench. All of these factors contribute to the representation of subway as a space of sensory challenges, and the complaint letters demonstrate increasing fragility in passengers' techniques of containment and protection against stimuli. At the same time, the proliferating conflicts between passengers reported in the letters serve as evidence of the system's decline. The frequency of assault, theft,

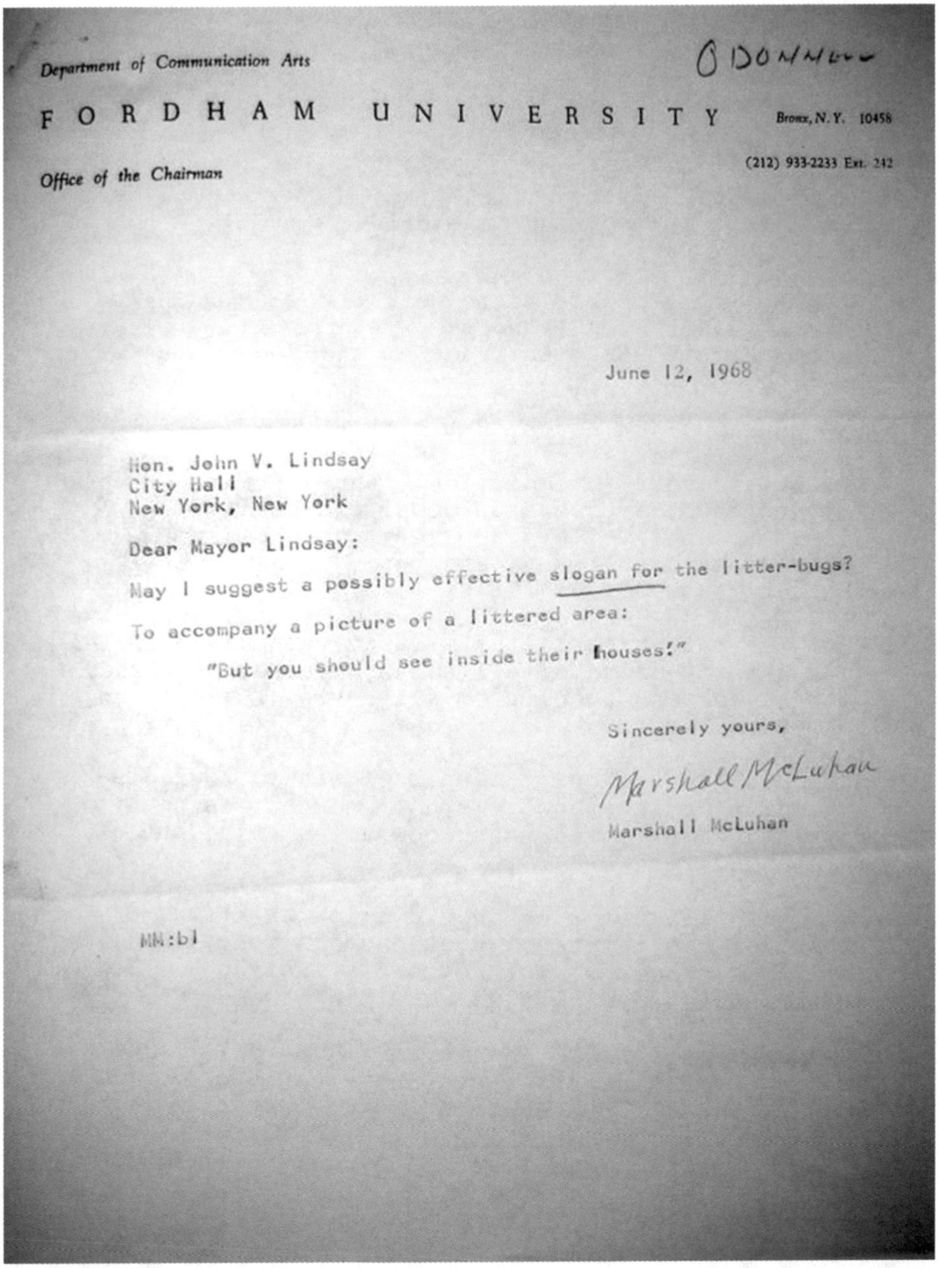

ODONNELL

Department of Communication Arts

FORDHAM UNIVERSITY Bronx, N.Y. 10458

Office of the Chairman (212) 933-2233 Ext. 242

June 12, 1968

Hon. John V. Lindsay
City Hall
New York, New York

Dear Mayor Lindsay:

May I suggest a possibly effective slogan for the litter-bugs?

To accompany a picture of a littered area:

"But you should see inside their houses!"

Sincerely yours,

Marshall McLuhan

Marshall McLuhan

MM:bl

Figure 5.1 Marshall McLuhan to Mayor Lindsay, June 12, 1968. Courtesy of New York Transit Museum. Author's photo.

60 First Avenue
New York 10009
November 29, 1964

√ 3286

New York City Transit Authority
370 Jay Street
New York City

Gentlemen:

I am one of a group of 1200 families that have
moved into the housing development known as
"Village View."

Most of these people go to work via the Independent
Line -- Sixth Avenue Subway -- and board the station
at the First Avenue entrance of the Second Avenue
station.

You are no doubt aware that this exit (or entrance)
is the derelict's toilet. The derelicts, of course
are not in your jurisdiction. The police are helpful and each day round up people to keep them away
from the subway. They are very humanitarian in
their work, I must say, and try very hard to keep
the derelicts off of the streets.

But, the subway is filthy! It is repugnant, filled
with latrine odors of all sorts, generously spread
with liquor and puddles of all sorts of heaved-up,
undigested matter.

A simple dispatch of a couple of men to pour dis-
infectant on the subway is not enough. We need to
have that station scrubbed down regularly, and
maintained regularly.

I do not think that we should allow shortages of
staff, or complaints of "not enough money" enter
into this. Somebody of authority must look into this
matter, and personally do something.

As a matter of fact, and in addition to the above
matter, I must report that I have noticed a general
decline in the appearance of our subway cars. I do
not know what is happening. I used to feel so proud
of our subways. Now I see dirty cars with slashed
seats.

I am active in the cooperative association at

Figure 5.2 Cecile G. Cutler to the Transit Authority, November 29, 1964. Courtesy of New York Transit Museum. Author's photo.

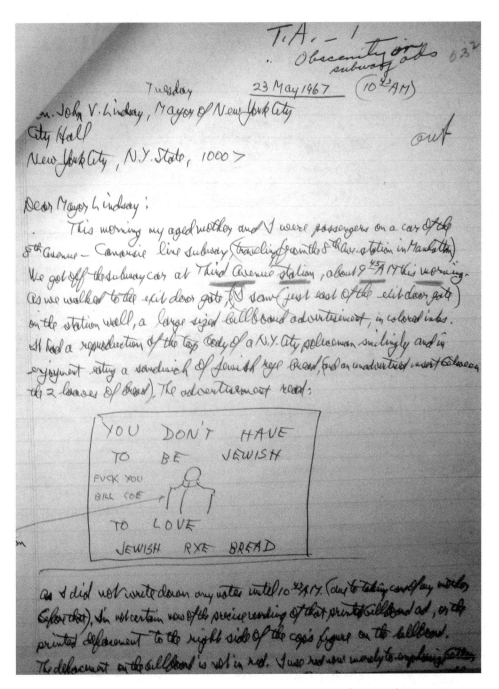

Figure 5.3 Sam W. Liske to Mayor Lindsay, May 23, 1967. Courtesy of New York Transit Museum. Author's photo.

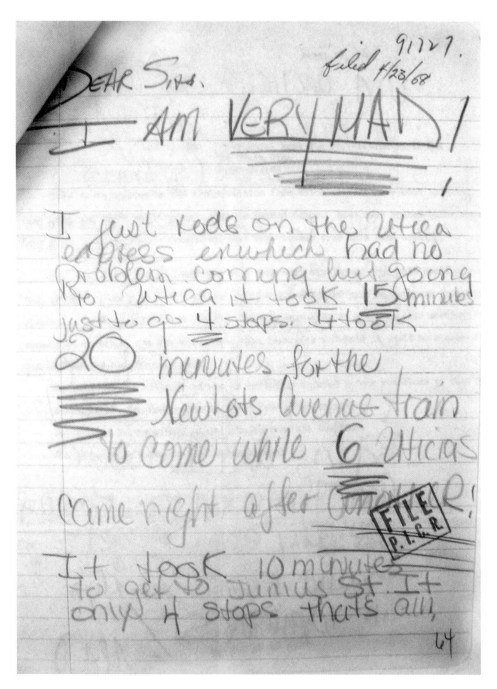

Figure 5.4 Anonymous to the Transit Authority, April 23, 1967. Courtesy of New York Transit Museum. Author's photo.

sexual harassment, and violent altercations rose exponentially. Especially in the 1960s, the subject culture of the subway was marked by crisis. And yet, their complaint letters can also be read as attempts to emancipate themselves, claiming individuality and autonomy by injecting their own experiences and demands into the administrative discourse built up around the subway.

Given the volume of complaints and responses, as well as the difficulties involved in interpreting and evaluating any historical source documents, a full account of this material would warrant its own study. Nonetheless, in the following, we will be able to identify and discuss some salient aspects of the complaints and the answers they provoked. The letters show a trend of increasing fragmentation that emerged toward the end of the machine age, profoundly altering the once relatively homogeneous subject culture of mass transit. Between 1954 and 1968, the subway was transformed into a place of intense fear, violence, and denunciation, with struggles playing out largely along the lines of race, class, age, and gender. There were signs of crisis to be found on all of these fronts during the time period in question: denunciation of poor people, exclusion of beggars and homeless people; vandalism by adolescents and school children; violence against women, homosexuals, and Jews; massive racism, and systematic discrimination against African Americans and Latinxs. Additionally, as the subway's technologies of supervision became increasingly militarized, there were more and more complaints regarding poorly trained police officers and the brutality and bureaucratic whims of security enforcement.

While the previous decades had seen some leveling of racial and class differences among individual subjects through the phenomenon of massification, attempts at group identification and distinction picked up with new momentum as of the 1950s. The increase in conflict and violence in the subway also changed the demographics of passenger crowds. While the middle class largely abandoned the system in favor of the automobile, the segment of the population that continued to take the subway consisted of those who had no alternative form of transit. By the mid-1970s, with political elites continuing to neglect system maintenance, the New York subway had the dubious reputation of being by far the most dangerous transit system in the world.

As valuable as this source material is for tracing the crisis in passenger culture as it unfolds, several problems arise for any analysis of structures of subjectivation through documents featuring self-disclosure, outrage, requests, and threats. To evaluate these materials, we must ask whether they are representative, how they are presented, where to situate them historically, and how they relate to other materials and methods considered elsewhere in this study. In the end, they reveal the methodological limits and gray zones in the historical reconstruction of modes of subjectivity.

Before discussing these problems and looking at possible solutions, we will look into the sparse research on the phenomenon of the complaint in order to more precisely describe and evaluate the source material at hand. We will then look into these particular complaints, investigating how they relate to the violation of explicit and implicit norms in the subway. Primary focus will be on letters that call for the exclusion or punishment of individual passengers or specific groups—often not shying from full-blown denunciation. We will also look at how the subway was discursively framed as a place of fear, as the increase in sexual harassment and vandalism became major topics of complaint. Such elements of violence contributed to the destabilization of cultural techniques of isolation and protection against stimuli that passengers had gone to such great lengths to cultivate. The drastic increase in violent incidents forced authorities to expand security regimes, which in turn led to more complaints of bureaucratic arbitrariness and police brutality. Finally, we will look at the strategies and rhetorical forms that passengers used to position themselves vis-à-vis those in power and lend weight to their concerns. We will analyze how complaint letters were processed and evaluated in bureaucratic terms, examining the complex negotiation of subject positions that plays out in the exchange between letter writers and public authorities. As we will see, these negotiations ultimately had less to do with eliminating the cause for complaint than with achieving mutual recognition of the needs of the passenger and the position of authority. In the processing of complaint letters, moments of fear and helplessness were disambiguated and transferred into discursive apparatuses in a way that parallels the subjectivation of passengers.

WHAT IS THE COMPLAINT?

Looking to history, sociology, or cultural studies for inspiration on how to evaluate these documents, it turns out that complaint letters are largely neglected as a type of source material. There is almost no literature on the subject to be found in English.[19] The small amount of historical work that has been done in this field in the German context concerns the process of submitting letters to public officials and party leaders in the former German Democratic Republic, including petitions sent to Communist leaders, requests and suggestions submitted to public administration, and denuncia-tions of alleged dissenters and class enemies.[20] In French scholarship, Luc Boltanski's work on late twentieth-century letters to the editor of the daily newspaper *Le Monde* offers valuable ideas for evaluating and presenting this type of source material.[21] The *lettres de cachet* of the French ancien régime, as edited and commented on by Michel Foucault and Arlette Farge, also suggest useful strategies for interpretation.[22]

English-language research on the phenomenon of denunciation also provides use-ful background for understanding complaint letters, especially the work of histori-ans Sheila Fitzpatrick and Robert Gellately.[23] It is to their credit that after decades of widespread disinterest, recent years have seen more research on historical practices of denunciation. Fitzpatrick and Gellately's analyses focus primarily on the totalitarian regimes of German National Socialism, the German Democratic Republic, and the Soviet Union under Stalin. Nonetheless, their insights have been useful here for inter-preting complaint letters written to subway authorities, particularly those calling for the exclusion and punishment of certain individuals or groups of passengers.[24]

While the relative lack of research in this field already complicates the evaluation of these documents, the task is made even more difficult by the fact that the material at hand comes from incomplete and poorly organized holdings. Not all letters and responses have survived in the New York Transit Museum Archives. Complaint let-ters in the archive date back to the time of the unification of the subway system and the formation of the Transit Authority, but we can assume that passengers also wrote letters to subway operators and authorities before this point. Furthermore, even the letters and bureaucratic files dating from 1954 to 1968 were archived halfheartedly at best.[25] For unknown reasons, files for the period between July 1958 and November 1962 are missing entirely, and there are several other gaps as well. In many cases, only the original submission remains, without the accompanying bureaucratic response or internal documentation. The opposite is also frequently the case: internal communi-cations and responses are present in the archive, but the original letter has been lost.

Thus, we are often dealing with fragmentary, obscure traces of long-past encounters between passengers and authorities. Similarly to Foucault's treatment of the "infamous men" of eighteenth-century France, we have only "fragments of discourse trailing the fragments of a reality in which they take part."[26]

Nonetheless, the corpus of source material at our disposal includes thousands of pages of letters, protocols, replies, evaluations, and internal reports. The fragmented state of the material makes it difficult to evaluate in quantitative terms, but we can observe that the number of letters and responses tended to increase over the years. Until the mid-1950s, the authorities received twenty to thirty letters a month. By the late 1960s, the average had reached one hundred letters per month (figure 5.5).[27]

Despite gaps in the transmission of documents, it is possible to identify dominant themes, from security issues and threats to complaints of police brutality and inappropriate passenger behavior. Some historical events caused a temporary surge in correspondence on particular topics. For example, the exceptionally high submission rate of letters in March and April 1965 was a reaction to the murder of Andrew Mormile in early March 1965, which resulted in a flood of complaints and demands for more security in the subway. While some topics were only relevant in a given moment, the call for more security and control remained a dominant theme. Often composing their letters in the aftermath of some frightening ordeal, many passengers asked the public authorities to exercise their monopoly on the use of force in the subway. Letter writers frequently called for the authorities to expel wrongdoers from the system and punish them. As to the identity of these wrongdoers and the implicit or explicit nature of their wrongdoings, passenger allegations varied widely.

While the Transit Authority was responsible for collecting and processing the letters, they frequently cooperated with the New York City Transit Police Department in investigating passenger complaints. Formed in 1953 when the three separate subway systems were united, the Transit Police became a separate department two years later, distinctly different from the NYPD.[28] It was initially staffed with five hundred officers, some of whom were women. In uniform or plain clothes, they patrolled the subway, fined and summoned wrongdoers, and arrested delinquent passengers. From the very beginning, a major aspect of their work was to register and following up on passenger complaints. With the number of complaints growing along with the rate of

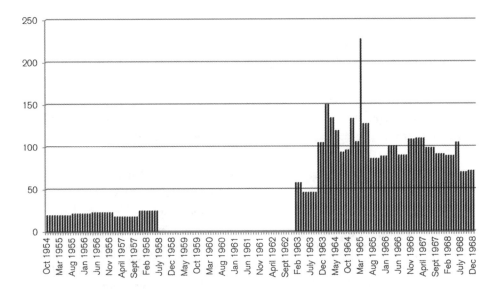

Figure 5.5 This diagram shows the number of archived documents per month, based on a survey conducted by the author and the archivist of the New York Transit Museum, Carey Stumm, in October 2011.

offenses, another 352 officers were added to the squad only a few months later.[29] As the rate of crime continued to rise and public pressure grew, the size of the department continued to expand, along with its powers.[30] In 1966 the department had 2,272 employees, and 3,600 by 1975.[31] In the late 1970s, the New York City Transit Police had become the fifth largest police force in the United States.[32] The need for such a high number of employees was due in part to the monthly inundation of passenger complaints; in order to investigate these complaints in close cooperation with the Transit Authority, it became necessary to mobilize an astonishing level of organizational and personnel resources.

The letters that reached the authorities between 1954 and 1968 differed greatly not only in content and frequency, but also in form and material. Some were quickly scribbled notes and angry outbursts, while others contained pages of detailed description of assaults or threatening situations, along with specific denunciations of alleged perpetrators. Some were apparently written in the heat of the moment, with wild typography, numerous exclamation marks, or hasty scrawling, whereas others included long lists, diagrams, or detailed sketches. Most of the letters were not only typed on typewriters, but also submitted on personalized stationary or the letterhead of the company where the writer was employed. It is obvious that much effort went into the visual appearance of letters in order to lend more weight to complaints.

Before delving into the rhetorical strategies employed to legitimize complaints, we will take a first look at the social background of the letter writers. While recognizing the limitations of generalization, the form and content of most letters suggest that it was primarily middle-class passengers who used this instrument to communicate their experiences and concerns. The majority of them also seem to have been white, as indicated by a variety of factors, including the large number of references to "negro passengers." From the mid-1960s onward, however, an increasing number of letters came from passengers identifying themselves as people of color, turning to the authorities to report scandalous police brutality or other experiences of violence and discrimination.

Men and women appear to have written an almost equal number of letters. However, the often drastic descriptions of sexual violence submitted by female passengers point to a central aspect of the discursive limits of the complaint: certain speaker positions and discursive fields yield recognition from bureaucratic apparatus, whereas others are marginalized or excluded. Foucault emphasizes the significance of this phenomenon when he notes in his definition of the dispositif that it includes "the said as much as the unsaid."[33] As we will see, the implicit rules of the discourse of passenger complaints did not allow for the direct naming of acts of sexual violence, or the repetition of the exact content of obscene graffiti.[34]

As of the 1960s, an increasing number of complaint letters involved police misconduct, although the authorities made efforts to render such incidents harmless or to avoid addressing them altogether. There were also many homophobic, anti-Semitic, and racist accusations and denunciations. Passengers largely adhered to historical convention in expressing their emotions and describing their experiences. The criteria for legitimate forms of complaint were often implicit. By reconstructing historical shifts in the letters themselves and their bureaucratic framing, it is possible to identify conjunctures and ruptures, not only in the content of the complaints, but also in the strategies used by passengers to position themselves as subjects in exchange with authority.

In addition to keeping in mind the discursive structures of what could and could not be said, it is also important to avoid two misunderstandings when analyzing these sources. On the one hand, we must be careful not to interpret these letters from passengers as testimony of "authentic" subjectivity. On the other hand, we must be cautious not to sever these sources from their historical situation within dispositifs of power. Philosopher Judith Butler also warns of naiveté in analyzing self-testimony.[35] She underscores that self-testimony does not provide privileged access to a subjectivity that is somehow true, pure, or coherent. In other words, these letters cannot be taken as testimony of the undisguised inner "self" of passengers, nor as objective descriptions of the contemporary condition of the subway. Nonetheless, these self-drafted documents do allow us to draw some conclusions about the personal circumstances and situations of the writers at the time they were written. Anonymous letters in particular tell us a great deal about how passengers saw themselves. In one way or another, writing a letter is always connected to an act of reflexive subjectivation that reveals one's one position as a speaker. In the narrative development of complaints, writers inevitably position themselves in relation to a system of norms, thereby making these norms explicit. Judith Butler describes how the narrative of the self is linked to defining one's place in society: "When the 'I' seeks to give an account of itself, it can start with itself, but it will find that this self is already implicated in a social temporality that exceeds its own capacities for narration; indeed, when the 'I' seeks to give an account of itself, an account that must include the conditions of its own emergence, it must, as a matter of necessity, become a social theorist."[36]

Historians Mary Fulbrook and Ulinka Rublack also point out the danger of isolating a subject's speaker position from the circumstances that produce this position. They stress that it is essential to contextualize documents of this type by comparing them with other historical sources and research.[37] With recourse to a decentered concept of subjectivity, as advocated by Foucault, Butler, and others, we must interpret the self-definitions presented in these letters as dynamic, in-situ inventions of self that are highly relational in their operation.[38] In other words, subjective testimony given by passengers is often incongruent and inconsistent—a fact lamented time and again by the officials whose task it was to process complaints.

We must also take into account that while complaints provide insight into the self-images of passengers, they were also written with tactical subtlety. In contrast to most diary entries, for example, their primary purpose is not self-reflection but the appeal to authority. They are meditated and calculated representations of a particular person and that person's particular experiences in the subway. Above all, passengers aimed to gain acknowledgment of their experiences and the legitimacy of their demands. Thus, these letters were far more than an instrument of catharsis for dealing with psychological stress. They have to be understood as a form of direct appeal to a higher power, bearing the desire for affirmation of subjective experience and stabilization of a subject position. One indication of this is the fact that many letters had multiple addressees, including not only the Transit Authority, but also the mayor, congressional representatives, and other higher authorities. It was also not unusual for passengers to address their complaints directly to the president of the United States.

Our analysis must take all of these factors into consideration, exercising precision and care in the interpretation of the letters as well as their bureaucratic treatment. Nonetheless, these documents are an exceptionally valuable source for the historical reconstruction of the forms of subjectivity found in the New York subway from the early 1950s to the late 1960s. Fulbrook and Rublack stress the particular value of such documents: "We can thus gain access to subjective experiences without trying to access any particular individual self, while retaining a very lively sense of the individual personalities who not only experienced and witnessed, but explicitly bore witness to the events of their times and the effects these had on people's perceptions and responses."[39]

While exploring the bureaucratic processing of complaints as part of a historically contingent infrastructural dispositif, we must also see it in relation to the political economy and cultural dynamics of the times. During the decades in question, both the city and the subway were undergoing considerable transformation: the subway was deteriorating from neglect, poverty was spreading throughout the inner city, passengers and residents were fleeing for cars and suburbs, and the crisis of the Fordist regime of accumulation was becoming ever more apparent. All of these circumstances and many others found their way into complaint letters and their bureaucratic processing.

NORMS AND TRANSGRESSIONS

This is not just a complaint letter. This is an accusation against the city of New York, which allows such lawlessness to exist in their subways.

—*Abe Goldberg to the Transit Authority, October 29, 1963*[40]

Sorting the letters according to dominant themes, one of the first clear categories is formed by the many complaints related to the decline of a code of conduct that had been so painstakingly established by passengers in the subway. These letters addressed a wide range of actions and behaviors that letter writers saw as transgressions of norm systems, from inappropriate bodily practices and violations of the imperative of emotional containment to harassment and violence. A large number of letters complained about the increasingly desolate condition of the material elements of the subway, disruptions in systems of visual order, and a general disregard for hygiene policies. Over the years, the Transit Police received plenty of complaints regarding the scandalous state of subway cars and stations, with graffiti on the walls, torn seats, and piles of trash everywhere.[41] As maintenance and repair were neglected, the subway's once modern and impeccable hygienic condition began to erode. The lack of regular cleaning, combined with the resurgence of habits like spitting, smoking, littering, and urinating, allowed the system to fall into disrepair, especially as of the 1960s. This was the topic of many outraged complaints, for example: "But, the subway is filthy! It is repugnant, filled with latrine odors of all sorts, generously spread with liquor and puddles of all sorts of heaved up, undigested matter."[42]

Although hefty fines were issued for violating sanitary regulations, deplorable conditions continued to spread throughout the system.[43] And if we believe the complaints, these transgressive acts were performed not only by the subway's paying customers, but also by police officers and subway employees.[44] Coded as disgusting and unhygienic, the spread of these practices was viewed as a violation of civilizing norms, as demonstrated by the degree of rage expressed by passengers in their letters. Many complaints had harsh words for the culprits: "It is most unsanitary to have spit and other filth disseminated in the closed areas of the subways by these human scavengers."[45] One letter from a new resident written in 1965 demonstrates the speed with which the subway went from being seen as a mundane infrastructure to symbolizing the decline of civilization: "This a totally new experience and a most disgusting one. Never have I met in the world such a group of discourteous and inconsiderate people. They are akin to pigs in front of a feed bin."[46] Over the course of the machine age, people had developed complex codes of behavior to cope with the new and unfamiliar situation of underground transit, but now, more and more cracks were opening up in the carefully established social structures of passenger culture.

In addition to outrage regarding the lack of civility among passengers and the dilapidation of the system, disruptions to the sign systems were another frequent cause of complaint. According to letters, schoolchildren were in the habit of switching or defacing signs, which frequently led to widespread confusion.[47] Passengers also protested against inappropriate advertisements, or political and religious agitation in the subway. Above all, they complained about the signs and symbols written by other passengers on the walls of the subway stations and cars, which only proliferated with time.[48] Sprayed graffiti did not appear until the 1970s, but when it did, globally circulated photographs of spray-painted subway cars became iconic images of New York's downfall.[49] Long before this, passenger scribblings had already begun to undermine the subway's hegemonic system of words, pictures, and symbols.[50] These marks elicited a level of outrage similar to the anger evoked by the appearance of billboards and advertisements more than half a century earlier.

Some of the oldest letters in the archives contain denunciations of passengers who covered walls and advertisements with drawings and slogans of their own.[51] In the eyes of passengers and subway operators alike, these often hastily scrawled messages and pictures were grounds for severe punishment. Messages with sexual or racist content were mentioned most often in complaint letters. These complaints bring a remarkable phenomenon to light: while there are very many letters in this category, few writers actually spell out the exact content or wording used in the graffiti. By highlighting the unsayable, these complaints underscore the extent to which these scribbles were experienced as threatening and intimidating. The contents are so unmentionable that, for many years, only a handful of passengers dared to reproduce them, and then only in an altered form. A letter from 1958 demonstrates the taboo:

> I noticed a drawing over an advertisement on the back of the motorman's cab, depicting the male and female's private parts and the words "Boy, girl, F————." How can these filthy, obscene and indecent writings, drawings, etc. go unnoticed by your many employees, supervisors and such?[52]

Besides unpresentable marks such as this, many letters also call for the removal of racist and anti-Semitic content. Slogans like "Kill all Kikes"[53] and "Gas all Jews Now"[54] (often with swastikas)[55] could be found throughout the system, arousing outrage on the part of individual passengers, authorities, and organizations.[56] Another aspect of visual obscenity derived from the literature sold by salesmen and at countless kiosks in the subway. Pressured by scores of complaints and public protest organized by the antipornography movement, in 1965 the Transit Authority began a campaign to eliminate printed material classified as highly dangerous.[57] Officials focused especially on magazines featuring half-naked women and articles about offensive and unlawful forms of love, sex, and violence. Magazines with titles such as *Dude*, *Inmate*, *Police Detectives*, and *Street Confessions* were confiscated as part of this campaign.[58] They were banned on the grounds that their elimination would reduce crime in the subway.[59] The subway was also a contested space when it came to political and religious propaganda. Several letter writers complained about religious agitation displayed on placards, preached in sermons, and distributed in pamphlets.[60] As a result, proselytizing passengers were fined and removed from the system.[61]

Another common complaint was that some passengers entered the system illegally by circumventing or outsmarting the gate system. After 1953, when the Transit Authority introduced purchasable tokens to be dropped into the turnstiles instead of coins, the number of attempts to get around or cheat the system increased. These subversive practices, almost unknown during the machine age, became more and more common with each fare hike. Passengers were creative: they used metal rods to force barriers out of their fittings, forged tokens and master keys, or twisted their bodies to get through the turnstiles without paying. Complex microeconomies sprung up after the introduction of tokens, with organized counterfeit rings and black-market dealers.[62]

Besides violations of official regulations, there were many complaints of failure to comply with unspoken rules of subway courtesy. Some complained of men who spread their legs too far apart or refused to offer elderly or pregnant women a seat.[63] Breaking the unspoken agreement not to put one's feet on the seat was also cause for outrage.[64] It was also tacitly forbidden to fall sleep on the subway. Many letters report such incidents and insist that the guards should wake and remove such "sleepers" and slap them with fines.[65] Some letters call for greater security measures to prevent, detect, and punish such behavior, with suggestions ranging from installing cameras to bringing in guard dogs.[66]

Another common characteristic of complaints regarding the violation of norms by other passengers is the subject position that writers assigned to themselves. While reporting the scandalous behavior of others, complainants frequently described themselves as decent, law-abiding citizens who know how to behave: "We dress nicely, we do not push or shove, we get off in time for our exit, we sit on one seat and not spread out on two. I only mention these rules of 'public transportation etiquette' which perhaps you could print and post up along with the ads."[67]

The majority of complainants mobilized such strategies, reinforcing their speaker position by pointing to conformity with the subway's implicit and explicit norms. This sometimes resulted in long imaginary instructions for other passengers, as demonstrated by the following letter, which incidentally contains an almost complete list of transgressions detected in the subway:

> Don't cut up subway seats or break or loosen subway bulbs where it [is] dark in cars. THINK before you do these awful things. Don't spit or throw papers on subway cars. It's unhealthy. Don't smoke. Signs are there for you to read and obey the law. Be a good CITIZEN, respect your law officer. Don't hold doors of subway cars. Leave it alone. Keep away from trouble makers. Don't bother good people. Act your age. JAILS are full of bad people. Give an old age person your seat. Someday you may be old yourself. Don't write on subway cars or busses. KEEP IT CLEAN, it's for you to ride and enjoy clean subways and busses . . . If cars are crowded don't go in and push. Wait for another train, avoid TROUBLEMAKERS . . . WOMEN keep away from fresh guys who push you and get near you . . . Move away from that kind old man . . . BE A GOOD CITIZEN NOT A BAD ONE. THINK, THINK, THINK . . .[68]

Just how widespread such deviant behavior had become can also be seen in the rigor with which the authorities punished infringements and violations of the rules, as demonstrated by the following example. One apparently very frightened woman piled her personal belongings onto the seat next to her, defending the space for herself even though the car was overcrowded. After witnessing this, another female passenger submitted an angry complaint.[69] A few days later, a police officer recognized the woman on the train and tried to take a seat next to her; she reacted aggressively and was immediately removed from the system, arrested, and taken to a psychiatric clinic.[70] People with apparent mental health issues, as well as people with intellectual or physical disabilities were the object of a massive number of complaints.[71] Although letter writers occasionally demonstrated some awareness of suffering and despair, they nonetheless denounced other passengers and called for their exclusion, often in radical tones. One passenger wrote: "I am hopeful that your organization will get rid of this animal so that the subways will be safe to ride as usual."[72]

These demands for the exclusion and punishment of transgressive subjects exposed the subway as a fragile territory, deeply dependent on conformity to norms. But the call to exclude individual passengers from the subway when they violated official or implicit norms was not the only form of exclusion supported by many complaint letters. There were also repeated appeals to exclude entire groups from the system regardless of their behavior. Above all, the targets of such denunciations were beggars, teenagers, and homeless people, as well as African Americans, Latinxs, and Jews.

<h2>DENUNCIATIONS</h2>

What makes a complaint a denunciation? Following historians Sheila Fitzpatrick and Robert Gellately, denunciation can be characterized very generally as a spontaneous communication from a private citizen to the state or other authorities (churches, administrative agencies, and so on), making an accusation against another person, group, or institution.[73] Denouncers also call for some form of punishment, either implicitly or explicitly. They usually invoke generally acknowledged values, norms, or laws, and deny personal interest in punishing the person they are denouncing. They justify the act of reporting their accusations based on the fulfillment of civil duty and an interest in the common good.

Not all passenger complaints fit into this category by far, but a large number of letters do fulfill the criteria identified by Fitzpatrick and Gellately. Besides passengers with mental illnesses and passengers acting confused, beggars and peddlers were high on the list of denounced persons. Although begging for money, peddling, and performing in the subway had been prohibited by law since the 1930s, these practices never disappeared entirely. When the economic crisis started to become manifest in New York in the late 1950s, the number of such incidents rose rapidly.[74] The Transit Authority regularly received complaints about beggars, peddlers, and members of charity organizations whose activities bothered passengers and compelled them to write:[75] "Why cannot passengers ride the Subways without being harassed and badgered by these professional beggars?"[76] Roaming peddlers selling flowers, newspapers, or snacks angered passengers and store owners alike, the latter fearing a loss of income from such activity.[77] Writers often legitimized their denunciations with the fear that these activities might attract rats as well as other vermin and threaten subway hygiene.[78] For example, one passenger denounced a man selling pastries: "This is a very congested station and this vendor takes up some space. It is a sanitary violation as the pretzels are not covered and take on dust and odors."[79]

This example shows that denunciation is often not easy to distinguish from general complaint, even for those doing the writing. In their research, Fitzpatrick and Gellately show that this was also true for denouncers in totalitarian regimes: "They would mentally categorize their communications as 'letters' or 'complaints,' avoiding any pejorative term associated with 'denunciation.' They were unlikely to feel that such an act was shameful, and the idea that it might make them—or make them appear— collaborators with the regime probably did not even occur to them."[80]

This seems to also apply to the complaints of many New York subway passengers, demonstrating that denunciation can be an important form of communication between individual subjects and bureaucratic authorities even in nontotalitarian contexts. Until now, denunciation has mainly been analyzed within the context of absolutist rule and totalitarian regimes; this is certainly due to its interpretation as a channel of discourse between the powerful apparatus of the state "above" and the people "below" who have very few other ways to participate in government.[81] However, even in the governmental context of democracies and functional constitutions, authorities sometimes explicitly encourage denunciations. This holds true in city subways as well, where passengers are sometimes promised rewards for information that leads to the arrest of delinquent or noncompliant passengers.[82]

Many denunciatory letters portrayed the activities of beggars and vendors as drastic disturbances that were impossible to fend off, even by exercising established techniques of containerization and protection against stimuli. One passenger complained of a salesperson: "His ear-splitting sales talk goes on for about six or seven minutes and if you have the misfortune to be near it is simply impossible to read a paper or even think."[83] Some passengers experienced artistic performances in the subway as such a disturbance—or even threat—that these also warranted denunciation. One letter writer complained of a group of young people making music and dancing: "It is offensive to transit patrons to have to suffer their noise, their urgent solicitation, their tasteless gyrations."[84] In order to put an end to begging and peddling, passengers suggested more severe police enforcement, or the installation of signs in every subway car prohibiting the sale of wares and the giving of alms.[85] This indicates that faith in the disciplinary power of the visual regime remained intact among passengers, despite increasing symptoms of crisis in the system.

Denunciatory letters were also often aimed at homeless people. The city's economic crisis had caused an increase in the number of people without homes, and many sought refuge in the subway. Their presence elicited so much rage from other passengers that many wrote letters demanding severe punishment and the exclusion of homeless people from the system.[86] However, passengers did also express concern for the many homeless people, insisting that the Transit Authority and other city authorities should do something to relieve their misery.[87] Some letters even accused the Transit Police of pursuing an almost fascist routine in the subway, treating homeless people as well as other vulnerable passengers with brutality, excluding them, and infringing on their constitutional rights.[88] Such letters were rare, however. Most complaint letters contain clear distinctions between their largely middle-class authors and poor and homeless people. Some writers even suggested additional steep hikes in the already continually rising price of tickets in order to keep them out of the subway.[89]

The contempt for the city's marginalized groups expressed in these denunciatory letters was underscored by the desire to battle and punish idleness and loitering. Many denunciations concerned idle police officers and subway employees: "I wish to ask you for your cooperation and rid the subway of loafers, lazy employees, and work dodgers who refuse to give the City a fair shake for a day's pay."[90] Others complained of people loitering in stations and entranceways, ignoring the hectic daily bustle that governed the lives of most passengers. As we saw in chapter 1, government authorities began targeting the practice of loitering already in the nineteenth century. A century later, loitering had come to be seen as something so improper and threatening that even the Transit Police took increased action, displaying warnings and sending people on their way. These strategies represent a form of governmentality that sought to ensure the productivity of subjects, pigeonholing homeless people, street peddlers, and loiterers as suspicious and potentially parasitic.

However, the conflicts that proliferated in the subway during the 1950s and 1960s did not always play out along the class lines of socio-economic status. Conflicts between passengers of different races were equally prevalent. Relevant literature from the first half of the twentieth century describes a racially diverse "salt and pepper crowd" fairly free of conflict and relatively harmonious.[91] Complaint letters from subsequent decades present an entirely different picture.

Anti-Semitism also played a strong role in many conflicts, displayed in graffiti and expressed through insults and attacks by passengers and even subway personnel.[92] The complaint letters further testify to ubiquitous racism against people of color. Although New York had never been legally segregated, in contrast to the southern US states, letters provide documentation of racist assaults, humiliation, and police brutality as daily occurrences in the subway. They speak to a climate of violence, discrimination, and exclusion that was dominant in the United States in the mid-1960s, leading to widespread protest and "racial unrest" in cities such as Chicago, Los Angeles, and New York.[93] In the New York subway around this time, violent disputes between white passengers and passengers of color were frequent occurrences. Media reports of the day often contributed to an atmosphere of fear and distrust among passengers. In 1964, when the United States Congress passed the Civil Rights Act officially ending segregation, massive protest broke out all over the country, including New York City. The subway factored into this conflict through the argument that segregation was urgently needed in the system in order to protect white passengers.

A growing number of complaint letters claimed that the subway was a privileged space of violence against white people. An anonymous letter from 1965 states: "The negroes have a hate for the white and they have been admitting it on television, in the newspaper, and they [have] proven it by stabbing us like as though they were killing jack-rabbits. We are living in a city of fear."[94] Many other letters report attacks and intimidation by passengers of color, and demand that the police take action.[95] The majority of the letters do not refer to specific incidents, instead serving to vent general suspicion and racist denunciations. There are often sentences such as: "the Negros in the city are getting too much education,"[96] or: "this hostile race both men + women cause trouble all the time, killing poor law-abiding people, just because they happen to be White."[97]

What becomes apparent here is that the culture of denunciation related to the subway differs from traditional patterns of denunciation in at least one way. Foucault and Farge as well as Gellately emphasize that, contrary to the altruistic rhetoric used by denouncers, denunciation often aims at some sort of advantage for oneself, such as eliminating a competitor.[98] This motive is largely absent from these passenger letters. In their own eyes, complainants may well have been seeking comfortable transit free of disturbances. However, their letters suggest a number of other motives, including personal gratification, compensation for perceived damage, and a demonstration of loyalty to the mayor or other authorities. The mainsprings of these complaints are contempt for certain groups of passengers and the desire to gain official recognition for one's own personal opinion and experiences. As research on the history of denunciation has repeatedly shown, defamation and stigmatization of one kind or another are elements of every society. Specifically, in Western societies in particular, there is a long tradition of denunciations against Jews, homosexuals, and people who are not white. But the primary purpose of denunciation is political on the surface only. As a performative practice, denunciation is likewise a powerful social and cultural instrument. This certainly applies to the denunciations of New York subway passengers, which served to negotiate questions of loyalty, solidarity, and identity.

As the civil rights movement gained momentum, denunciation was also employed as a means of drawing attention to racist or anti-Semitic attacks. Especially from 1965 onward, there was an increase in complaints from both white passengers and passengers of color regarding racist hostility and abuse by the Transit Police and subway personnel. These complaints cover a range of topics, such as racist advertising,[99] a conductor failing to intervene when a woman displayed an offensive sign protesting "Negros" in the subway,[100] or the severe mistreatment of an African American man by a group of police officers.[101] While rare, such acts of solidarity among various groups of passengers indicate that despite the fragmentation of the masses, there were still moments of collective bonding and solidarity.

In complaint letters, a number of the mechanisms that create social hierarchy and enable the exercise of power became part of the discourse of the day. Examination of this process reveals the ongoing struggle to gain control of, and access to, the territories of urban transit. Alongside race and class, gender constitutes another central category of denunciation. Important work from the context of historical gender research has often found that when established subject orders degenerate, gender-specific allocations of space become an object of intense debate.[102] This was also true of passenger culture in the New York of the 1950s and 1960s. Complaints often addressed implicit expectations for women, women's understanding of themselves, and the demands placed on women. Correspondingly, the coding of the subway as a "space of fear"[103] and a privileged setting for sexual violence points to more than just the erosion of passenger culture. It also highlights the implicit rules of discourse that governed complaints related to anxiety and violence in the subway.

DISCOURSES OF FEAR

Especially after the end of the machine age, the subway was transformed into a terrain of danger and fear for many passengers. Complaint letters registered a perceived permanent risk of attack. More and more shop owners sent letters expressing their frustration at losing loyal customers, who no longer came in because they were afraid to take the subway.[104] Additionally, an increasing number of passengers wrote that they avoided the subway after undergoing traumatic experiences: "I am sending you this letter because I feel very disturbed by this incident. So many of my friends have ceased attending the theatre and concerts in the evenings because of their fear of traveling in the subways."[105] It was mostly female passengers who submitted vivid descriptions of violent attacks and sexual harassment, demanding more police protection.[106] These letters made it clear that the women of New York had been victims of such harassment for their entire lives, but these problems had rarely been the topic of public debate:

> I wish to report a situation in the New York Subways which I have not read about in any newspaper. I was recently approached for the third time by a sexual pervert on the D-train of the IND line. The first two did not attempt to touch me but the third one did. These incidents all happened during the day and there was no Police officer around. I am old enough now to understand that these men are sick, but the first time it happened, I was only fifteen and was quite frightened by the experience.[107]

A number of women described similar experiences of being threatened by men, the thought of which still frightened them years later.[108] Some proudly reported what they did to protect themselves, like blowing whistles or wielding a hairpin.[109] While full of gaps, Transit Police records do show that over the years, cases of violence by male passengers against female passengers steadily grew in number.[110] As the subway gained the reputation of life-threatening danger, businesses in New York City began hiring private security guards to escort their female employees right into the trains.[111] More and more women demanded cars reserved for them—a feature that had already been experimented with and dismissed during the early years of the subway.[112]

These reports also reveal a discursive norm that did not allow victims to describe an incident in detail, to make a scandal out of it, or to file charges. For example, after reporting that a group of male passengers had sexually harassed her, one woman wrote of herself in the third person: "She did not scream and did not want to make a commotion."[113]

The taboos shaping the discourse around sexualized violence were not limited to complaint letters. Until the late 1960s, judiciary and bureaucratic institutions paid hardly any attention to assault of this kind. It was not until 1967 that incidents of such violence were legally coded and punished as criminal offenses. Only as of 1975 were these violent offenses recorded statistically and categorized more precisely.[114] The 1980 study "Sex Crimes in the Subway" by Anne Beller, Sanford Garelik, and Sydney Cooper provided the first detailed investigation of these phenomena.[115] In their work, they show that although practices like exhibitionism and frottage (in this context, pressing one's genitals against a stranger's body) had been around since ancient times, they did not become widespread until human bodies were concentrated in the dense spaces of twentieth-century metropolises. The first efforts to register and categorize such incidents—psychologically, legally, and bureaucratically—did not come about until after the Second World War.

The centerpiece of Beller, Garelik, and Cooper's study is a detailed evaluation of Transit Police records compiled after 1975. In their analysis, the authors show that most incidents of indecent exposure, harassment, and other types of improper sexual behavior in public took place in the New York subway system. Puzzled by the revelation that this element of infrastructure apparently presents a privileged location for such offenses, the authors speculate as to why this might be the case: "But something peculiar to the physical setting of the subway seems to make it a preferred site for both these types of sex offenders, and metropolitan subway systems appear to have become a sort of endemic focus or reservoir for such behaviors in modern urban life."[116]

The territory of the subway, with its dark, confusing areas and the particular social situation of anonymous masses of bodies, seems to constitute the central reason why—then as now—more than three-fourths of all sexual assault victims in New York are women on the subway.[117] This phenomenon became more and more visible in complaint letters as of the 1960s, not only because of the striking rise in the number of incidents, but also because of the creeping erosion of the discursive taboo around the experience of sexual violence. As the women's rights movement gained in strength, public awareness of the implicit sexual power relations structuring urban space also grew.[118] As a result, more victims found the courage to report what had happened to them, and to demand more police presence in the system. At the same time, they also emphasized how much more dangerous the system had become, especially for female passengers: "The Subways were a marvelous thing when they were built, but now they are used in every way possible to endanger the life and threaten the respect of every woman riding in them."[119]

As these conflicts were heating up in the 1960s, the general rate of crime on the subway was also rapidly increasing. Despite the exodus of many passengers, overcrowded subway cars continued to offer ideal conditions for pickpockets, along with more and more aggressive confrontations between passengers over attempted theft, unintentional jostling, and so on.[120] Complaint letters describing such incidents or calling for greater police presence make up a large part of this corpus of archive material. Media reports on these crimes also increased during this time, as did the degree of panic in the tone of reporting.

TERROR IN THE SUBWAY

In the course of the 1960s, one central area of passenger conflict as reported in complaint letters was juvenile vandalism and transgressive behavior.[121] According to the Transit Authority, delight in destruction was mostly directed at material elements of the subway.[122] Teenagers damaged car furnishings, turnstiles, signs, and vending machines. They also severely disrupted subway operation by blocking doors and pulling emergency brakes, leading to investigation and significant punishment.[123] Another massive problem was the removal of lightbulbs for use as missiles, which resulted in damage to surrounding buildings as well as serious injury to other passengers.[124] According to many letters, it added to passenger discomfort and fear to find bulbs destroyed in cars and stations that were already deemed insufficiently illuminated. As one woman wrote: "The subway stations are dangerously dark. Personally, I don't have to read on the trains, but they are too dark for safety in a town as lacking in Police as New York."[125]

Vandalism by young passengers was only one of the grievances that piled up over the years. According to Transit Authority reports, teenagers were the primary perpetrators of rowdiness, theft, and fare dodging.[126] According to passenger letters, young people were taking part in brawls and all-out fistfights.[127] One anonymous letter signed "High School Rowdy" from 1962 serves as a particularly noteworthy piece of evidence (figure 5.6). In it, the writer criticizes the system but also seeks attention from the authorities, demonstrating the inherent ambivalence of the instrument of complaint

An internal Transit Authority memo regarding this letter noted: "By his own admission, the writer of this anonymous letter is not very intelligent. Apparently he tried to remove the extinguisher without breaking the glass. If the glass is broken by any easily available object, such as a shoe, the extinguisher can be removed in a few seconds."[128]

As the frequency of such incidents rose in the 1960s, newspapers began reporting them more often, regularly printing angry letters to editors about subway juvenile delinquency.[129] By this point, rowdies had apparently gone from attacking passengers with stink bombs and itching powder to mugging and seriously injuring them.[130] Furthermore, young passengers now also attacked police officers and subway employees, as well as banding together to hold up kiosks.[131] Not all of these delinquents were male. Many letters reported groups of schoolgirls who not only fought violently among themselves, but also brutally attacked other passengers.[132] A number of parents wrote that their children had been attacked on the subway, or that they were worried about

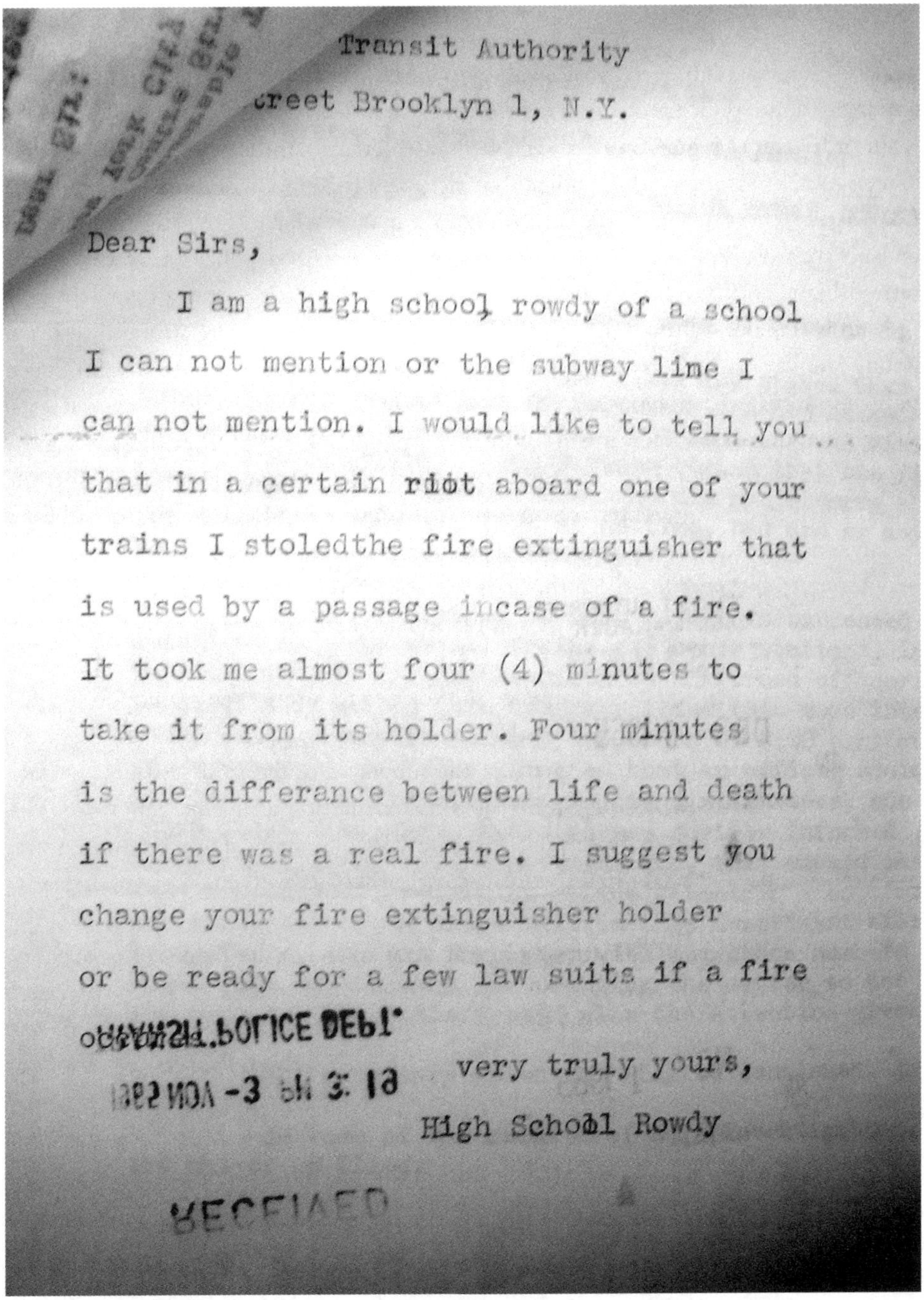

Transit Authority
...reet Brooklyn 1, N.Y.

Dear Sirs,

I am a high school rowdy of a school
I can not mention or the subway line I
can not mention. I would like to tell you
that in a certain riot aboard one of your
trains I stoledthe fire extinguisher that
is used by a passage incase of a fire.
It took me almost four (4) minutes to
take it from its holder. Four minutes
is the differance between life and death
if there was a real fire. I suggest you
change your fire extinguisher holder
or be ready for a few law suits if a fire

very truly yours,
High School Rowdy

Figure 5.6 Anonymous letter signed "High School Rowdy" to the Transit Police, November 3, 1962. Courtesy of New York Transit Museum. Author's photo.

safety.[133] Some parents even threatened to organize patrols and militias to protect their children, should police protection continue to prove lacking.[134]

In order to get a grip on these problems, the Transit Authority arranged various cooperation programs with school authorities, parent councils, and similar groups.[135] Likewise, teachers turned to the authorities in search of help, acknowledging that they had no control over the situation and requesting increased police presence.[136] One retired New York teacher wrote that "'Progressive Education' was a complete failure and was making little 'Hitlers' out of teen agers."[137] There were attempts to reach young subway passengers through education, appeals, and even specially produced radio programs, with dubious success.[138] Despite increased police raids and mass arrests, the number of incidents continued to rise, resulting in a flurry of letters describing the subway as a space of violence and fear.[139] Especially after 17-year-old Andrew Mormile was stabbed to death by other teenagers on the subway in March 1965, the call for a massive increase in police presence could no longer be ignored. In their letters, many people questioned whether the authorities were doing their job, demanding that a state of emergency be declared and that the National Guard be deployed in the subway system.[140]

This crisis in legitimation compelled authorities to assert their monopoly on the use of force. In early April, New York Mayor Robert F. Wagner (1910–1991) gave a prime-time televised speech on safety in the subway. Printed in newspapers nationwide, the speech announced the immediate implementation of an emergency program to combat catastrophic conditions in the subway. Wagner came down hard on groups of delinquent passengers, or in his words, "roughnecks and wolf packs of young brutes and sadists who have terrorized and tormented subway riders."[141]

To restore the subway's security regime and put an end to the exodus of passengers, Mayor Wagner promised to install various technological innovations, such as alarm systems, radio communication between train conductors and police stations, and even surveillance cameras. One idea was to close off particularly dark areas of stations at night. Another idea was to put a stop to mugging by encouraging further crowding, with half of the subway cars locked after dark to force passengers into closer quarters. Thus, the overcrowding of the system—otherwise a logistical problem—would be employed to enhance security. This plan was scrapped, however, when trains became so overloaded that police patrols were impossible.[142] The mayor also announced that all fire fighters, police officers, and correction officers would be permitted to use the system free of charge; they were meant to intervene whenever necessary in order to make the system more safe. In fact, the mayor appealed to all passengers to step up: "It is not enough to complain. You, too, must help advance the cause of safety in the subways and in the city at large."[143] But while passengers were called upon to actively produce their own security, it was unclear exactly how they were supposed to manage this.

The most important part of the emergency program was the immediate expansion of the Transit Police. Within weeks following the mayor's speech, more than one hundred new recruits were hastily trained, and existing patrols were reinforced with officers from other departments. Now, police officers were assigned to every train from 8 p.m. to 4 a.m. Many passengers welcomed this measure in their letters.[144] At the same time, the increase in police presence also led to more police brutality, abuse of power, and arbitrariness. Beginning in 1966, the authorities received letters almost every day complaining of inappropriate, violent, and even criminal behavior on the part of the police. Passengers complained of bullying, unwarranted reprimands, and other unfriendly and ruthless behavior.[145] There were also increasing complaints of sexual assault and brutal mistreatment by police officers.[146] To express what they found scandalous about such incidents, an astonishing number of passengers made reference to dictatorial regimes, drawing comparisons with the Gestapo and Stalin's secret police.[147]

The authorities were unruffled by this storm of accusations and indignation. Complaints rarely led to disciplinary measures or consequences for personnel. Most officers and subway employees who were questioned in the course of internal investigation of these events said that they could not remember what had happened.[148] Prosecution was pursued only when there was unequivocal evidence that a police officer had committed a serious crime, such as robbing a passenger.[149]

While the emergency program did lead to a reduction in the rate of crime in the New York subway, along with a minor ebb in the tide of complaint letters, these effects were short lived. In the following years, levels of subway crime and passenger complaints rose again. Complaint letters were no longer archived after 1968, but this conclusion to the paper trail does not by any means mark the end of the system's decline. Indeed, the downward spiral of material neglect, urban economic crisis, and the exodus of subway passengers continued in the decades to come. Meanwhile, passenger testimonies from the years leading up to 1968 constitute a powerful image of the increasing fragility of subject culture in the subway.

FRAGILE SUBJECTS

The experience was frightening and made me very nervous and upset.

—*Letter from Rica Whelehan to the Transit Authority, January 29, 1964.*[150]

Portraying the subway as an increasingly threatening and contested territory, complaint letters also testify to how radically public opinion of the subway had deteriorated within a matter of decades. Once celebrated as a great achievement of modern civilization, the subway was subsequently normalized as a reliable urban convenience, before degenerating to the point where it came to embody all of the evils and barbarisms of urban life. This is particularly apparent in the preferred metaphors employed in the letters, with passengers describing the subway as hell, an inferno, or death row.[151] The jungle served as another recurrent image.[152] In keeping with these organicist metaphors, authors of complaint letters frequently referred to other passengers as animals.[153]

Letters from passengers indicate that the deterioration of the subway not only contributed to the exodus of passengers, but also to the loss of population for the city of New York. In a letter from 1956 with the heading, "Why you are not only losing me but six of my friends," one passenger of many years complained of several incidents of harassment that she had experienced while station guards failed to act: "As to your employees, they are the laziest, nastiest bunch of punks that I have ever met."[154] At the end of the letter, she swore never to set foot in the subway again.

This was not an isolated case. Over the years, New York gained the reputation of being a "rotten city," the seat of the downfall of civilization and morals, definitely a place worth leaving.[155] In remarkably drastic terms, passengers described their states of psychological and physical distress or sensory overload. Regularly, there were detailed reports of trauma following attacks and frightening situations that caused lasting damage and required medical treatment.[156] Experiences of panic and fear led to countless statements such as: "I find that the fright destroys me more than any possible loss."[157] Both men and women diagnosed themselves with shock, and some ended their descriptions of traumatic situations in the subway with sentence like, "The shakeup and assault caused nervous and mental instability to my person,"[158] or, "I went home, sobbing and near physical breakdown."[159]

At the same time, as truthful and sincere as these descriptions may appear, it is important to keep in mind that such testimonies of the fragile subjectivity of passengers were often calculated attempts to lend weight to their complaints. The letters doubtlessly demonstrate the immense uncertainty experienced by passengers of every age, gender, ethnic background, and social status. Despite their many differences, most shared a desire for validation of their experiences by institutions of authority. Letter writers mobilized many different rhetorical strategies to underscore the legitimacy of their complaints, simultaneously articulating the deepening crisis in the social order of the subway.

How to Address Complaint to Authority?

I am human—if you went through it—you wouldn't like it one bit—and there are thousands of people who feel the same way—of course, of this, I am sure you are also aware.

—*Sydelle G. Carlton to the Transit Authority, December 14, 1956.*[160]

Which rhetorical strategies are successful when composing a complaint letter? How does one compel authorities to recognize and acknowledge one's accusations and requests? In order to answer these questions and understand the function of these letters, it is necessary to consider the meaning that passengers themselves ascribed to their complaints. In 1965 a teacher from New York attached the following note to a bundle of papers that her students had written to the Transit Authority: "These letters that have been written to you are their first efforts to make their voices heard; to communicate, as citizens, with the authorities that shape their world."[161]

These few lines demonstrate several of the key elements involved in submitting complaints: the belief that the letter will actually be read and grievances heard; the legitimation of the writer's position as a citizen of the United States; and the acknowledgment of the power of the addressee. However, the teacher does not express any awareness of this dual function. A complaint questions authority but also affirms it. It usually first highlights some mistake or offense on the part of the authority, but then proceeds to suggest that the institution in question has both the competence and the resources to solve the problem. This tension between questioning and acknowledging authority reveals the contradictory nature of the instrument of complaint. Indignation at the authority's failure to solve a problem involves admitting that, unlike the complainant, the authority is in a position to change things. In their letters, passengers often addressed this predicament quite candidly, calling out the hegemony of the governmental dispositif: "You have the power to correct this evil situation."[162]

Thus, the complaint letters of subway passengers often served to maintain social hierarchies rather than undermining them. In this way, these letters fulfill a similar function to the *lettres de cachet*, petitions written by citizens of eighteenth-century Paris to the king's chancellery, or the police, requesting that other people be seized and locked away. These petitions were often attempts to rid oneself of family members, at least temporarily. Through these letters, Michel Foucault and Arlette Farge describe the relationship between absolutist authority and subjects:

> What an astonishing process this appeal to the king was; passing through him as an intermediary meant interesting (inflecting) his will, directing (catching) his gaze, which was ordinarily turned toward matters of state, existing in his eyes, requesting that it might please him to linger over the details of lives that normally would have every reason to remain submerged in the opaqueness of the multitude. To write to the king, to oblige his hand, was to introduce oneself into History, and to compensate in a spectacular manner for one's social status.[163]

The appeals of subway passengers to New York authorities had an analogous form. Most letter writers took care to address the authorities with the utmost respect, often mobilizing an elaborate choreography of rhetorical persuasion. They often began with a gesture of gratitude or respect, praising the accomplishments of the Transit Authority or the mayor, and offering assurance that the writer basically thinks well of the addressee. This was usually followed by personal information aimed at demonstrating social standing, for example by pointing out that the writer had been riding the subway for many years, or had a respectable occupation. Some letters mentioned personal acquaintance with leaders from politics or business or even Transit Authority's managers.[164]

Other letters were written by groups of passengers expressing solidarity with one another, submitting petitions with long lists of signatures in the hope of lending more weight to their calls for more police presence or their demands for better service.[165] Frequently, passengers sent their letters not only to the Transit Authority or the mayor's office, but also to senators, members of congress, and local newspapers.[166] Especially in these letters, passengers regularly mentioned their political party affiliation or the candidate they had voted for in the last election.[167]

After thus positioning themselves, writers normally went on to present the actual complaint in a subservient manner, operating with the assumption that the writer and the authority had common interests and were on the same side. For example, one letter rationalized the significance of the complaint by pointing out that the deficiency in question would also make a bad impression on visitors and tourists.[168] Similar arguments were used in denunciations or requests for the exclusion of certain groups of people from the subway, especially homosexuals, beggars, and people of color. With formulations such as, "Please make the subway safe for white middle-class taxpayers,"[169] letter writers aimed to demonstrate that they belonged to the most legitimate or relevant portion of the population. Surely public authorities would be favorably disposed to such complaints!

Some passengers used the same strategy in an attempt to stave off fines for fare dodging, smoking in the subway, and other kinds of misconduct. Sometimes they wrote long and very personal letters explaining the unusual circumstances surrounding an action, expressing regret or swearing innocence. Parents of children caught vandalizing or dodging fares also wrote at length. They tried to circumvent fines by vowing that an important lesson had already been learned by their remorseful off-spring, or by enclosing the unpaid fare in the envelope, albeit to no avail.[170]

In the 1960s, as the system continued to deteriorate and the rate of crime rose, a growing number of writers began to deviate from this formula of subservience, instead striking a demanding, accusatory, or threatening tone. A sense of helplessness and being at the mercy of the system often led to outrage and insults: "Has every-body in N.Y.'s officialdom cracked up?"[171] or "Instead of Transit Authority, the name should be Transit Perverts."[172] Some writers underscored their anger and frustration by threatening to sue or boycott the system, or share their stories with the press.[173] Faced with what they considered an insufficient number of officers on patrol, several passen-gers also threatened to carry guns and purge the system of delinquents themselves.[174] Such threats were particularly frequent in anonymous letters, which constitute a spe-cial category. While the roughly 200 anonymous letters make up only a small part of the corpus of material in the archives, they reveal a great deal about how letter writers defined their own subject positions and supported the legitimacy of their complaints.

The first thing to note is that anonymous writers almost always signed their letters, not with their names, but with the description of a position that would lend weight to their concerns without revealing their identity. Some designations, like "night worker,"[175] or "an honest paying passenger,"[176] came across as neutral, whereas others were charged with emotion, such as "worried mother,"[177] "a frightened victimized cit-izen,"[178] and "terrified woman rider."[179] The majority of anonymous complaints, how-ever, bore the signature of "citizen" or "taxpayer." These particular designations of subject position apparently represented gravity and recognition for many writers seek-ing to underscore the legitimacy of their complaints. At the same time, these signatures also reflect historical dynamics. In the 1960s, as the number of anonymous complaint letters increased, there was a general shift in the frequency of these self-designations, with reference to national citizenship giving way to legitimation through the payment of taxes. Rather than a subject form with specific rights guaranteed by national law, passengers pointed to their economic participation in order to add to the efficacy of their complaints.

At the same time, a number of anonymous letters were signed in ways that delib-erately inserted distance between the person writing and the issue of complaint ("No name because I am a coward").[180] One complaint about intoxicated "negros" in the subway was signed "One million middle-class whites who paid their taxes of New York."[181] Other anonymous authors signed their letters in ways that emphasized their belonging to a certain collective, such as "Subway Riders on the Fourth Avenue Local Trains"[182] or "A Member of the Veterans of Foreign Wars."[183]

This strategy of refusing to identify oneself and claiming to represent a collective subject position frequently created serious bureaucratic problems for the Transit Police. Internal department memos expressed frustration that it was not possible to investigate complaints when writers did not give their names, or when they gave false names and addresses.[184] The inability of the administrative apparatus to deal with this blurring of the identity of subjects issuing appeals was based in large part on the proliferation of complaint writers, and the lack of clarity in what they represented. When there was no clear author behind a complaint, or when the author was obscured behind a general category like citizen or taxpayer, it could have been written by almost anyone. This intentional strategy allowed passengers to keep themselves out of the bureaucratic spotlight, while at the same time granting themselves the power to speak for others.

The authorities evaluated the claims made in anonymous complaints, but generally pursued them no further. All other complaints were subject to complex administrative processing. The passengers who submitted complaints under their real names were identified, questioned, and examined though a process meant to determine their subject status. Underlying this process is an extensive microphysics of power that demonstrates the often difficult and contradictory relationship between subjects and bureaucratic dispositifs.

VALIDATION . . . FOR THE FILES

Put your complaints in order, state the most important first and then the others in descending order, then perhaps you won't even need to mention most of them.[185]

—*Franz Kafka*

Tracing the complex path of these letters as they made their way through the administrative workings of the Transit Authority and Transit Police, it quickly becomes clear that an immense amount of effort went into processing passenger complaints. Complaints were dated, stamped as received, and then evaluated in terms of possible official reactions. Accusations against Transit Authority or Transit Police employees floated around the department until one staffer or another declared him- or herself willing to look into it. The first step on the part of the Transit Authority was usually to examine the validity of the complaint, determine the writer's place of residence, and explore ways to get in touch with him or her. In many cases, this involved cooperation with the Transit Police, who left no stone unturned to find the author of a given complaint. They went through files at registration offices, contacted possible neighbors and relatives, and even made multiple visits to the place of residence. By looking up the names in police databases, they were also able to determine whether an individual had a criminal record or might be on the run from the law.[186] In the process, they frequently identified notorious complainants whose accusations apparently never proved true and therefore warranted no reply.[187]

When a letter was written anonymously or under a false name, as discussed above, this did not mean that it was dropped immediately. A clerk still wrote up a report, noted any results, and asked a superior for permission to archive the folder.[188] If the identity of the author was verified, which was possible for the majority of the letters, the real processing of the complaint began. Considerable effort went into arranging an appointment to discuss the contents of the letter with its author, either at his or her home or place of work.[189]

Passengers reacted quite differently to Transit Police requests to discuss the matter in person. They were often surprised to discover that their letter had been taken seriously, and immediately agreed to meet. Others, however, wanted to avoid any direct conversation: they denied having written the letter, or claimed that they had written it in the heat of the moment and the contents were unfounded. In some cases, they were not home when the police came at the arranged time, or they had asked neighbors to say that they were not home. Perhaps they rightly suspected that the purpose of this procedure was to test them, to critically evaluate their complaints, and to inform them of the many security measures already in place.

Whenever officers did actually meet with the author of a complaint, the encounter was carefully protocolled. These records often provide valuable insight into the motives of both passengers and officials. In an astounding number of cases, complainants admitted that their descriptions of events were exaggerated or entirely fictitious. Officers frequently added their own personal assessment of a passenger to the report. For example, after visiting a man named Mr. Lazar at his home (Mr. Lazar had complained of the lack of police on the subway), one officer noted: "The investigating officer found Mr. Lazar to be an over-zealous and anxious individual who expressed great concern in the subject matter but who was unable to discuss the topic to any intelligible extent."[190]

Another striking aspect of these encounters is how many passengers were unable to provide any concrete examples of the generally horrible and dangerous conditions they complained of in the subway.[191] Officers took this as an indication of the complaint's lack of credibility. Looking at the internal evaluations of people who had submitted complaints, it is clear that it was part of the officer's duty to assess whether a person could be considered normal and sane. These judgments of a subject's normalcy were especially important for evaluating a complaint that involved a direct accusation leveled at a specific police officer or another official.[192] These procedures offer a vivid example of Judith Butler's thesis, informed by Foucault, that regimes of truth are constitutive of the processes of subjectivation.[193]

In most cases, the complaints expressed in conversation with the investigating officers did largely match those expressed in the letter. Once this had been established, officers proceeded to inform the passenger at length about how seriously the authorities took complaints, and how they would go to great lengths to redress the issue. Most of the protocols end with the following statement: "The complainants were informed that the Transit System is patrolled on a 24 hour basis . . . They were assured that special attention will be given to the subject area . . . Both expressed satisfaction with the attention given to this matter . . . Recommended that all papers be filed."[194]

These few lines hint at a central function of the processing of complaints: once passengers had thanked officers for the attention paid to their concerns and problems, both parties offered gestures of mutual recognition and assurance. For both parties, this act of acknowledgment was apparently much more important than eliminating the grounds for complaint. Passengers did hope that their words would lead to action and results, but rarely did a letter of complaint ever leave the realm of discourse. Some complaints led to further investigation and accused officers were questioned by their superiors, but usually that was all. Often the officers in question replied that they had no recollection of the incident, or they described the circumstances entirely differently.[195] When complaints contained accusations against officers, there seems to have been a tendency to believe the testimony of colleagues more than the accounts presented by passengers. Thus many reports ended with the statement: "No dereliction noted on the part of any member of this department,"[196] or "Investigation has failed to substantiate the complainant's allegations."[197]

The fact that authorities made relatively little effort to eliminate the cause of complaint, instead mobilizing considerable resources to track down and assuage complainants, sheds light on the significance of these letters for the authorities. On the one hand, passenger concerns appear to have been largely trivial in their eyes. On the other hand, accusations seem to have roused some vague danger that needed to be thwarted. Generally, unacknowledged complaints threaten to create wider ripples, discrediting the reputation of the authorities; the authorities rely on the public for legitimation. In their study of petitions submitted to the monarchy of the ancien régime, Farge and Foucault underscore the reciprocity involved in calling upon power and the pursuit of mutual acknowledgment: "A petition not only made it possible to avoid losing honor, but granted upon the person who wrote it the pride of being recognized by the most important state figure. Was this not, moreover, a dual movement?"[198]

The procedures used to process subway complaints can be interpreted as a similar kind of dual subjectivation—that of the passengers on the one hand, and subway authorities on the other. What is primarily negotiated in these complaints is the validity and objectivity of subjective perceptions and feelings. In the course of bureaucratic assessment, passenger accounts went through various processes of translation: first they were registered, then checked, verified, or falsified, and then either filed away in archives, or used as a basis for concrete actions.

If successful, the processing of complaints brought about synchronization between the subjectivation of passengers and that of subway authorities. In other words, a reciprocal attribution of subject positions took place, stabilizing subject orders as well as trust in the infrastructure itself and the legitimacy of its operators. This process can be understood as a means of successfully disciplining and solidifying the passenger's subjectivity, as well as an instrument that upheld social order in the subway.

Perhaps this explains why bureaucrats took note of even the most trivial and insignificant complaints. Many passengers wrote letters of appreciation once the police had visited them and heard them out. The brief encounter with an institution of power appears to have provided a somewhat uplifting moment for many. Knowing that their individual concerns were important enough to be heard by authorities gave complainants a feeling of triumph, awe, and loyalty. The procedure was very effective at restoring passengers' belief in social order and the authority of public institutions. The procedures were in fact so successful at reassuring and stabilizing passengers that some begged the same for others. For example, one passenger wrote: "My mother for 20 years has been riding the subway to see her brother in the Bronx. She lives in Brooklyn. She is now afraid to ride the subway for fear of her life with all the terrible things that are going on. Could you please write her and tell her that everything is going to be alright and that she has nothing to fear?"[199] And indeed, a police officer was actually sent to visit both the writer and his mother at her home, assuring them of the police presence in the system, which they appreciated.[200]

When we look at the complex procedures used by New York authorities to handle passenger complaints, we find a subjectifying force at work that inverts Althusser's model of interpellation in a certain sense. In his iconic text *On the Reproduction of Capitalism: Ideology and Ideological State Apparatuses*, Althusser depicts a kind of primal scene in which he compares the process of becoming a subject with a police officer's call to someone on the street: "Hey, you there!" The person summoned turns around, and it is this act of responding that makes him a subject, because he "has recognized that the hail 'really' was addressed to him and that 'it really was he who was hailed.'"[201] Subway passenger complaint letters demonstrate the opposite movement, so to speak. Here, subjects appeal to authority, demanding confirmation and recognition of their subjectivity. Historians Alois Hahn and Volker Kapp have made the controversial claim that Western modernity is marked by an increase in the intensity of social control involving refined techniques of self-observation, as well as other forms of voluntary and involuntary subject commitment.[202] In a certain sense, this applies to many aspects of the complaint regime of the New York subway. The subjectivity of both passengers and the authorities appears in a decidedly reciprocal relationship. Complaint letters and the bureaucratic responses they elicited show that becoming a subject involved not only subjugation, but also empowerment, as evidenced by the claims and demands of passengers. By demanding that their experiences be acknowledged, and thereby making their experiences part of bureaucratic discourse, passengers insisted on their individuality and constituted themselves as independent people with opinions deserving of respect. By doing so in writing, they defined themselves as participants in official discourse, with a "body made of paper" and a "heart made of writing."[203] Thus, passengers compelled the administrative dispositif of the subway to align itself with them.

The way that the authorities interpreted and evaluated complaints also reveals the intricate interplay and microphysics of power that existed between passengers and bureaucratic apparatuses. Accusations brought forth by passengers created relationships among a number of actors, including complainant, addressee, accused, and possible witnesses. In processing these complaints, authorities strove to negotiate between these actors, to produce moments of reciprocal recognition, and to maintain social order and control. For these reasons, complaint letter were also powerful instruments for subjectifying passengers through infrastructure.

The archive does not hold complaint letters written after 1968, but the end of the paper trail for this study does not coincide with any let up in the decline of the subway system. On the contrary, the erosion that had begun in the 1950s accelerated in the late 1960s and early 1970s, bringing the system to the verge of collapse. According to contemporary sources, both the city and the subway proceeded to slide into the worst crisis in their history—a crisis that would last for almost twenty years.

Limits of Containment

From Vigilantes to Observers

According to media reports and statistics from the time, 1982 was the worst year in the history of the New York subway.[1] In the years after the Second World War, the subway transported almost seven million passengers every day; by 1982 ridership had dropped by more than half.[2] In October of that year, the number of passengers barely reached 1917 levels.[3] The subway was in a miserable state, with accidents, staff shortages, and faulty equipment causing increasingly frequent breakdowns.[4] By this point, a trip that had taken ten minutes in 1910 now often lasted more than forty minutes.[5] The system also made a catastrophic impression on passengers, as described in a *New York Times* article from January 1982: "The subway is frightful looking. It has paint and signatures all over its aged face. It has been vandalized from end to end. It smells so hideous you want to put a clothespin on your nose, and it is so noisy the sound actually hurts. Is it dangerous? Ask anyone, and, without thinking, he will tell you there must be about two murders a day on the subway."[6]

Although the rate of murder in the subway was much lower than hysterical media reporting suggested, it had nonetheless hit a record high of thirteen murders per year.[7] Meanwhile, the rate of other crimes in the subway likewise broke one record after another. The Transit Police registered more than 250 offenses per week, much more than any other transit system in the world.[8] Faced with these abysmal conditions, more and more New Yorkers preferred to commute by car. The logistics experts of the Metropolitan Transportation Authority calculated in panic that if the current trend were to continue, the last passenger would turn his or her back on the subway in 2002.[9]

As control slipped from the hands of subway operators over the years, petty criminals and so-called youth gangs began to represent more and more of a counterforce to the institutional regime. At the same time, homeless people, beggars, and substance users increasingly found their way into the system. Photographs from the period convey an almost post-apocalyptical view of the subway with its dilapidated stations, broken lights, and growing piles of trash. Works by New York artist Bruce Davidson (figure 6.1)[10] and magazine features by war photographer Christopher Morris (figure 6.2)[11] provide glimpses of the subway as a subterranean landscape of ruins, where stations and cars covered with rust, filth, and graffiti suggest decaying monuments to an unfulfilled promise of the future.

Although the system is said to have hit absolute rock bottom during these years, it did not fully collapse. For millions of people in New York, the subway continued to be an indispensable part of their daily routine; they endured its discomforts and dangers day in and day out. The subway had become such a constitutive element of New York that its collapse would have meant the breakdown of urban order in general. In 1982 the city had come closer to this breaking point than ever before, begging the questions of what led to this crisis, and how it was resolved.

In the late 1960s, although crisis conditions were already creeping into the subway, the system had experienced one more peak of standardization and modernization. In 1968 the subway was united with all of New York's busses and ferries under the umbrella of the recently founded Metropolitan Transportation Authority (MTA). With more than 60,000 employees, the MTA became one of the largest public institutions in the United States. Indeed, it prides itself on being one of the largest regional public transportation providers in the Western hemisphere. Shortly after it was constituted, the MTA was able to secure an enormous contribution of 600 million dollars from the federal government of the United States in order to prevent further deterioration of the system. However, the MTA invested most of this money in ambitious construction and development projects, many of which were either entirely abandoned or massively delayed—like the Second Avenue line.[12]

Figure 6.1   Photo: Bruce Davidson, subway, New York City, USA, 1980. © Bruce Davidson/Magnum Photos.

Figure 6.2 Photo: Christopher Morris, discarded newspapers and trash on the floor of a subway car, 1981.

While passenger numbers continued to decline at an accelerating rate, in the 1970s subway operators finally implemented a uniform system of signs for all lines and stations, with a clear formal design that is still considered exemplary today.[13] Meanwhile, however, both the MTA and the city above ground faced increasing fiscal distress. Industry and commerce steadily moved away from the city, leading to ever deeper crisis for an economy modeled on Fordism. In 1975 the city's failure to generate tax revenue led to an official declaration of bankruptcy. By then, almost a million New Yorkers—most of them from the white middle class—had left the city over the previous years.[14]

The phenomenon of "white flight," along with the continued rapid increase in poverty and crime, soon gave New York the reputation of being a lawless metropolis beyond hope of salvation. The media propagated an exaggerated image of the city as a place where civilization itself was in a state of emergency. More than any other image, the rundown, graffiti-covered subway became a symbol of New York's demise, making headlines in international newspapers and on television. During the same period, the subway also became a symbol of the deterioration of moral order on the silver screen. Successful contemporary films like *The Taking of Pelham One Two Three* (1974) and *The Warriors* (1979) portray the New York subway as a lawless territory ruled by marauding gangs.

These themes also feature centrally in *Death Wish*, a New York drama of vigilante justice from 1974. In one iconic scene, the protagonist played by action star Charles Bronson is attacked at night on the subway by two criminals and threatened with a knife. He pulls out a revolver, shoots the attackers, and then stoically executes them both as they lie already injured on the ground. This cinematic fulfillment of the revenge fantasies of many New York passengers yielded high box-office results and several sequels. Additionally, it served as inspiration for some passengers to follow Bronson's example in real life.

In 1977 a group of passengers calling themselves the Magnificent 13 formed an organization that soon became known worldwide as the Guardian Angels.[15] By the early 1980s, this civilian watch had grown to include 220 men and women who patrolled the subway in self-styled uniform day and night, performing citizen's arrests, and enacting a new ideal of the passenger.[16] Taking matters into their own hands, passengers no longer appeared as frightened subjects at the mercy of their surroundings, but as empowered individuals. As militant producers of security and law, they were the opposite of the fragile passenger-subjects discussed in chapter 5. News of these subway vigilantes traveled fast, gaining widespread media attention and the enthusiastic support of large portions of the population. The group was portrayed by among others photographer Bruce Davidson, who had accompanied the Guardian Angels for several months in 1979 (figure 6.3).

Thanks to the group's media success, additional groups of Guardian Angels soon began to crop up in other urban underground systems, for example in Boston, Washington, DC, and London.[17] The crisis in the Fordist accumulation regime that had already shaken New York for years now began to impact other urban centers in North America and Europe, also negatively affecting their public transit infrastructures.[18]

In New York and elsewhere, these passenger watch groups were unable to curb the sharply rising rates of crime in the subway.[19] Although skeptical of the group, the Transit Authority initially allowed the Guardian Angels to carry out their patrols, which provided much needed assistance even as they undermined the official monopoly on the use of force in the subway. Soon enough, however, there was real cause for concern regarding self-empowered passenger-subjects willing to defend the order of the system by force. On December 22, 1984, an incident occurred that remains a matter of heated controversy to this day. According to legal documents, on this day in the early afternoon, thirty-two-year-old white electrical engineer Bernhard Goetz was attacked by four African American passengers in a train below Manhattan. He defended himself with several shots from an illegally purchased revolver, seriously injuring his attackers. Although Goetz initially fled to evade responsibility for his actions, only turning himself in more than a week later, his actions were met with widespread public approval. In special broadcasts and on countless cover pages, the media celebrated Goetz as a heroic "subway vigilante."[20] In light of the unacceptable conditions in the subway and the perceived incompetency of the police, many passengers viewed Goetz's deeds as exemplary, donating thousands of dollars via the Guardian Angels to post his bail. At the same time, critical voices deplored his actions as brutal and disproportionate. In court an almost exclusively white jury exonerated him on all counts, and he was sentenced to only eight months in custody for illegal possession of a weapon. Deemed

Figure 6.3 Bruce Davidson, 1980, *Members of the passenger watch group Guardian Angels in 1979.* © Bruce Davidson/Magnum Photos.

racist by many, this verdict prompted angry protest. Goetz became the most famous and controversial New York subway passenger of the day, and an icon for those who dreamed of defending themselves against subway crime with armed violence.

It was only as the city began recovering economically in the 1980s that the downward slide of the subway began to reverse course, with more money flowing into the coffers of the MTA. It was finally possible to carry out long overdue repairs, and New Yorkers increasingly returned to the subway. Through massive police intervention and new chemical methods, subway operators were also able to get a handle on the "graffiti war." A decrease in the rate of crime beginning in the 1990s has been attributed to Mayor Rudolph Guiliani's controversial "zero tolerance" policy, along with the recovery of the economy and a reduction in unemployment.[21] With the subsequent boom of the financial industry, New York advanced to become a leading global city.[22]

Although the reliability and safety of the subway system began to improve, it was slow to shake off the reputation of being a highly dangerous territory. In the 1990s, when New York reinvented itself as a city of tourism and finance, the subway began to regain importance.[23] The system's traditional hours of operation—around the clock since the day of its inception—proved highly compatible with the new rhythms of flexible work organization in a post-Fordist urban economy. The subway thus supported the restructuring of New York's economy, now oriented toward tourists, service workers, and those belonging to the "creative classes."

As public authorities reestablished control over order in the subways, the MTA also registered a strong decline in crime rates. This trend was promptly instrumentalized for a far-reaching marketing campaign propagating the renaissance of the system with slogans like: "The Subway. We're coming back, so you come back."[24] By the late 1990s, the volume of passengers had almost climbed back to the levels from shortly after the Second World War, and the subway's comeback was undeniable.

In the past three decades, and over the turn of the twenty-first century, the subway has once again become a mundane part of everyday life for the majority of New Yorkers. Little about it reminds people of the dread and discomfort suffered in times of crisis. The control regime has been successfully reestablished using previously proven methods of surveillance and discipline. Expertise gained over decades of controlling passengers has been routinely and successfully implemented as the MTA's financial and human resources continue to grow. Existing control methods have been augmented by camera surveillance and stricter prosecution of petty offenses in accordance with the controversial model of "broken windows" policing.[25]

As a result of the attacks on the World Trade Center on September 11, 2001, and the subsequent increasing militarization of urban space, new ideals of the subway passenger have once again emerged in New York. American flags in all stations and every subway car now remind passengers of their patriotic allegiance. There has been an increase in identity checks, bag searches, camera surveillance, and other security measures in order to immediately identify passengers deemed dangerous and remove them from the system. In light of such procedures, from the perspective of law enforcement authorities, the passenger appears as a potential terrorist threat to be barred from the system at all costs. At the same time, however, passengers are increasingly regarded as crucial actors in the fight against terrorism and the maintenance of public order.[26] In late 2002, as part of a broad security campaign, the MTA put up thousands of posters with the message: "If You See Something, Say Something!" This campaign aimed to encourage subway riders to report suspicious or nonconformist passengers to the authorities. This call to denounce other passengers revives procedures from the complaint regime. More than ever before, the responsibility for maintaining order in the system is delegated to passengers themselves, who are prompted to suspiciously observe and control one another. Communicating to passengers that their own actions are crucial for sustaining safety and security in the subway, authorities indirectly admit the limits of the institutional capacity to produce order. This "responsibilization" of passengers can also be interpreted as acceptance of the desire of passengers to empower themselves and take control of security in the system, as the Guardian Angels did of their own accord. Following political scientist Susanne Krasmann, these appeals can be seen as paradigmatic of neoliberal techniques of subjectivation: "The strategy of 'responsibilization' is the subjective pendant to the state delegation of responsibility: While the paradigm of solidarity with the policies of the welfare state recedes, the political rationality of neoliberalism advances the principle of the responsibility of atomized individuals."[27]

Underground in New York City, actual subway passengers seem largely unimpressed by these imperatives.[28] While over five million passengers rode the subway daily in 2008, the MTA registered only seven calls per day related to such reporting. Yet widespread despite criticism and the dubious efficacy of the campaign, in 2010 the United States Department of Homeland Security made "If You See Something, Say Something" the center of its nationwide public awareness campaign. By now, it is also used by authorities in parts of Australia and Canada.[29]

Beyond new techniques of subjectivation, the New York subway continues to pave the way in other areas as well. The technological innovations and methods of passenger control devised for the New York system became exemplary for more than 150 other underground trains around the world.[30] Its design and organization also provide models for more than thirty systems currently under construction. In the future, an increasing number of people in the growing cities of China, India, and Iran, and elsewhere will undergo the transformation into subterranean passengers, demonstrating the ongoing need for such infrastructures. From the historical perspective of *longue durée*,[31] with respect to the 150-year history of the New York subway, the crisis of the system in the 1970s and 1980s represents a relatively short regressive phase in an otherwise remarkable success story that continues to this day.

A CENTURY OF INFRASTRUCTURAL SUBJECTIVITY

This history of passengers of the New York City subway has shown that strategies of shaping its users are not limited to the territories of the subway itself. The dynamics of passenger culture are closely tied to historical, social, and economic circumstances in the city, as well as overarching developments in society. This study has also shown also that while the dispositifs of the subway have been effective in determining the subjectivity of its passengers, these subject forms remain highly dynamic and contested. As Reckwitz has pointed out, subject forms stake out "a field for struggles over cultural differences that determine what constitutes a subject and how it can transform itself."[32] Consequently, the kinds of passengers discussed in this study represent snapshots taken under concrete historical conditions. Nonetheless, it is possible to outline a number of general findings:

1. The analysis has shown that many of the phenomena and conflicts that emerged within the context of New York subway passenger culture occurred simultaneously

or with some delay in other parts of society as well. More than just one setting among many, the subway often served as a kind of laboratory for the testing of new governmental technologies and social practices. This applies particularly to the spread of machines into all areas of society around 1900, and to the immense crowding of human bodies into the new and ambivalent collective subject of the masses. We have seen that the subway provided an early field for the application of methods from logistics to regulate social situations that arose with mass circulation. Engineering principles such as standardization, rationalization, modularization, and the maximization of efficiency were transferred onto passengers, establishing a new dispositif of knowledge that was in turn inscribed into material elements of the subway such as turnstiles, seating arrangements, and station interiors. By confronting passengers with specific behavioral demands, regulations, and norms through sign systems and technical artefacts, processes of inscription produced social order and determined subjectivation. These machine codes were effective not only at the level of discursive attribution, they also structured passenger perceptions, emotions, and bodily practices. This process goes hand in hand with an understanding of the individual as a subject steered by external signals, as demonstrated by the visual strategies used throughout the subway system to shape behavior as well as the subject's self-image as a patriot, tax payer, or consumer.

2. The underground railway systems of the twentieth century were crucial to the establishment of liberal forms of urban governing. By allowing for the reliable, orderly circulation of the city's population, they promoted Fordist—and post-Fordist—modes of production and consumption. This infrastructure proved itself as an elementary background for everyday urban life with its specific forms of work, leisure, and family life. Coinciding with economic crisis in New York, the decline of the subway and the exodus of passengers in the 1960s and 1970s reflected the loss of the primary function of local public transportation, namely, to keep the workforce moving.

3. The subway was also an early field of application for new biopolitical models of measurement and control. Breaking passenger behavior down into a series of spatially and temporally defined actions to be counted and predicted made it possible to consider passengers as uniform and comparable units. Intertwining social norms with technical norms, this process facilitated the development of a container ethics. At the same time,

the creation of social order based on logistical and technological methods of governing also had liberating effects. When passengers are viewed as standardized, black-boxed entities, other attributes such as gender, race, or class lose their relevance—at least temporarily—offering opportunities for new freedoms.

4. Many conflicts developed among subway passengers over the years, often along the lines of gender, age, race, and class. Especially for female passengers, the subway constituted a potentially threatening territory. The same held for people of color. In contrast to other modes of public transportation in the United States, New York's subway was never segregated, but racist discrimination and violence was widespread nonetheless. Dynamics of inclusion and exclusion also played out in efforts to exclude beggars and homeless passengers from the subway. The banality of transit brought various social tensions to light. At the same time, these struggles fostered moments of deviance, complicity, and appropriation that led to the development of transgressive practices and modes of subjectivity.

5. Not only related to the overall dynamics of society, the emergence of new forms of passengers was often a response to methods of subjectivation already implemented in the subway. For example, the emergence of the subway vigilante can be understood as a reaction to the erosion of modes of order and control that had been so painstakingly established in the system over the course of decades. By the 1940s at the latest, the subway's mechanisms of subjectivation had advanced to the point that masses of passengers were seen as iconic embodiments of "one-dimensional man" rather than as barbaric hordes. Radical criticism of the commodifying power of mass culture in rationalist modernity testified to the first signs of a fundamental change in society. The rise of counterculture and the revolts of the late 1960s fed the development of newly awakened desires for self-actualization. This transformation of the idea of the individual subject also found its expression in growing enthusiasm for new ways of living made possible by the affordability of automobiles and suburban single-family homes.[33] The solidification of hedonistic consumer culture in the 1960s and 1970s, and the related claim to individuality, also left its mark on New York subway passengers. Their rejection of uniformity and conformity manifested itself as graffiti on station walls and subway car interiors, for example. Philosopher Jean Baudrillard has called graffiti a veritable "insurrection of signs,"[34] attesting to desire of young New Yorkers

in particular to aesthetically appropriate their surroundings. Even the militant passengers of youth gangs or the Guardian Angels can be seen as signaling an emancipatory impulse aimed at establishing order beyond governmental organs of control. Yet the subjectivation of passengers remains dynamic and adaptive, as shown by the skepticism of subway operators with respect to such forms of self-empowerment, along with official efforts to repress these impulses.

UNRULY CONTAINERS

With these perspectives on the ambivalent dynamics of infrastructured subjectivity, this study opens up paths for further exploration. Comparable historical studies of other urban transit systems and passenger cultures are called for to determine how the subject forms of the New York subway resemble or differ from other subway passengers. This would also allow us to gain a better historical understanding of how the socio-technological dispostifs inscribed into infrastructure can spark societal change.

As this study has shown, the impact of infrastructure on subjectivity becomes especially apparent when we take stock of how containerization has affected society and culture in the twentieth century. Methods of containerization were particularly productive in controlling New York subway passengers, but this can also be seen as a paradigm for the organization of modern technological society in general. The second half of the twentieth century saw the emergence of a distinct social dispositif shaped by ideas of the container and containment, with the New York subway as one of its primary areas of application.[35]

Across the wide range of passenger subject forms found in the New York subway throughout the twentieth century, all are shaped in one way or another by strategies of containerization and containment. These strategies played a role in the logistical standardization and black-boxing of passengers, and they contributed to diagnoses of isolation and conformity in sociology and cultural theory as the machine age drew to a close. They also fit into models of perception inscribed into the visual regimes used to address passengers via external signals, and they were at work in the way passengers compartmentalized their emotions and contained themselves by separating their inner lives from the infrastructural surroundings. Complaint letters highlighted this aspect of passenger experience, accusing others of transgressing the code of conduct that dictated emotional containment. Finally, more recent calls to passengers to be observant and take responsibility for their own security exemplify the ideal of containerized subjects who mutually control and govern one another by eliminating disturbances and deviant behavior. Efforts by management to delegate aspects of system control to subway riders may reveal a new ideal of the passenger as a "smart" container.

Not only constitutive of the culture of New York subway passengers, these various methods of containerization and containment can be found in many other areas of society as well. In the ongoing progression of the containerization of the subject, architectural critic Lieven de Cauter sees the "genesis of a capsular civilization."[36] For him, the massive proliferation of technologies of mobile isolation and exclusion marks the dawn of dystopian social order, heralded by an increase in control and social fragmentation as well as the ubiquitous spread of neoliberal utilization and exploitation.[37]

In his book *The Practice of Everyday Life*, in a brief chapter titled "Railway Navigation and Incarceration," philosopher Michel de Certeau arrives at a similarly pessimistic take on containerization.[38] As the chapter title suggests, de Certeau characterizes the passenger as a helpless prisoner, entering the infrastructure of the container only to find him- or herself at the mercy of the "gridwork of technocratic discipline."[39] "The unchanging traveller is pigeonholed, numbered, and regulated in the grid of the railway car, which is a perfect actualization of the rational utopia."[40] Whereas for Foucault, a ship was still a "heterotopia *par excellence*,"[41] a moving place that turned social power relations upside down in many respects, for de Certeau, the railway car marks the authoritarian manifestation of Hegel's rational state. As an inescapable "rationalized cell,"[42] the subway car constitutes the final stage of subjectivation through infrastructures of transit. In turn, the complete subjugation of passengers to the rational and technological principles of containerization makes the system itself fade into the background as a black box. For de Certeau, these moments of social and cognitive relief are all that makes room for a remnant of passenger subjectivity and autonomy. Fleeting moments of contemplation and daydreaming allow the last "Robinson Crusoe adventure of the travelling noble soul,"[43] amidst the totalitarian transit apparatus of control and discipline.

We could close the book with this melancholic image of the passenger. But de Cauter and de Certeau's pessimistic predictions ignore the fact that regimes of subjectivation can also create new freedoms and new forms of agency. The history of the New York subway passenger underscores the ambivalence between the many demands made on passengers and the potential for emancipation established by the dispositif of transit. Reconstructing the events that led to building the subway in the first place, for example, revealed the enormous hopes and expectations that escorted the future passenger into being. The realization of these hopes and expectations may have been ambivalent, but again and again, the use of this infrastructure led to moments of autonomy and solidarity. The proliferation of critical accounts of passenger culture as of the 1950s, highlighting isolation, alienation, and a sense of being at the mercy of the transit apparatus, can be seen as symptomatic of a dynamic pattern of alternating decontainerization and recontainerization. Tracing this dynamic, Alexander Klose writes: "Every effort at codifying and establishing specific standards for passenger transportation—velocity, space per passenger, behavioral regulation, the logistics of entering and exiting, and even the goals and purposes of transportation—is followed by a movement of de-containerization, when passengers either turn to other means of transportation or regain their individual elbowroom through deviant behavior."[44]

As we have seen again and again, subway passengers' insistence on self-determination emerges under circumstances marked by inertia and stubbornness, when subjects dodge the demands and impositions of the system. In the eyes of engineers, logistics experts, and authorities, these passengers appear primarily as sources of disruption, troublemakers, or unpredictable elements of an already fragile technical system, but emancipation may reside precisely in these features. Situations involving deviant behavior point to moments of passenger tenacity and autonomy that lie beyond the control and subjectifying powers of urban infrastructure. Staying on this track, it is unlikely that passenger individuality will ever be entirely obliterated in the way that Foucault prophesies for the image of the human, vanishing from the modern human sciences "like a face drawn in sand at the edge of the sea."[45] As the history of New York subway passengers teaches us, people will always find ways to throw sand in the gears.

Notes

Preface to the English Edition

1. Jeffrey E. Harris, "The Subways Seeded the Massive Coronavirus Epidemic in New York City," working paper series, no. 27021, National Bureau of Economic Research: Cambridge, MA, April 2020, http://www.nber.org/papers/w27021.

2. Kyle Smith, quoted in Laura Bliss, "The New York Subway Got Caught in the Coronavirus Culture War," *City Lab*, April 21, 2020, https://www.citylab.com/transportation/2020/04/coronavirus -cases-new-york-subway-infection-riders-mta/610159/.

Introduction

1. "Clamor for Tickets for Subway Opening," *New York Times*, October 26, 1904.

2. "Our Subway Open: 150,000 Try It," *New York Times*, October 28, 1904.

3. "Exercises in City Hall," *New York Times*, October 28, 1904.

4. Brian J. Cudahy, *Under the Sidewalks of New York: The Story of the Greatest Subway System in the World* (New York: Fordham University Press, 1995), 2.

5. The mayor appears to have been so captivated by the first subway that the *New York Times* devoted a detailed story to the moment. "McClellan Motorman of First Subway Train," *New York Times*, October 28, 1904. Subway historians never tire of relating the circumstances. See, for instance, Clifton Hood, *722 Miles: The Building of the Subways and How They Transformed New York* (Baltimore: John Hopkins University Press, 2004), 91.

6. "Our Subway Open: 150,000 Try It."

7. For instance, "Finish Plans for Subway Celebration," *New York Times*, October 18, 1904.

8. "Our Subway Open: 150,000 Try It"; "Subway Opening To-day with Simple Ceremony," *New York Times*, October 27, 1904.

9. "Our Subway Open: 150,000 Try It."

10. Michael Brooks, *Subway City: Riding the Trains, Reading New York* (New Brunswick, NJ: Rutgers University Press, 1977), 37.

11. Andreas Bernard, *Lifted: A Cultural History of the Elevator* (New York: NYU Press, 2014); Stephen Graham, *Vertical: The City from Satellites to Bunkers* (London: Verso, 2017).

12. Helmut Lethen, *Cool Conduct: The Culture of Distance in Weimar Germany*, trans. Don Reneu (Berkeley: University of California Press, 2002), 21–32.

13. Ian Hacking, "Making Up People," in *Reconstructing Individualism: Autonomy, Individuality, and the Self in Western Thought*, ed. Thomas C. Heller, Morton Sosna, and David E. Wellberg (Stanford: Stanford University Press, 1986), 222.

14. Karl Marx, *Grundrisse: Foundations of the Critique of Political Economy* (London: Penguin UK, 2005), 92.

15. A passenger could be of any gender. As we will see, transit encouraged the development of numerous complex attributes and conflicts related to the gender identity of its passengers.

16. Among others, this has been noted by German historian Dirk van Laak, "Infra-Strukturgeschichte," *Geschichte und Gesellschaft* 27, no. 3 (September 2001): 367–393. See also Wolfgang Essbach, "Antitechnische und antiästhetische Haltungen in der soziologischen Theorie," in *Technologien als Diskurse*, ed. Andreas Lösch (Heidelberg: Synchron Wissenschaftsverlag der Autoren, 2001), 123–136; Susan Star Leigh, "The Ethnography of Infrastructure," *American Behavioral Scientist* 43, no. 3 (November 1999): 377–391.

17. John Urry, *Mobilities* (Cambridge: Polity Press, 2007); John Urry, *Tourist Gaze* (London: Sage, 1990); Richard Sennett, *The Corrosion of Character: The Personal Consequences of Work in the New Capitalism* (New York: W. W. Norton, 1998). See also contributions to the following anthologies: Markus Schroer and Stephan Moebius, *Diven, Hacker, Spekulanten: Sozialfiguren der Gegenwart* (Frankfurt am Main: Suhrkamp, 2010); Peter-Ulrich Merz-Benz and Gerhard Wagner, eds., *Der Fremde als sozialer Typus* (Konstanz: UVK Verlagsgesellschaft, 2007).

18. British history of transportation does include some attempts to encourage study of this field, especially in connection with the publications *Transfers* and *Journal for Transport History*. See, for instance, Barbara Schmucki, "On the Trams: Women, Men and Urban Public Transport in Germany," *Journal of Transport History* 23, no. 1 (March 2002): 60–72; Christopher Kopper, "Mobile Exceptionalism? Passenger Transport in Interwar Germany," *Transfers* 3, no. 2 (June 2013): 89–107.

19. For an impressive study of this issue in the nineteenth century, see Wolfgang Schivelbusch, *The Railway Journey: The Industrialization of Time and Space in the Nineteenth Century* (New York: Urizen Books, 1979).

20. Dirk Van Laak, "Infrastruktur und Macht," in *Umwelt und Herrschaft in der Geschichte*, ed. Françoise Duceppe-Lamarre and Jens Ivo Engels (Munich: R. Oldenbourg, 2008), 106–114.

21. On the idea of the built environment as a social medium, see the inspiring work of Heike Delitz, *Gebaute Gesellschaft: Architektur als Medium des Sozialen* (Frankfurt am Main: Campus, 2010).

22. Louis Althusser discusses the concept of interpellation in detail in: *On the Reproduction of Capitalism: Ideology and Ideological State Apparatuses* (London: Verso, 2014), 261–270. Further reading: Ulrich Bröckling, *The Entrepreneurial Self: Fabricating a New Type of Subject* (London: Sage, 2015), 1–19; Andrea Allerkamp, *Anruf, Adresse, Appell: Figurationen der Kommunikation in Philosophie und Literatur* (Bielefeld: transcript, 2005). We will return to Althusser's model of interpellation in chapter 5.

23. Steve Woolgar, "Configuring the User: The Case of Usability Trials," in *A Sociology of Monsters: Essays on Power, Technology, and Domination*, ed. John Law (London: Routledge, 1991), 57–99. See also Nelly Oudshoorn and Trevor Pinch, "Introduction: How Users and Non-Users Matter," in *How Users Matter: The Co-Construction of Users and Technology*, ed. Nelly Oudshoorn and Trevor Pinch (Cambridge, MA: MIT Press, 2005), 1–25.

24. In addition, then as now, New York leads in the duration of subway rides. An average commuter ride on the subway to work lasts 39.4 minutes, adding up to almost 80 minutes of work-related transit daily, not including trips for leisure. About 5.6 percent of New Yorkers spend three hours a day on the subway going to and from work. See Patrick McGeehan, "Mass Transit Grows as Commuters' Trip of Choice," *New York Times*, September 2, 2006 and American Community Survey 2002, *US Census Bureau*, factfinder.census.gov, as well as statistics variable on the webpage of the Metropolitan Transportation Authority of New York: web.mta.info/nyct/facts/ridership/.

25. Before the construction of underground train systems, metropolises already had transit systems such as horse-drawn busses and elevated railways. These systems were not only much slower and smaller, they moved a different kind of passenger. For more detail, see chapter 1.

26. Passengers went unmentioned in the *Encyclopedia Britannica*, *Collier's Encyclopedia*, and *Webster's Dictionary*. German lexica from the late nineteenth century such as *Meyers Konversations-Lexicon* or *Brockhaus* defined passengers as "travelers [. . .], especially those carried by wagons, railways, or steamships." *Brockhaus Konversations-Lexikon*, 14th ed. (Leipzig, Berlin & Vienna: F.A. Brockhaus, 1894), s.v. "Passagier," translated by the author.

27. *Oxford Universal Dictionary*, vol. 2 (Oxford: Clarendon, 1959), 1443.

28. Schivelbusch, *The Railway Journey*. A similar study was published by Jo Guldi, *Roads to Power: Britain invents the Infrastructure State* (Cambridge, MA: Harvard University Press, 2012).

29. Besides de Certeau's work in *The Practice of Everyday Life* (Berkeley: University of California Press, 1984), special attention should be given to his work on Jules Verne, *Heterologies: Discourse on the Other* (Minneapolis: University of Minnesota Press, 1986).

30. Leo Marx, *The Pilot and the Passenger: Essays on Literature, Technology, and Culture in the United States* (New York: Oxford University Press, 1988). See also Simon Ward, "The Passenger as Flaneur? Railway Networks in German-language Fiction since 1945," *Modern Language Review* 100, no. 2 (April 2005): 412–428.

31. Bernhard Siegert, *Passagiere und Papiere: Schreibakte auf der Schwelle zwischen Spanien und Amerika* (Munich: Wilhelm Fink Verlag, 2006).

32. See the following essays by Paul Virilio: *Negative Horizon: An Essay in Dromoscopy* (London: Continuum, 2005); *Speed and Politics: An Essay on Dromology* (New York: Semiotext(e), 1977); *Negative Horizon: An Essay in Dromoscopy* (London: Continuum, 2005).

As original as Virilios's thoughts and prognoses often are, they also contain strong traces of cultural pessimism and anti-feminism. For a critique of these aspects of his work, see Verena A. Conley, "The Passenger: Paul Virilio and Feminism," in *Paul Virilio: From Modernism to Hypermodernism and Beyond*, ed. John Armitage (London: Sage, 2000), 201–215.

33. One exception to this rule is the work of Marc Augé, *In the Metro* (Minneapolis: University of Minnesota Press, 2001); as well as Marc Augé, *Non-Places: An Introduction to Supermodernity* (London: Verso, 2008).

34. See Urry, *Mobilities*, 91.

35. For instance, Benson Bobrick, *Labyrinths of Iron: Subways in History, Myth, Art, Technology, and War* (New York: Henry Holtand Company, Inc., 1994); Philip Ashforth Coppola, *Silver Connections: A Fresh Perspective on the New York Area Subway Systems*, vol. 3 (Maplewood: Four Oceans Press, 1988); Brian J. Cudahy, *Cash, Tokens, and Transfers: A History of Urban Mass Transit in North America* (New York: Fordham University Press, 1990); Brian J. Cudahy, *A Century of Subways: Celebrating 100 Years of New York's Underground Railways* (New York: Fordham University Press, 2003); Lorraine B. Diehl, *Subways: The Tracks That Built New York City* (New York: Clarkson Potter, 2004); Jim Dwyer, *Subway Lives: 24 Hours in the Life of the New York City Subway* (New York: Crown, 1991); New York Transit Museum, *Subway Style: 100 Years of Architecture and Design in the New York City Subway* (New York: Stewart, Tabori & Chang, 2004); Gene Sansone, *New York Subways: An Illustrated History of New York City's Transit Cars* (Baltimore: John Hopkins University Press, 1997); James Blaine Walker, *Fifty Years of Rapid Transit, 1864–1917* (North Stratford: Ayer Publishing, 1918). For an overview of subway research, see also the book review of Darius Sollohub, "The Machine in Society," *Journal of Urban History* 34, no. 4 (May 2008): 532–540.

36. Historian Donald F. Davis passes harsh judgment on such approaches: "[M]ost of the books written about transit history have been narrowly internalist, even antiquarian, and that their failure to place technological change in its wider, socio-cultural context sharply limits their usefulness." Donald F. Davis, "North American Urban Mass Transit, 1890–1950: What if We Thought about It as a Type of Technology?," *History and Technology* 12, no. 4 (Winter 1995): 309.

37. Brooks, *Subway City*; Tracy Fitzpatrick, *Art and the Subway: New York Underground* (New Brunswick, NJ: Rutgers University Press, 2009).

38. Brian J. Cudahy, *Under the Sidewalks of New York*; Hood, *722 Miles*; Clifton Hood, "Changing Perceptions of Public Space on the New York Rapid Transit System," *Journal of Urban History* 22, no. 3 (March 1996): 308–331; Clifton Hood, "The Impact of the IRT on New York City," in *Historical American Engineering Record: Interborough Rapid Transit Subway (Original Line) NY-122* (New York, 1979): 145–206.

39. For more on this approach, see Philipp Felsch, "Merves Lachen," *Zeitschrift für Ideengeschichte* 2, no. 4 (November 2008): 11–30; Steve Woolgar, "Reflexivity is the Ethnographer of the Text," in *Knowledge and Reflexivity: New Frontiers in the Sociology of Knowledge*, ed. Steve Woolgar (London: Sage, 1988), 14–36.

40. Niklas Luhmann, "Die Tücke des Subjekts und die Frage nach dem Menschen," in *Der Mensch, das Medium der Gesellschaft?*, ed. Peter Fuchs and Andreas Göbel (Frankfurt am Main: Suhrkamp, 1994), 40–56. This is also prominently discussed in Slavoj Žižek, *The Ticklish Subject: The Absent Centre of Political Ontology* (London: Verso, 1999).

41. See Brigitte Kible, "Subjekt," in *Historisches Wörterbuch der Philosophie* 10, ed. Joachim Ritter and Karfried Gründer (Stuttgart: Schwabe & Co. AG, 1998), 373–400; Andreas Reckwitz, *Subjekt* (Bielefeld: transcript, 2008), 9–21.

42. Foucault also took interest in the state of being a passenger and the role it plays for social and cultural orders, albeit in an entirely different historical context: *Madness and Civilization: A History of Insanity in the Age of Reason* (New York: Random House, 1988), 6–17.

43. Michel Foucault, "The Subject and Power," *Critical Inquiry* 8, no. 4 (Summer 1982): 777–795. Following Foucault, over the past twenty years a large number of researchers have grappled with the same question. The following studies offer a good introduction to this field of research: Judith Butler, *The Psychic Life of Power: Theories in Subjection* (Stanford, CA: Stanford University Press, 1997); Nikolas S. Rose, *Governing the Soul: The Shaping of the Private Self* (London: Free Association Books, 1999); Nikolas S. Rose and Peter Miller, *Governing the Present: Administering Economic, Social, and Personal Life* (New York: John Wiley & Sons, 2013). Also, German sociologists Andreas Reckwitz and Ulrich Bröckling have written inspiring analyses on historical and current subject forms.

44. For more on these fields of study, individual works, and the genesis of Foucault's work as a whole, see Ulrich Johannes Schneider, *Michel Foucault* (Darmstadt: Primus, 2006).

45. Foucault, "The Subject and Power," 781.

46. See van Laak, "Infra-Stukturgeschichte," 377.

47. Unfortunately, Foucault rarely wrote about the material side of subjectivation; his focus is primarily on how discourse, forms of knowledge, and the use of power objectivize subjects, not on the physical and technical factors involved. Foucault admitted this himself. See among others Michel Foucault, "Technologies of the Self," in *Technologies of the Self: A Seminar with Michel Foucault*, ed. Luther H. Martin, Huck Gutman, and Patrick H. Hutton (Amherst, MA: University of Massachusetts Press, 1988), 18. When

Foucault does deal with the material dimensions of objectivizing the subject, as he does in his study of the controlling power of Jeremy Bentham's panopticon, the result is an analysis that has become iconic. See Michel Foucault, *Discipline and Punish: The Birth of the Prison* (New York: Random House, 1995).

48. Dirk van Laak, "Der Begriff 'Infrastruktur' und was er vor seiner Erfindung besagte," *Archiv für Begriffsgeschichte* 41 (1999): 280–299.

49. On the central role of infrastructure in colonization, particularly in the eighteenth and nineteenth centuries, see Michael Adas, *Machines as the Measure of Men: Science, Technology, and Ideologies of Western Dominance* (Ithaca: Cornell University Press, 1989); Maria Paula Diogo and Dirk van Laak, *Europeans Globalizing: Mapping, Exploiting, Exchanging* (London: Palgrave Macmillan, 2016); Dirk van Laak, *Imperiale Infrastruktur: Deutsche Planungen für die Erschliessung Afrikas 1880–1960* (Paderborn: Schöningh, 2004).

50. Walter Buhr, "What Is Infrastructure?" (discussion paper, Volkswirtschaftliche Diskussionsbeiträge—Discussion Paper no. 107-03, Department for Economic Science, University of Siegen, Siegen, 2003).

51. Christopher G. Boone and Ali Modarres, *City and Environment* (Philadelphia: Temple University Press, 2006), 96.

52. Benjamin Lee and Edward LiPuma, "Cultures of Circulation: The Imaginations of Modernity," *Public Culture* 14, no. 1 (January 2002): 191–213. See also Alexandra Boutros and Will Straw, "Introduction," in *Circulation and the City*, ed. Alexandra Boutros and Will Straw (Montreal: McGill-Queen's University Press, 2010), 3–22. Circulation will be discussed at length in chapters 1 and 2.

53. Michel Foucault, "The Confession of the Flesh," in *Power/Knowledge: Selected Interviews and Other Writings 1972–1977*, ed. Colin Gordon (New York: Pantheon Books, 1980), 194–228.

54. The same holds for the theory of assemblage (*agencement*) developed by Gilles Deleuze and Felix Guattari, which has numerous similarities with Foucault's concept of the *dispositif*. See Gilles Deleuze and Félix Guattari, *A Thousand Plateaus: Capitalism and Schizophrenia* (London: Continuum, 1988), 303–305; Gilles Deleuze, "What Is a Dispositif?," In *Michel Foucault Philosopher*, trans. Timothy J. Armstrong (New York: Routledge, 1992). However, Deleuze and Guattari stress that their understanding of assemblage is not primarily shaped by power, but rather by desire. For more on this distinction, see also Gilles Deleuze, *Foucault* (Minnesota: University of Minnesota Press, 1988). Guattari too developed the concept further in his writing. See, for example, Felix Guattari, *Chaosmosis: An Ethico-Aesthetic Paradigm* (Bloomington: Indiana University Press, 1995). See also Giorgio Agamben, "What Is an Apparatus?" in *What Is an Apparatus? And Other Essays* (Stanford: Stanford University Press 2009); Andreas D. Bührmann and Werner Schneider, *Vom Diskurs zum Dispositiv: Eine Einführung in die Dispositivanalyse* (Bielefeld: transcript, 2008).

55. For further detail see Deleuze and Guattari, *A Thousand Plateaus*. For more on the concept of assemblage, especially in urban and architecture studies see also Ash Amin and Nigel Thrift, *Cities: Reimagining the Urban* (Cambridge: Polity Press, 2001), 78–83; Delitz, *Gebaute Gesellschaft*, 126–136; George

E. Marcus and Erkan Saka, "Assemblage," *Theory, Culture & Society* 23, no. 2–3 (May 2006): 101–106; J. Macgregor Wise, "Assemblage," in *Gilles Deleuze: Key Concepts*, ed. Charles J. Strivale (Montreal: McGill-Queen's University Press, 2005), 77–86.

56. Deleuze and Guattari develop this understanding of machines in *Anti-Oedipus*, with respect to Mumford's philosophy of technology: Deleuze, Gilles and Félix Guattari, *Anti-Oedipus: Capitalism and Schizophrenia* (Minneapolis: University of Minnesota Press, 1983), 141. On Mumford and the concept of the subway as a megamachine, see also chapter 4.

57. Hughes developed his approach in a series of publications, especially Thomas P. Hughes, "The Evolution of Large Technological Systems," in *The Social Construction of Technological Systems: New Directions in the Sociology and History of Technology*, ed. Wiebe E. Bijker, Thomas P. Hughes, and Trevor J. Pinch (Cambridge, MA: MIT Press, 1989), 51–82.

58. See Hartmut Böhme, Peter Matussek, and Lothar Müller, *Orientierung Kulturwissenschaft: Was sie kann, was sie will* (Reinbek bei Hamburg: Rowohlt, 2000), 164.

59. For more detail, see Martina Hessler, "Ansätze und Methoden der Technikgeschichtsschreibung (Zusatztexte im Internet)," in *Kulturgeschichte der Technik* (Frankfurt am Main and New York: Campus Verlag, 2010), 6–7.

60. See Ignacio Farias, "Introduction: Decentering the Object of Urban Studies," in *Urban Assemblages: How Actor-Network Theory Changes Urban Studies*, ed. Thomas Bender and Ignacio Farias (New York: Routledge, 2010), 1–24; Stefan Höhne and Rene Umlauf, "Die Akteur-Netzwerk Theorie: Zur Vernetzung und Entgrenzung des Sozialen," in *Theorien in der Raum- und Stadtforschung: Einführungen*, ed. Jürgen Ossenbrügge and Anne Vogelpohl (Münster: Westfälisches Dampfboot, 2015), 195–214.

61. Böhme, Matussek, and Müller, *Orientierung Kulturwissenschaft*, 178, translated by the author.

62. Madeleine Akrich, "The De-Scription of Technical Objects," in *Shaping Technology/Building Society: Studies in Sociotechnical Change*, ed. Wiebe J. Bijker and John Law (Cambridge, MA: MIT Press, 1992); Madeleine Akrich and Bruno Latour, "A Summary of a Convenient Vocabulary for the Semiotics of Human and Nonhuman Assemblies," in *Shaping Technology/Building Society*; Bruno Latour, "Technology Is Society Made Durable," *Sociological Review* 38, no. 1 suppl. (May 1990): 103–131.

63. Bruno Latour famously uses the example of the so-called Berlin key to explain how such scripts work. Designed especially for the main door of tenements in the city, the key had two bits; after opening a door, in order to remove the key, it had to be pushed entirely through the lock, thus locking the door once again from the inside. The key was only used during the night. In the morning, the caretaker used a smooth key to activate a mechanism within the lock that allowed tenants to come and go without pushing the key entirely through the lock. This simple element of design kept the main entrances of West Berlin tenements locked all night, but it also led to the development of a number of subversive practices, including filing down the second bit. See Bruno Latour, "The Berlin Key, or How to do Things with Words," in *Matter, Materiality and Modern Culture*, ed. P. M. Graves-Brown (London: Routledge, 2000).

64. Akrich, "The De-Scription of Technical Objects," 208.

65. Akrich and Latour, "A Summary of a Convenient Vocabulary," 261.

66. See van Laak, "Infra-Strukturgeschichte"; Stephen Graham and Simon Marvin, *Splintering Urbanism: Networked Infrastructures. Technological Mobilities and the Urban Condition* (London: Routledge, 2001).

67. See Andreas Reckwitz, *Das hybride Subjekt: Eine Theorie der Subjektkulturen von der bügerlichen Moderne zur Postmoderne* (Weilerswist: Velbrück Wissenschaft, 2010), 14.

68. Reckwitz, *Subjekt*, 13.

69. In a comprehensive study titled *Das hybride Subjekt*, Andreas Reckwitz has explored the value of the concept of subject forms for cultural and historical analysis.

70. Michel Foucault, "Self Writing," in *Ethics: Subjectivity and Truth*, ed. Paul Rabinow (New York: The New Press, 1997), 221.

71. The term "container subject" was coined by Alexander Klose, "Who do you want to be today? Annäherungen an eine Theorie des Container-Subjekts," in *Das Motiv der Kästchenwahl: Container in Psychoanalyse, Kunst, Kultur*, ed. Insa Härtel and Olaf Knellessen (Göttingen: Vandenhoeck & Ruprecht, 2012), 21–38.

72. Michel Foucault, "The Confession of the Flesh," in *Power/Knowledge: Selected Interviews and other Writings 1972–1977*, ed. Colin Gordon (New York: Pantheon Books, 1980), 195.

CHAPTER I

1. Quoted in Irving L. Allen, *City in Slang: New York Life and Popular Speech* (New York: Oxford University Press, 1995), 93.

2. John B. McDonald, "The Man That Built the Subway," interview by Kate Carew, *New York World*, October 23, 1904, quoted in Michael Brooks, *Subway City: Riding the Trains, Reading New York* (New Brunswick, NJ: Rutgers University Press, 1977), 64.

3. See James Blaine Walker, *Fifty Years of Rapid Transit, 1864–1917* (North Stratford, NH: Ayer Publishing, 1918), 66.

4. Clifton Hood, "The Impact of the IRT on New York City," in *Historical American Engineering Record: Interborough Rapid Transit Subway (Original Line) NY-122* (New York, 1979), 145–206.

5. "Exercises in City Hall," *New York Times*, October 28, 1904.

6. See Robert F. Wesser, "McClellan, George Brinton," in *The Encyclopedia of New York City*, ed. Kenneth T. Jackson (New Haven: Yale University Press, 1995), 704.

7. "Exercises in City Hall."

8. For more detail, see Walker, *Fifty Years of Rapid Transit*, 176–181.

9. "Exercises in City Hall."

10. For more detail, see Maria Kaika and Erik Swyngedouw, "Fetishizing the Modern City: The Phantasmagoria of Urban Technological Networks," *International Journal of Urban and Regional Research* 24, no. 1 (March 2000): 120–138. See also Jürgen Osterhammel, *Transformation of the World: A Global History of the Nineteenth Century* (Princeton, NJ: Princeton University Press, 2015); Stephen Kern, *The Culture of Time and Space, 1880–1918* (Cambridge, MA: Harvard University Press, 2003); Thomas P. Hughes, *American Genesis: A Century of Invention and Technological Enthusiasm, 1870–1970* (Chicago: University of Chicago Press, 1989).

11. "Exercises in City Hall."

12. See Brian J. Cudahy, *Under the Sidewalks of New York: The Story of the Greatest Subway System in the World* (New York: Fordham University Press, 1995), 2–3.

13. "Modern Martyrdom," *New York Herald*, October 2, 1864, quoted in Walker, *Fifty Years of Rapid Transit*, 7.

14. These and the following statistics on New York's population development are taken from Nathan Kantrowitz, "Population," in *Encyclopedia of New York City*, 920–923.

15. For a good survey of these huge waves of migration, see Carol Groneman and David M. Reimers, "Immigration," in *Encyclopedia of New York City*, 581–589. For more detail, see Robert Ernst, *Immigrant Life in New York City: 1825–1863* (Syracuse: Syracuse University Press, 1994).

16. Long past the mid-nineteenth century, New York did not include the five districts of Manhattan, the Bronx, Queens, Brooklyn, and Staten Island, but simply Manhattan Island. The shape of the city as we know it today did not begin forming until 1874, when parts of the Bronx were annexed, and 1898, when formerly independent parts of the city were aggregated.

17. See Edward K. Spann, "Grid Plan," in *Encyclopedia of New York City*, 510.

18. See Brooks, *Subway City*, 8.

19. Around 1900 the number of residents in New York City equaled that of the entire combined population of all other North American cities in 1850. See Paul S. Boyer, *Urban Masses and Moral Order in America, 1820–1920* (Cambridge, MA: Harvard University Press, 1992), 123.

20. Cudahy, *Under the Sidewalks of New York*, 2.

21. For more on the development of traffic regulation and modalities in the United States, see Clay McShane, *Down the Asphalt Path: The Automobile and the American City* (New York: Columbia University Press, 1995), esp. 173–202.

22. Walter Benjamin, "The Paris of the Second Empire in Baudelaire," in *The Writer of Modern Life: Essays on Charles Baudelaire*, ed. Michael W. Jennings, trans. Howard Eiland et al. (Cambridge, MA: Belknap Press of Harvard University Press, 2006), 83–88.

23. "The natural suitability of the flaneur to the culture of New York was evident in the degree to which he was associated with and easily assimilated into the culture of spectacle that was as prominent in New York as it was in any city of Europe." Dana Brand, *The Spectator and the City in Nineteenth-Century American Literature* (Cambridge: Cambridge University Press, 1991), 74.

24. Walt Whitman, "A Broadway Pageant," *Leaves of Grass* (New York: William E. Chapin Printers, 1867), 193.

25. See Susan Buck-Morss, "The Flaneur, the Sandwichman and the Whore: The Politics of Loitering," *New German Critique* 39 (Autumn 1986): 99–140. See also Susan Buck-Morss, *The Dialectics of Seeing: Walter Benjamin and the Arcades Project* (Cambridge, MA: MIT Press, 1991).

26. See also chapter 5 for more on fighting idleness and loitering in the subway.

27. According to Benjamin, this placed the flaneur in the proximity of prostitution. See Walter Benjamin, "Paris, the Capital of the Nineteenth Century," in *The Writer of Modern Life*, 41.

28. William Foote Whyte, *Street Corner Society: The Social Structure of an Italian Slum*, 4th ed. (Chicago: University of Chicago Press, 1993).

29. Asa Greene, *A Glance at New York* (New York: Craighead & Allen, 1837), 5. Quoted in Clifton Hood, *722 Miles: The Building of the Subways and How They Transformed New York* (Baltimore: John Hopkins University Press, 2004), 41.

30. For details see Hood, *722 Miles*, 37–38.

31. As Hood has carefully calculated, the city had 94 miles of elevated railways, 265 miles of rails for horse-drawn streetcars, 137 miles of routes for horse-drawn omnibuses, and a cable car to bring commuters over the newly erected Brooklyn Bridge. In 1860 fourteen horse-drawn streetcar companies competed for the business of carrying more than 38 million passengers annually. Twenty-one horse-drawn omnibus lines altogether were served by 671 vehicles that made ten tours uptown and downtown every day. See Hood, *722 Miles*, 25.

32. See Mark H. Rose and Vincent Seyfried, "Streetcars," in *Encyclopedia of New York City*, 1127–1128.

33. See Glen E. Holt, "The Changing Perception of Urban Pathology: An Essay on the Development of Mass Transit in the United States," in *Cities in American History*, ed. Kenneth T. Jackson and Stanley K. Schultz (New York: Alfred A. Knopf, 1972), 324–343.

34. Quoted in Holt, "The Changing Perception of Urban Pathology," 392.

35. The *New York Herald* journalist continues: "The discomforts, inconveniences, and annoyances of a trip in one of these vehicles are almost intolerable. From the beginning to the end of the journey a constant quarrel is progressing. The driver quarrels with the passengers and the passengers quarrel with the driver. There are quarrels about getting out and quarrels about getting in. There are quarrels about change and quarrels about the ticket swindle." Quoted in Holt, "The Changing Perception of Urban Pathology," 392.

36. In addition, as Holt details in "The Changing Perception of Urban Pathology," it was customary in all US cities for workers to ride on the platform outside the wagon. In St. Louis and Philadelphia, African Americans were prohibited from using the transit system at all, and workers were barred from using it on Sundays because their noise disgruntled service-goers in the churches along the route.

37. Samuel R. Curtis, "Omnibuses," *New York Times*, May 26, 1860.

38. For more on the elevated trains and their meaning for the city and its residents, see Tracy Fitzpatrick, *Art and the Subway: New York Underground* (Piscataway: Rutgers University Press, 2009), 11–12; "East Side Rapid Transit," *New York Times*, July 2, 1878.

39. Hood, *722 Miles*, 49.

40. Erica Judge, Vincent Seyfried, and Andrew Sparberg, "Elevated Railways," in *Encyclopedia of New York City*, 368–370.

41. See chapter 2 for more on introducing the vertical perspective of the city at the turn of the century.

42. Brooks, *Subway City*, 36–37.

43. "East Side Rapid Transit."

44. "East Side Rapid Transit."

45. Medical discourse from the period shows that elevated trains were considered a danger to health not only because of exhaust and acoustic pollution, but because many doctors feared that the wear of friction between rails and wheels released fine particles of metal that could cause eye damage and inflammation. Brooks, *Subway City*, 35.

46. Vincent Seyfried, "Blizzard of 1888," in *Encyclopedia of New York City*, 118. See also Hood, *722 Miles*, 26.

47. Michael Brooks provides a collection of such ideas in *Subway City*, 10–17.

48. For more detail on Beach's idea, see Brooks, *Subway City*, 20–27, and Hood, *722 Miles*, 42–48.

49. Brooks, *Subway City*, 12–13.

50. A fine collection of numerous large projects for New York that never materialized can be found in Rebecca Read Shanor, *The City That Never Was: Two Hundred Years of Fantastic and Fascinating Plans That Might Have Changed the Face of New York City* (New York: Viking, 1988). In *Weiße Elefanten: Anspruch und Scheitern technischer Großprojekte im 20. Jahrhundert* (Munich: Deutsche Verlags-Anstalt, 1999), Dirk van Laak presents an inspiring study of failed large-scale infrastructure projects in the twentieth century.

51. Peter Derrick, *Tunneling the Future: The Story of the Great Subway Expansion That Saved New York* (New York: NYU Press, 2001), 13.

52. Marx himself pointed out the importance of mobility: "In the production of commodities, circulation is as necessary as production itself, so that agents are just as much needed in circulation as in

production." Karl Marx, *Capital: A Critique of Political Economy. Volume II: The Process of Circulation of Capital*, trans. Ernest Untermann (Chicago: Charles H. Kerr, 1910), 144.

53. Schivelbusch, *The Railway Journey*, 188–192. For more on the crisis of circulation around the turn of the century, see James Beniger, *Control Revolution: Technological and Economic Origins of the Information Society* (Cambridge, MA: Harvard University Press, 2009).

54. Around 1900 New York's harbor was the busiest in the United States. See Norman J. Brouwer, "Port of New York," in *Encyclopedia of New York City*, 927–929.

55. See Brouwer, "Port of New York," 14. See also Donald R. Stabile, "New York Chamber of Commerce and Industry," in *Encyclopedia of New York City*, 825–826.

56. Chamber of Commerce, *Thirty-Sixth Annual Report* (New York: Press of the Chamber of Commerce, 1894), 91, quoted in Hood, *722 Miles*, 63.

57. For a good introduction to the complex history of the chamber, see Donald R. Stabile, "New York Chamber of Commerce and Industry," in *Encyclopedia of New York City*, 825–826.

58. For a comprehensive study on the idea of circulation pursued in the eighteenth century, see Albrecht Koschorke, *Körperströme und Schriftverkehr: Mediologie des 18. Jahrhunderts* (Munich: Fink, 2003), 112–129. See also Philipp Sarasin and Andreas Kilcher, "Editorial," in *Nach Feierabend: Zürcher Jahrbuch für Wissensgeschichte: Zirkulationen* 7, ed. David Gugerli et al. (Zurich: diaphanes, 2001), 8–10.

59. Physician Andreas Cesalpino is considered the first scientist to have used the word *circulatio* in a physiological and medical context. See Georg Töpfer, *Historisches Wörterbuch der Biologie: Geschichte und Theorie der biologischen Grundbegriffe* 2 (Stuttgart: Metzler, 2011), 303.

60. For further detail, see Richard Sennett, *Flesh and Stone: The Body and the City in Western Civilization* (New York: W. W. Norton, 1996), 261–266.

61. See Michel Foucault, *The Order of Things* (New York: Pantheon Books, 1971), 125; Töpfer, *Historisches Wörterbuch der Biologie*, 302–320. See also Alfred P. Fishman and Dickinson W. Richards, *Circulation of the Blood: Men and Ideas* (New York: Oxford University Press, 1964).

62. As science historian Georg Töpfer points out, circulation also played a primary role for Georg Simmel, who took reciprocity as a condition of circulation from biology and applied it to fundamental functions of society, ultimately determining the unity of all that is social. See Töpfer, *Historisches Wörterbuch der Biologie*, 310.

63. Another idea that nineteenth-century biology contributed to the explanation of social processes was that of metabolism. The question discussed by contemporary urban planners of how to organize functional differentiation in a city and how to diffuse its inhabitants was steeped in metaphors of metabolism. See Sabine Barles, "Urban Metabolism of Paris and Its Region," *Journal of Industrial Ecology* 13, no. 6 (December 2009): 898–913; David Wachsmuth, "Three Ecologies: Urban Metabolism and the Society-Nature Opposition," *Sociological Quarterly* 53, no. 4 (Autumn 2012): 506–523.

64. Schivelbusch, *The Railway Journey*, 194–197.

65. Michel Foucault, *Security, Territory, Population: Lectures at the College De France, 1977–78*, ed. Michel Senellart, trans. Graham Burchell (Basingstoke: Palgrave Macmillan, 2014), 34.

66. For further detail on how New York's city government was organized, see Charles Brecher, "City Council," in *Encyclopedia of New York City*, 229–230.

67. See Boyer's excellent study, *Urban Masses and Moral Order in America*.

68. Boyer, *Urban Masses and Moral Order in America*, 124.

69. Boyer, *Urban Masses and Moral Order in America*, 127. See also Groneman and Reimers, "Immigration."

70. Ernst, *Immigrant Life in New York City*, 135–139. On the American Tract Society, see Elizabeth Twaddell, "The American Tract Society, 1814–1860," *Church History* 15, no. 2 (1946): 116.

71. Alan Trachtenberg, *The Incorporation of America: Culture and Society in the Gilded Age* (New York: Hill and Wang, 2007); Robert H. Wiebe, *The Search for Order, 1877–1920* (New York: Hill and Wang, 1967).

72. Richard Hofstadter, *Age of Reform: From Bryan to F.D.R.* (New York: Vintage Books, 1971), remains the standard work on these efforts as of 1890.

73. Foucault, *Security, Territory, Population*, 29.

74. Patrick Joyce, *Rule of Freedom: Liberalism and the Modern City* (London: Verso, 2003).

75. Boyer, *Urban Masses and Moral Order in America*, 175. In the United States, these ideas were advocated particularly within the City Beautiful movement. See Michele Helene Bogart, *Public Sculpture and the Civic Ideal in New York City, 1890–1930* (Washington, DC: Smithsonian Institution Press, 1997).

76. For further detail, see the excellent study by Wolfgang Schivelbusch, *Disenchanted Night: The Industrialization of Light in the Nineteenth Century* (Berkeley: University of California Press, 1998).

77. Joyce, *Rule of Freedom*, 144–182. On the role that sewage networks played in creating urban order, see also Martin Melosi, *The Sanitary City: Urban Infrastructure in America from Colonial Times to the Present* (Baltimore: John Hopkins University Press, 2000).

78. Joyce, *Rule of Freedom*, 11.

79. The notion of the city as an apparatus of circulation was not new, of course. As Foucault has shown, the idea began emerging in city plans from the sixteenth and seventeenth centuries. See Foucault, *Security, Territory, Population*, 33–38. At that time, streets were channeled through the center of town to enable a more efficient diffusion of goods and people. This was not simply an economic necessity, but also an effective way to improve urban hygiene and political order.

80. For more on the introduction of urban water systems, see John Duffy, *A History of Public Health in New York City, 1625–1866* (New York: Russell Sage Foundation, 1968), 391–404; Steven H. Corey, "Sanitation," in *Encyclopedia of New York City*, 1041–1043.

81. "Loving Cup to Belmont Given at Subway Feast; Chief Engineer Parsons Says City Should Be Satisfied," *New York Times*, October 28, 1904.

82. Mary Douglas, *Purity and Danger: An Analysis of Concepts of Pollution and Taboo* (London: Routledge, 2002), 2.

83. Roy Lubove, *The Progressives and the Slums: Tenement House Reform in New York City, 1890–1917* (Pittsburgh: University of Pittsburgh Press, 1974).

84. Daniel T. Rogers, *Atlantic Crossings: Social Politics in a Progressive Age* (Cambridge, MA: Harvard University Press, 1998).

85. From the extensive literature on the progressive movement in the United States, the following serve as good introductions to the topic: John Whiteclay Chambers, *Tyranny of Change: America in the Progressive Era, 1890–1920* (New Brunswick, NJ: Rutgers University Press, 2000); Michael El McGerr, *A Fierce Discontent: Rise and Fall of the Progressive Movement in America, 1870–1920* (New York: Oxford University Press, 2005); Hood, *722 Miles*, 126.

86. Lubove, *The Progressives and the Slums*.

87. Alexander Alland, *Jacob A. Riis: Photographer & Citizen* (New York: Aperture, 1993).

88. Jacob A. Riis, *How the Other Half Lives* (New York: Barnes & Noble, 2004); Tyler Anbinder, *Five Points: The 19th-Century New York City Neighborhood that Invented Tap Dance, Stole Elections, and Became the World's Most Notorious Slum* (New York: Blume, 2002).

89. From the extensive literature on New York's housing in the nineteenth century, see Richard A. Pluntz, "On the Uses and Abuses of Air: Perfecting the New York Tenement, 1850–1901," in *Like and Unlike: Essays on Architecture and Art from 1870 to the Present*, ed. Josef P. Kleihues and Christina Rathgeber (New York: Rizzoli, 1993), 159–179.

90. Riis, *How the Other Half Lives*, 2.

91. Douglas, *Purity and Danger*, 2.

92. Derrick, *Tunneling the Future*, 18–19.

93. Riis, *How the Other Half Lives*, 89.

94. Riis, *How the Other Half Lives*, 89.

95. John M. Jordan, *Machine-Age Ideology: Social Engineering and American Liberalism, 1911–1939* (Chapel Hill, NC: University of North Carolina Press, 1994).

96. On New York, see Lubove, *The Progressives and the Slums;* on Paris, see Donald Reid, *Paris Sewer and Sewermen: Realities and Representations* (Cambridge, MA: Harvard University Press, 1991).

97. Adna F. Weber, *The Growth of Cities in the Nineteenth Century: A Study in Statistics* (New York: Macmillan, 1899), 157.

98. Adna F. Weber, "Rapid Transit and the Housing Problem," *Municipal Affairs* 6 (June 1902): 408–417, 11. See also Hood, "The Impact of the IRT on New York City," 155.

99. Kenneth T. Jackson, *Crabgrass Frontier: The Suburbanization of the United States* (New York: Oxford University Press, 1987), 116–118.

100. Hood, "The Impact of the IRT on New York City," 155–156.

101. Lubove, *The Progressives and the Slums*, esp. 117–166.

102. Weber, "Rapid Transit and the Housing Problem."

103. Hood, *722 Miles*, 127.

104. Hood, "The Impact of the IRT on New York City," 154.

105. Hood, "The Impact of the IRT on New York City," 159.

106. Weber, "Rapid Transit and the Housing Problem," 412.

107. Hood, "The Impact of the IRT on New York City," 157.

108. Walker, *Fifty Years of Rapid Transit*, 66.

109. Brooks, *Subway City*, 30.

110. "Exercises in City Hall."

111. On the idea of infrastructure constituting an element of welfare (*Daseinsfürsorge*), see Dirk van Laak, "Infra-Strukturgeschichte," *Geschichte und Gesellschaft* 27, no. 3 (Autumn 2001): 367–393.

112. "Exercises in City Hall."

113. This coding of the subway is an example of "normative universalism" as described by Jürgen Habermas in *The Inclusion of the Other: Studies in Political Theory*, ed. Ciaran P. Cronin and Pablo De Greiff (Cambridge, MA: MIT Press, 1998).

114. The dynamics that ultimately led to construction contracts for the first stretch of the subway were both complicated and controversial. As they had little effect on the definition of the passenger, we do not need to deal with them in depth here. For details, see Derrick, *Tunneling the Future*.

115. Derrick, *Tunneling the Future*, 40–44.

116. Cudahy, *Under the Sidewalks of New York*, 20.

117. That station was closed on December 31, 1945. See Walker, *Fifty Years of Rapid Transit*, 162–164. Alternative plans for the subway can be found in Shanor, *The City That Never Was*.

118. Hood, "The Impact of the IRT on New York City," 150.

119. "Rapid Transit Tunnel Begun—Ground Officially Broken," *New York Times*, March 25, 1900.

120. For more detail, see Interborough Rapid Transit Company, *New York Subway: Its Construction and Equipment* (New York: McGraw, 1904).

121. For more on construction of the subway see Stan Fischler, *Uptown Downtown: A Trip through Time on New York's Subways* (New York: Hawthorn Books, 1976), 33–40.

122. Cudahy, *Under the Sidewalks of New York*, 22.

123. Cudahy, *Under the Sidewalks of New York*, 24.

124. This opinion was held, for instance, by John B. McDonald. See "The Man that Built the Subway," interview by Kate Carew, *New York World*, October 23, 1904. Quoted in Brooks, *Subway City*, 64.

125. There is almost as much literature on the history of the London Underground as there is on the history of the New York subway. For a particularly good introduction, see Christian Wolmar, *The Subterranean Railway: How the London Underground Was Built and How It Changed the City Forever* (London: Atlantic Books, 2005).

126. Hood, *722 Miles*, 91–94.

127. James E. Mooney, "Sage, Russell," in *Encyclopedia of New York City*, 1032.

128. Quoted in Hood, *722 Miles*, 92.

129. Norbert Elias, *The Civilizing Process: Sociogenic and Psychogenetic Investigations* (Malden, MA: Blackwell, 2000), 443.

130. This was particularly true for the Manhattan lines, which were almost entirely subterranean. Later, in other city districts, elevated train installations were often used for subways. Today, two-thirds of the subway network run underground.

131. For further detail, see David L. Pike, *Subterranean Cities: The World beneath Paris and London, 1800–1945* (Ithaca: Cornell University Press, 2005); David L. Pike, *Metropolis on the Styx: The Underworlds of Modern Urban Culture, 1800–2001* (Ithaca: Cornell University Press, 2007).

132. Alex Marshall, *Beneath the Metropolis: The Secret Lives of Cities* (New York City: Caroll & Graf, 2006), 4–9.

133. See Rosalind Williams, *Notes on the Underground: An Essay on Technology, Society, and the Imagination* (Cambridge, MA: MIT Press, 2008).

134. Gaston Bachelard, *The Poetics of Space* (Boston: Beacon Press, 1994), 17–22.

135. For a study on this process in Paris, see Reid, *Paris Sewer and Sewermen*.

136. Pike, *Metropolis on the Styx*, 33.

137. W. J. Passingham, *The Romance of London's Underground* (London: Sampson, Low, Marston, 1932), II. Quoted in Pike, *Subterranean Cities*, 33.

138. Brooks, *Subway City*, 3. In chapter 3, we will return to the experience of the subway as hell or Hades.

139. Christopher Gray, "Streetscapes: New York's Subway; That Engineering Marvel Also Had Architects," *New York Times*, October 10, 2004; Marjorie Pearson, "Heins and La Farge," in *Encyclopedia of New York City*, 537.

140. Schivelbusch, *The Railway Journey*, 171–177.

141. Quoted in Jeffrey Richards and John M. MacKenzie, *The Railway Station: A Social History* (London: Faber & Faber, 2010), 3.

142. New York Transit Museum, *Subway Style: 100 Years of Architecture and Design in the New York City Subway* (New York: Stewart, Tabori & Chang, 2004), 4–9.

143. Fitzpatrick, *Art and the Subway*, 30.

144. The article is mentioned without a source in Brooks, *Subway City*, 70; and in Christopher Cumo, *Science and Technology in Twentieth Century American Life* (Westport, CT: Greenwood Publishing Group, 2007), 56. See also Holt, "The Changing Perception of Urban Pathology."

145. Steven Johnson, *Invention of Air: A Story of Science, Faith, Revolution, and the Birth of America* (New York: Riverhead Books, 2008).

146. On the idea of the laboratory and mechanical objectivity, see Lorraine Daston and Peter Galison, *Objectivity* (New York: Zone Books, 2007), 115–124; see also Bruno Latour, *We Have Never Been Modern* (Cambridge, MA: Harvard University Press, 1993), 15–20.

147. On Chandler's method and findings, see "Subway Air Good—Chandler," *New York Times*, November 19, 1904. The authorities also later published a pamphlet on Chandler's findings, titled: *Subway Air Pure as in Your Own Home.*

148. John W. Alexander, "Portrait of Dr. Chandler," in *Charles Frederick Chandler Testimonial Supplement,* ed. Fannie Casseday Duncan (New York: Columbia University Press, 1910), 6.

149. Fritjof Capra, *Web of Life: A New Scientific Understanding of Living Systems* (New York: Anchor, 1997).

150. Matthew Gandy, "Cyborg Urbanization: Complexity and Monstrosity in the Contemporary City," *International Journal of Urban and Regional Research* 29, no. 1 (March 2005): 26–49, esp. 28–30.

151. Wachsmuth, "Three Ecologies."

152. "Clamor for Tickets for Subway Opening," *New York Times*, October 26, 1904.

153. Fitzpatrick, *Art and the Subway*, 36.

154. Directives were implemented in stations and wagon interiors, for example, in the way that seating was arranged, and in the use of turnstiles, symbols, and lighting. See chapter 3 for more detail.

155. "Some Ifs and Don'ts," *New York Times*, October 27, 1904.

156. One question related to physical health is striking: "Q: Is subway travel injurious to the eyes? A: A well-known oculist says that looking at the rows of white columns is very straining. Therefore, don't look at them." "Some Ifs and Don'ts."

157. "Schedule of Trains for the Subway Out," *New York Times*, October 25, 1904.

158. "Subway Opening To-day with Simple Ceremony," *New York Times*, October 27, 1904.

159. "Subway Opening To-day with Simple Ceremony."

160. "Finish Plans for Subway Celebration," *New York Times*, October 18, 1904.

161. "Schedule of Trains for the Subway Out."

162. The *New York Times* wrote: "The rush for tickets to the opening continued unabated yesterday, and scores of demands had to be refused, with the result that the applicants went away declaring they had been slighted." "Clamor for Tickets for Subway Opening."

163. "Our Subway Open: 150,000 Try It."

164. "McClellan Motorman of First Subway Train," *New York Times*, October 28, 1904.

165. "Things Seen and Heard Along the Underground," *New York Times*, October 28, 1904.

166. "Our Subway Open: 150,000 Try It."

167. Hood, *722 Miles*, 95.

168. Hood, *722 Miles*, 95.

169. Hood, *722 Miles*, 96.

170. "Things Seen and Heard Along the Underground."

171. "Rush Hour Blockade Jams Subway Crowds," *New York Times*, October 29, 1904.

172. "Loving Cup to Belmont Given at Subway Feast; Chief Engineer Parsons Says City Should Be Satisfied."

173. "Rush Hour Blockade Jams Subway Crowds."

174. "Visiting Crowds Swamp Subway Service," *New York Times*, October 31, 1904.

175. Hood, *722 Miles*, 95.

176. "Our Subway Open: 150,000 Try It."

177. "Our Subway Open: 150,000 Try It."

178. Earl Mayo, "New York's Subway in Operation," *Outlook Magazine* (November 1904): 563–568.

179. "Our Subway Open: 150,000 Try It."

180. "Rush Hour Blockade Jams Subway Crowds."

181. "Our Subway Open: 150,000 Try It."

182. "Our Subway Open: 150,000 Try It."

183. For more on the concept of liminality, see Victor Turner, *The Ritual Process: Structure and Anti-Structure* (Chicago: Aldine, 1995); as a historical source, see Arnold van Gennep, *The Rites of Passage* (London: Routledge, 2010).

184. Orvar Löfgren, "Motion and Emotion: Learning to be a Railway Traveler," *Mobilities* 3, no. 3 (October 2008): 331–351.

185. Georg Simmel, "The Metropolis and Mental Life," in *On Individuality and Social Forms: Selected Writings*, ed. Donald N. Levine (Chicago: University of Chicago Press, 1971); see also Schivelbusch, *The Railway Journey*, 124–128.

186. "Our Subway Open: 150,000 Try It."

187. Ulrich Bröckling, *The Entrepreneurial Self: Fabricating a New Type of Subject* (London: Sage, 2015), 13.

CHAPTER 2

1. Quoted in Neal Bascomb, *Higher: A Historic Race to the Sky and the Making of a City* (New York: Broadway Books, 2004), 13.

2. Andreas Reckwitz, *Das hybride Subjekt: Eine Theorie der Subjektkulturen von der bürgerlichen Moderne zur Postmoderne* (Weilerswist: Velbrück Wissenschaft, 2010), 275–288. See also David Frisby, *Fragments of Modernity: Theories of Modernity in the Work of Simmel, Kracauer, and Benjamin* (London: Routledge, 2013) for a discussion of how these diagnoses were addressed by Simmel, Kracauer, and Benjamin.

3. See, for example, Reyner Banham, *Theory and Design in the First Machine Age* (New York: Praeger, 1967); John M. Jordan, *Machine-Age Ideology: Social Engineering and American Liberalism, 1911–1939* (Chapel Hill, NC: University of North Carolina Press, 1994); Richard Guy Wilson, Dianne H. Pilgrim, and Dickran Tashijan, *The Machine Age in America: 1918–1941* (New York: Brooklyn Museum/Harry N. Abrams, 1986).

German pacifist Bertha von Suttner also used the term as early as 1899 for the title of a utopian novel, a critical evaluation of the nineteenth century, written as a fictional return from the future. Bertha von Suttner, *Das Maschinenzeitalter: Zukunftsvorlesungen über unsere Zeit* (Dresden: E. Pierson, 1899). In France and in the United States, the term became popular as of the 1920s, used for example by Le Corbusier and Clarence Arthur: Le Corbusier, *La Ville radieuse: Eléments d'une doctrine d'urbanisme pour l'équipment de la civilization machinist* (Boulogne-Sur-Seine: Editions de l'Architecture d'aujourd'hui, 1935); Clarence Arthur Perry, *Housing for the Machine Age* (New York: Russell Sage Foundation, 1939).

4. Stephan Kern, *The Culture of Time and Space, 1880–1918* (Cambridge, MA: Harvard University Press, 2003); Leo Marx, *The Machine in the Garden: Technology and the Pastoral Ideal in America* (New York: Oxford University Press, 1964).

5. Wilson, Pilgrim, and Tashjian, *The Machine Age in America*, 24.

6. Hughes, *American Genesis: A Century of Invention and Technological Enthusiasm, 1870–1970* (Chicago: University of Chicago Press, 1989), 295–303.

7. For more on transatlantic transfers and adaptions, see Anson Rabinbach, *The Human Motor: Energy, Fatigue, and the Origins of Modernity* (Berkeley: University of California Press, 1992); Daniel T. Rodgers, *Atlantic Crossings: Social Politics in a Progressive Age* (Cambridge, MA: Harvard University Press, 1998).

8. Hughes, *American Genesis*, 3.

9. See Reckwitz, *Das hybride Subjekt*, 275–288.

10. Reckwitz borrows the concept from Peter Wagner. See Reckwitz, *Das hybride Subjekt*, 283n10.

11. Reckwitz defines the period of organized modernity loosely from 1920 to 1970 (Reckwitz, *Das hybride Subjekt*, 336). In chapter 4 we will see that its features were visible much earlier and that it was already beset by crisis in the 1950s, at least in North American metropolises.

12. Reckwitz, *Das hybride Subjekt*, 298, translated by Sage Anderson.

13. See Michael Gamper, *Masse lesen, Masse schreiben: eine Diskurs- und Imaginationsgeschichte der Menschenmenge 1765–1930* (Munich: Fink, 2007).

14. José Ortega y Gasset, *The Revolt of the Masses* (New York: W. W. Norton, 1932), 11–12.

15. Sigmund Freud, "Group Psychology and the Analysis of the Ego," in *The Standard Edition of the Complete Psychological Works of Sigmund Freud*, volume 18 *(1920–1922): Beyond the Pleasure Principle, Group Psychology and Other Works* (London: The Hogarth Press, 1955). Siegfried Kracauer, "The Mass Ornament," in *The Mass Ornament: Weimar Essays*, trans. and ed. Thomas Y. Levi (Cambridge, MA: Harvard University Press, 1995): Robert Ezra Park, *The Crowd and the Public and Other Essays* (Chicago: University of Chicago Press, 1972).

16. For more on the crowd in American literature, see Mary Esteve, *Aesthetics and Politics of the Crowd in American Literature* (Cambridge: Cambridge University Press, 2003); Nicolaus Mills, *The Crowd in American Literature* (Baton Rouge: Louisiana State University Press, 1986).

17. Michael Makropoulos, *Theorie der Massenkultur* (Paderborn: Fink, 2008).

18. See Matthew Gandy, "Cyborg Urbanization: Complexity and Monstrosity in the Contemporary City," *International Journal of Urban and Regional Research* 29, no. 1 (March 2005): 26–49, esp. 28–29. There were, however, several trends opposing conceptualization of the city as a machine, for example, the Garden City movement and work by Frank Lloyd Wright. For more detail, see Peter Hall, *Cities of Tomorrow: An Intellectual History of Urban Planning and Design in the Twentieth Century* (Chichester: Wiley-Blackwell, 2002), 90–147; Harold L. Platt, "Planning Modernism: Growing the Organic City in the 20th Century," in *Thick Space: Approaches to Metropolitanism*, ed. Dorothee Brantz, Sasha Disko, and George Wagner-Kyora (Bielefeld: transcript, 2012), 165–212.

19. Quoted in Maurice Zolotow, "Manhattan's Daily Riot," *Saturday Evening Post*, March 10, 1945.

20. See David Ward and Oliver Zunz, "Between Rationalism and Pluralism: Creating the Modern City," in *The Landscape of Modernity: New York City, 1900–1940*, ed. David Ward and Oliver Zunz (Baltimore: John Hopkins University Press, 1992), 3–18. On the history of American port cities in the twentieth century, see Boris Vormann, *Global Port Cities in North America: Urbanization Processes and Global Production Networks* (London: Routledge, 2015).

21. See Janet L. Abu-Lughod, *New York, Chicago, Los Angeles: America's Global Cities* (Minneapolis: University of Minnesota Press, 1999); Saskia Sassen, *The Global City: New York, London, Tokyo* (Princeton,

NJ: Princeton University Press, 2001); Saskia Sassen, *Deciphering the Global: Its Scales, Spaces, and Subjects* (London: Routledge, 2007).

22. For more on the development of the skyscraper, see Bascomb, *Higher: A Historic Race to the Sky and the Making of a City*; George H. Douglas, *Skyscrapers: A Social History of the Very Tall Building in America* (Jefferson, NC: McFarland, 2004); Francisco Mújica, *History of the Skyscraper* (Cambridge, MA: Da Capo Press, 1977). On contemporary debates regarding skyscapers, see Roger Shepherd, ed., *Skyscraper: The Search for an American Style, 1891–1941* (New York: McGraw-Hill, 2003).

23. For more on the reorganization of the vertical city at the turn of the century, see David L. Pike, *Metropolis on the Styx: The Underworlds of Modern Urban Culture, 1800–2001* (Ithaca: Cornell University Press, 2007), 11–24.

24. For a history of the elevator, see Andreas Bernard, *Lifted: A Cultural History of the Elevator* (New York: NYU Press, 2014); Stephen Graham, *Vertical: The City from Satellites to Bunkers* (London: Verso, 2017).

25. See James E. Vance, Jr., *The Continuing City: Urban Morphology in Western Civilization* (Baltimore: John Hopkins University Press, 1990), 472–479.

26. Luxury New York hotels such as the Ritz Carlton were pioneers in this kind of reorganization by assigning the lower floors to the service personnel and reserving higher levels for exclusive guests. Historian Ward Morehouse has called this development the creation of a "vertical Beverly Hills." See Ward Morehouse III, *Waldorf-Astoria. America's Gilded Dream* (New York: 1991), 137. Quoted in Bernard, *Lifted*, 71.

27. At times, more than 9,000 people per hectare lived in this area in miserable conditions. See Brian J. Cudahy, *Under the Sidewalks of New York: The Story of the Greatest Subway System in the World* (New York: Fordham University Press, 1995), 2.

28. Further north, in the part of the Bronx that had largely resembled a village until then, population increased by 150 percent. See Clifton Hood, *722 Miles: The Building of the Subways and How They Transformed New York* (Baltimore: John Hopkins University Press, 2004), 113.

29. For more on the history of Harlem, see Jonathan Gill, *Harlem: The Four Hundred Year History from Dutch Village to Capital of Black America* (New York: Grove Press, 2001).

30. Hood, *722 Miles*, 114.

31. Clifton Hood, "The Impact of the IRT on New York City," *Historical American Engineering Record: Interborough Rapid Transit Subway (Original Line) NY-122* (New York City, 1979: 147–148.

32. Cudahy, *Under the Sidewalks of New York*, 37.

33. The negotiations that led to signing the dual contracts were so complicated that subway historian Peter Derrick has written an entire book about them: Peter Derrick, *Tunneling the Future: The Story of the Great Subway Expansion That Saved New York* (New York: NYU Press, 2001). While the judicial and administrative details of the contracts are less important for the present study of social change, they play a significant role in the history of subway technology and financing.

34. Hood, "The Impact of the IRT on New York City"; Hood, *722 Miles*, 174–180; Derrick, *Tunneling the Future*, 245–248.

35. Hood, *722 Miles*, 193.

36. Cudahy, *Under the Sidewalks of New York*, 85–89.

37. "Gay Midnight Crowds Ride First Trains in the New Subway," *New York Times*, September 10, 1932.

38. Hood, *722 Miles,* 183–184; Michael Brooks, *Subway City: Riding the Trains, Reading New York* (New Brunswick, NJ: Rutgers University Press, 1997), 114–115. For more on the impact of the automobile, see Stephen B. Goddard, *Getting There: The Epic Struggle between Road and Rail in the American Century* (Chicago: University of Chicago Press, 1996).

39. Hood, *722 Miles*, 239.

40. Despite growing numbers of passengers, the subway continued to lose money. New York's public purse subsidized every trip on the subway with fourteen cents per ride. After the Second World War and the city's greatest financial crisis up to that point, in 1947 Mayor LaGuardia doubled the fare to ten cents. It had remained the same for more than forty years, even in periods of inflation. When the New York City Transit Authority (NYCTA) was founded in 1953, the price was hiked up to fifteen cents. Hood, *722 Miles*, 239.

41. Hood, *722 Miles*, 254.

42. Kenneth T. Jackson, *Crabgrass Frontier: The Suburbanization of the United States* (New York: Oxford University Press, 1987).

43. See Hughes, *American Genesis*, 3.

44. Charles G. Poore, "Times Square Becomes Biggest Tube Station," *New York Times*, March 13, 1927. Also found in Brooks, *Subway City*, 109.

45. *Wörterbuch der Soziologie*, comp. Karl-Heinz Hillmann, 5th ed. (Stuttgart: Krömer, 2017), s.v. "Masse."

46. For example, Karl Marx wrote in 1847: "Economic conditions had first transformed the mass of the people of the country into workers. The domination of capital has created for this mass a common situation, common interests. This mass is thus already a class as against capital, but not yet for itself. In the struggle, of which we have pointed out only a few phases, this mass becomes united, and constitutes itself as a class for itself. The interests it defends become class interests. But the struggle of class against class is a political struggle." Karl Marx, "The Poverty of Philosophy," in *Marx-Engels Collected Works*, volume 6: *Marx and Engels, 1845–1848* (New York: International Publishers, 1976), 211.

47. Gustave Le Bon, *The Crowd: A Study of the Popular Mind* (Kitchener, Ont: Batoche, 2001).

48. See Susana Barrows, *Distorting Mirrors: Visions of the Crowd in Late Nineteenth-Century France* (New Haven: Yale University Press, 1981).

49. See Sigmund Freud, "Massenpsychologie und Ich-Analyse," in *Massenpsychologie und Ich-Analyse/Die Zukunft einer Illusion* (Frankfurt am Main: Fischer, 1993), 34. In the English translation, *Menschenhaufen* (heaps of people) was translated as "crowds of people," losing the somewhat derogatory association. See Freud, "Group Psychology and the Analysis of the Ego," 70.

50. Freud, "Group Psychology and the Analysis of the Ego," 77.

51. Le Bon, *The Crowd*, 19. With slightly different wording, Freud uses this quote from Le Bon in "Group Psychology and the Analysis of the Ego," 77.

52. Benjamin De Casseres, "Darwin Defied in Our Subways," *New York Times*, April 16, 1922.

53. See Christoph Asendorf, *Batteries of Life: On the History of Things and Their Perception in Modernity* (Berkeley: University of California Press, 1993), 105–111; Schivelbusch, *The Railway Journey: The Industrialization of Time and Space in the Nineteenth Century* (New York: Urizen Books, 1979), 77–78.

54. The first composers of subway songs avoided slang, idioms, and the rhythms of ragtime that were considered titillating and proletarian. Sunny Stalter, "The Subway Crush: Making Contact in New York City Subway Songs, 1904–1915," *Journal of American Culture* 34, no. 4 (December 2011): 321–331, esp. 322.

55. "The Subway Express" was originally a song from the musical called *Fascinating Flora* (1907), with music written by Jerome Kern and lyrics by James O'Dea. Lyrics quoted in Hood, *722 Miles*, 99–100.

56. Berman and Baumann were not the first to characterize modernity in terms of fluidity. Baudelaire's concept of modernity was already strongly characterized by the idea of movement: "Modernity is the transient, the fleeting, the contingent." Charles Baudelaire, "The Painter of Modern Life," in *The Painter of Modern Life and Other Essays*, ed. and trans. Jonathan Mayne (London: Phaidon Press, 1995), 7. See also Zygmunt Baumann, *Liquid Modernity* (Oxford: Polity Press, 2000); Marshall Berman, *All That Is Solid Melts into Air: The Experience of Modernity* (New York: Verso, 1991).

57. See Bernard, *Lifted*, 244.

58. Music and lyrics by Fred Fischer and Bob Emmerich. Quoted in Nancy Groce, *New York, Songs of the City* (New York: Billboard Books, 1999), 120. For more on this and other songs, see Tracy Fitzpatrick, *Art and the Subway: New York Underground* (New Brunswick, NJ: Rutgers University Press, 2009), 41.

59. On the concept of liminality, see Victor Turner, *The Ritual Process: Structure and Anti-Structure* (Chicago: Aldine, 1995).

60. Freud, "Group Psychology and the Analysis of the Ego," 88.

61. See also Fitzpatrick, *Art and the Subway*, 55.

62. Paul S. Boyer, *Urban Masses and Moral Order in America, 1820–1920* (Cambridge, MA: Harvard University Press, 1992), 252–260.

63. Sigmund Freud, "Three Essays on Sexuality," in *The Standard Edition of the Complete Psychological Works of Sigmund Freud*, volume 7 (1901–1905): *A Case of Hysteria, Three Essays on Sexuality and Other*

Works, trans. James Strachney (London: Hogarth Press, 1953), 201–202. See also Asendorf, *Batteries of Life*, 108.

64. Christopher Cumo, *Science and Technology in Twentieth Century American Life* (Westport, CT: Greenwood Publishing Group, 2007), 41; Fitzpatrick, *Art and the Subway*, 40.

65. *Blackwell Encyclopedia of Sociology*, comp. Clark McPhail (London: Blackwell, 2007), s.v. "Crowd Behavior."

66. See Reimut Reiche's introduction to Sigmund Freud, *Massenpsychologie und Ich-Analyse / Die Zukunft einer Illusion*, esp. 15–17.

67. Numerous authors have pointed out that the territories of world metropolises were anything but gender neutral spaces. For a good introduction, see Susanne Frank, *Stadtplanung im Geschlechterkampf: Stadt und Geschlecht in der Großstadtentwicklung des 19. und 20. Jahrhunderts* (Opladen: Leske + Budrich, 2003); Waltraud Ernst, "Möglichkeiten (in) der Stadt: Überlegungen zur Öffentlichkeit und Privatheit geschlechtlicher Raumordnungen," in *Street Harassment: Machtprozesse und Raumproduktion*, ed. Feministisches Frauenkollektiv (Vienna: Mandelbaum, 2008), 75–93; Franziska Roller, "Flaneurinnen, Straßenmädchen, Bürgerinnen: Öffentlicher Raum und gesellschaftliche Teilhabe von Frauen," in *Geschlechter-Räume: Konstruktionen von "gender" in Geschichte, Literatur und Alltag*, ed. Margarete Hubrath (Cologne: Böhlau, 2001), 251–265.

68. Brooks, *Subway City*, 173.

69. We will return to the experience of fear in chapter 5, reconstructing the experiences of female passengers with the help of complaint letters.

70. Brooks, *Subway City*, 173.

71. See Ellen Carol DuBois, "Equality League of Self-Supporting Women," in *Encyclopedia of New York City*, ed. Kenneth T. Jackson (New Haven: Yale University Press, 1995), 281.

72. Hood, *722 Miles*, 119.

73. "No Cars for Women Only," *New York Times*, August 4, 1909.

74. Brooks, *Subway City*, 179–180.

75. The employment policies of subway operators also discriminated against women. When there was a lack of labor during the First World War, women were employed at ticket counters and gates, but they lost their jobs immediately when the war was over. See Brooks, *Subway City*, 174.

76. Brooks, *Subway City*, 174.

77. The Transit Museum Archive stores a wealth of documents, particularly from the 1950s, testifying to both individual and structural racism. See chapter 5.

78. Catherine A. Barnes, *A Journey from Jim Crow: The Desegregation of Southern Transit* (New York: Columbia University Press, 1983).

79. They did not become inspectors, wagon conductors, or train engineers until the 1950s. See Brooks, *Subway City*, 183.

80. Until 1942 the employment policy of the operating companies prohibited African Americans from becoming bus drivers. See Brooks, *Subway City*, 183.

81. Hood, *722 Miles*, 94–95.

82. See Fitzpatrick, *Art and the Subway*, 81.

83. See Hood, *722 Miles*, 15.

84. Hood, *722 Miles*, 214.

85. See Fitzpatrick, *Art and the Subway*, 94–95.

86. In chapter 4, we will return to this shift in interpreting the role of the crowd, and to the critique of standardized mass society after the Second World War in particular.

87. Christopher Morley, *Christopher Morley's New York* (New York: Fordham University Press, 1988), 116.

88. See Bion J. Arnold, *Report No. 1–7 on the Subway of the Interborough Rapid Transit Company of New York City* (New York: Public Service Commission, 1909), Report No. 2, 13. These reports can be accessed at the New York Transit Museum Archives. See also Hood, *722 Miles*, 115.

89. Sigfried Giedion, *Mechanization Takes Command* (New York: W. W. Norton, 1948), 96.

90. See Reckwitz's analysis of what he calls the "code of socio-technology" in *Das hybride Subjekt*, 338–343.

91. Reckwitz, *Das hybride Subjekt*, 275.

92. On the concept of time-space compression, see David Harvey, *The Condition of Postmodernity* (Cambridge, MA: Blackwell, 1990), 260–283. See also Kern, *The Culture of Time and Space*.

93. Jordan, *Machine-Age Ideology*, 33–44.

94. For more on how powerful this notion was, see Jennifer Karns Alexander, *Mantra of Efficiency: From Waterwheel to Social Control* (Baltimore: John Hopkins University Press, 2008).

95. Jordan, *Machine-Age Ideology*. In 1890, however, Frank Bunker Gilbreth (1868–1924) undertook studies and experiments to increase efficiency in masonry that can be seen as landmarks in the transfer of the machine code to the realm of work organization. See Harvey, *The Condition of Postmodernity*, 125–126; Daniel Nelson, *Managers and Workers: Origins of the Twentieth-Century Factory System in the United States, 1880–1920* (Madison: University of Wisconsin Press, 1996), 65–78.

96. Hughes, *American Genesis*, 187.

97. See Harvey, *The Condition of Postmodernity*, 125–129.

98. Henry Ford, *My Live and Work* (New York: Doubleday, Page, 1922). On the impact of Ford's ideas, see Lindy Briggs, *The Rational Factory: Architecture, Technology, and Work in America's Age of Mass Production* (Baltimore: Johns Hopkins University Press, 1996); David Hounshell, *From the American System to Mass Production, 1800–1932: The Development of Manufacturing Technology in the United States* (Baltimore: Johns Hopkins University Press, 1985); David Gartman, *From Autos to Architecture: Fordism and Architectural Aesthetics in The Twentieth Century* (New York: Princeton Architectural Press, 2009).

99. Dirk van Laak, "Infra-Strukturgeschichte," *Geschichte und Gesellschaft* 27, no. 3 (Autumn 2001): 367–393, esp. 390.

100. Besides Taylor and Ford, Hughes takes a closer look especially at entrepreneur and inventor Samuel Insull (1859–1938), who implemented Chicago's first integrated electric power supply systems and decisively advanced the electric grid in the United States. See Hughes, *American Genesis*, 185–248; Thomas P. Hughes, "The Electrification of America: The System Builders," *Technology and Culture* 20, no. 1 (January 1979): 124. For more on Insull, see also John F. F. Wasik, *The Merchant of Power: Sam Insull, Thomas Edison, and the Creation of the Modern Metropolis* (New York: Palgrave Macmillan, 2008).

101. Paul Virilio, "Perception, Politics and the Intellectual: Interview with Niels Brügger," in *Virilio Live: Selected Interviews*, ed. John Armitage (London: Sage, 2001), 82–96, esp. 91. See also Paul Virilio, *Logistics of Perception* (New York: Verso, 1989).

102. For an introduction to the various kinds of knowledge that constitute logistics from the perspective of the history of science and culture, see James Beniger, *The Control Revolution: Technological and Economic Origins of the Information Society* (Cambridge, MA: Harvard University Press, 2009); Richard Vahrenkamp, *Die logistische Revolution: Der Aufstieg der Logistik in der Massenkonsumgesellschaft* (Frankfurt am Main: Campus Verlag, 2011); Christopher Jahns and Christine Schüffler, *Logistik: Von der Seidenstraße bis heute* (Wiesbaden: Springer Gabler Verlag, 2008). Alexander Klose's work on the conceptual and cultural history of the container is both inspiring and highly informative. See Alexander Klose, *The Container Principle: How a Box Changes the Way We Think* (Cambridge, MA: MIT Press, 2015). See Marc Levinson, *The Box: How the Shipping Container Made the World Smaller and the World Economy Bigger* (Princeton, NJ: Princeton University Press, 2006). Monika Dommann and Gabriele Schabacher have also written helpful histories of logistics. See Monika Dommann, "Handling, Flowcharts, Logistik. Zur Wissensgeschichte und Materialkultur von Warenflüssen," in *Nach Feierabend: Zürcher Jahrbuch für Wissensgeschichte: Zirkulationen* 7, ed. David Gugerli et al. (Zurich: diaphanes, 2011), 75–103; Gabriele Schabacher, "Raum-Zeit-Regime: Logistikgeschichte als Wissenszirkulation zwischen Medien, Verkehr und Ökonomie," *Archive für Mediengeschichte* 8 (2008): 135–148.

103. Klose, *The Container Principle*, 172–176. For more detail, see Bernhard Siegert, *Relais: Geschicke der Literatur als Epoche der Post 1751–1913* (Berlin: Brinkman u. Bose, 1993).

104. Klose, *The Container Principle*, 168–170. For more detail, see also the following standard sources: Oskar Morgenstern, "Note on the Formulation of the Theory of Logistics," *Naval Research Logistics Quarterly* 2, no. 3 (March 1955): 129–136; Martin L. van Creveld, *Supplying War: Logistics from Wallenstein to Patton* (Cambridge, MA: Cambridge University Press, 1980).

105. Giedion, *Mechanization Takes Command*, 209–246. See also Dorothee Brantz, *Slaughterhouse City: Paris, Berlin, and Chicago, 1780–1914* (Baltimore: Johns Hopkins University Press, forthcoming).

106. According to the *Oxford English Dictionary*, the expression "logistics" came into use in the late nineteenth century, having evolved from the French *logistique* (*loger*, meaning "to accommodate"), a term used mainly to designate the temporary provision of (sleeping) quarters to soldiers. Klose, however, traces the term back to the Greek verb *logizomai,* which means to calculate, think, or consider. In English and in German, the term did not gain widespread use until the 1920s, but then became the topic of countless treatises and books. For more on the history of the concept of logistics, see Klose, *The Container Principle*, 168–169; and Monika Dommann, "Material Manövrieren: Eine Begriffsgeschichte der Logistik," *Via Storia*, no. 2 (2009): 13–27. Deborah Cowens's work is also highly informative in this context: *The Deadly Life of Logistics: Mapping Violence in Global Trade* (Minneapolis: University of Minnesota Press, 2014), 23–43.

107. Philipp Sarasin and Andreas Kilcher, "Editorial," in *Nach Feierabend: Zürcher Jahrbuch für Wissensgeschichte: Zirkulationen 7*, ed. David Gugerli et al. (Zurich: diaphanes, 2011), 8–10.

108. Quoted by Dirk van Laak, "Infrastruktur und Macht," in *Umwelt und Herrschaft in der Geschichte*, ed. François Duceppe-Lamarre and Jens Ivo Engels (Munich: R. Oldenbourg, 2008), 106–114. The concept of infrastructure has had a similar fate. See Dirk van Laak, "Der Begriff 'Infrastruktur' und was er vor seiner Erfindung besagte," *Archiv für Begriffsgeschichte* 41 (1999): 280–299.

109. Dommann, "Material Manövrieren: Eine Begriffsgeschichte der Logistik"; Ake E. Andersson, "The Four Logistical Revolutions," *Papers in Regional Science* 59, no. 1 (January 1986): 1–12.

110. This is the first sentence of the introduction to the current standard work on passenger logistics: Paul Fawcett, *Managing Passenger Logistics: The Comprehensive Guide to People and Transport* (London: Kogan Page, 2000).

111. These and the following numbers are taken from Cudahy, *Under the Sidewalks of New York*, 31–34.

112. Besides the general volume of passengers being much higher than anticipated, predictions also miscalculated the distribution and capacity of express trains. New York subway riders often had to traverse long distances underground to get to their places of work. See Cudahy, *Under the Sidewalks of New York*, 31.

113. See chapter 4.

114. See Bion J. Arnold, *Report No. 1–7 on the Subway.*

115. As Arnold proudly reported, he was able to increase the speed of circulation during rush hour by almost ten percent by carefully reorganizing the signaling system. Using complex calculations, he also showed that slightly changing train braking systems would result in greater speeds, which individually only amounted to four or five seconds but taken in sum meant a significant increase in the volume of passengers circulated. Arnold, *Report No. 1–7 on the Subway*, No. 1, 33–39.

116. Michel Foucault, *Security, Territory, Population: Lectures at the Collège de France, 1977–78*, ed. Michel Senellart, trans. Graham Burchell (Basingstoke: Palgrave Macmillan, 2014); Michel Foucault, *The Birth of Biopolitics: Lectures at the Collège de France, 1978–1979*, ed. Michel Senellart, trans. Graham Burchell (New York: Palgrave Macmillan, 2008). For its first use, see Michel Foucault, *The History of Sexuality*, volume 1: *The Will to Knowledge* (New York: Pantheon, 1978).

117. Michel Foucault, "The Meshes of Power," in *Space, Knowledge, and Power: Foucault and Geography*, ed. Jeremy W. Crampton and Stuart Elden (Aldershot: Ashgate, 2007), 161.

118. Werner Sohn, "Bio-Macht und Normalisierungsgesellschaft: Versuch einer Annäherung," in *Normalität und Abweichung: Studien zur Theorie und Geschichte der Normalisierungsgesellschaft*, ed. Werner Sohn and Herbert Mehrtens (Opladen/Wiesbaden: Westdeutscher Verlag, 1999), 9–29, esp. 21.

119. Jürgen Link has laid out his concepts in a number of publications, most comprehensively in his main work, *Versuch über den Normalismus: wie Normalität produziert wird* (Opladen: Westdeutscher Verlag, 1999). Link's reception and interpretation of Foucault is not without controversy, however. See Christina Bartz and Marcus Krause, "Einleitung: Spektakel der Normalisierung," in *Spektakel der Normalisierung*, ed. Christina Bartz and Marcus Krause (Paderborn: Fink, 2007), 7–24.

120. Jürgen Link, "'Normativ' oder 'Normal'? Diskursgeschichtliches zur Sonderstellung der Industrienorm im Normalismus, mit einem Blick auf Walter Cannon," in *Normalität und Abweichung: Studien zur Theorie und Geschichte der Normalisierungsgesellschaft*, ed. Werner Sohn and Herbert Mehrtens, trans. Sage Anderson (Opladen/Wiesbaden: Westdeutscher Verlag, 1999), 30–44, esp. 41.

121. Since the 1990s, an enormous amount has been written on normalism and normalization, but good introductions can be found in Bartz and Krause, "Einleitung: Spektakel der Normalisierung," and Werner Sohn and Herbert Mehrtens, *Normalität und Abweichung: Studien zur Theorie und Geschichte der Normalisierungsgesellschaft* (Opladen/Wiesbaden: Westdeutscher Verlag, 1999).

122. Arnold thus discovered that December was the busiest month for the subway, and that cars were the least crowded in July. Sunday volumes were the lowest, while Monday was by far the weekday with the heaviest overload. Interestingly, Arnold attributed heavy transit loads on Mondays to the fact that people read advertisements in the papers on the weekends and went shopping on Mondays. See Arnold, *Report No. 1–7 on the Subway*, No. 6, 12.

123. Arnold, *Report No. 1–7 on the Subway*, No. 1, 5.

124. Arnold, *Report No. 1–7 on the Subway*, No. 1, 45.

125. Foucault, *The History of Sexuality*, 146.

126. Michel Foucault, *Discipline and Punish: The Birth of the Prison* (New York: Random House, 1995). See also Ulrich Johannes Schneider, *Michel Foucault* (Darmstadt: Primus, 2006), esp. 47–55 and 167–181.

127. Arnold, *Report No. 1–7 on the Subway*, No. 1, 6.

128. See chapter 5 for more on the Transit Police.

129. New York City Transit Authority, *The New York City Transit Police Department: History and Organization,* December 1990, internal report, NYTMA.

130. These phrases were soon seen on posters and postcards and turned up in many subway songs. See Stalter, "The Subway Crush," 5.

131. "Courtesy Phrases Standardized," *B.R.T. Monthly,* October 1917.

132. New York Transit Museum, *Subway Style: 100 Years of Architecture and Design in the New York City Subway* (New York: Stewart, Tabori & Chang, 2004), 4–5. See also chapter 1.

133. See Wilson, Pilgrim, and Tashjian, *The Machine Age in America,* 149–155.

134. New York Transit Museum, *Subway Style,* 5–7.

135. The last pavilion was torn down in 1967. See Brooks, *Subway City,* 67.

136. For a more detailed analysis, see Stefan Höhne, "Vereinzelungsanlagen: Die Genese des Drehkreuzes aus dem Geist automatischer Kontrolle," *Technikgeschichte* 83, no. 2 (2016): 103–124.

137. Michel de Certeau, *The Practice of Everyday Life* (Berkeley: University of California Press, 1984), 113.

138. Hood, *722 Miles,* 221–222.

139. "Rush Hour Blockade Jams Subway Crowds," *New York Times,* October 29, 1904.

140. State of New York Transit Commission, "First Annual Report" (April 25, 1921–December 31, 1921), 52. The annual reports are accessible in the NYTMA.

141. Advertisement of the Perey Manufacturing Company in *Electric Railway Journal* 58, no. 13 (November 1921): 169.

142. Subsequent models driven by compressed air pumps proved to be too loud, expensive, and high-maintenance. See the anonymous and undated manuscript *A History of Turnstiles in the New York City Subways,* accessible in the NYTMA.

143. State of New York Transit Commission, "First Annual Report," 52.

144. "I. R. T. Tests Turnstiles; Ticket Choppers May Be Eliminated if New Devices Work Well," *New York Times,* January 23, 1921.

145. In the 1920s, the real value of the nickel fare rarely equaled more than two and a half cents. Hood, *722 Miles,* 221.

146. Hood, *722 Miles,* 222.

147. See, for example, "96 Out of 100 Have Nickels Ready," *Electric Railway Journal* 59, no. 1 (January 1922): 56.

148. In the 1920s, the BMT installed 234 "Coinpassors." When it opened in 1932, the IND furnished all of its stations with the same model. Older equipment was regularly replaced by new models that were

smaller, faster, and lower-maintenance. A small number of turnstiles were even made for left-handed passengers. *A History of Turnstiles in the New York City Subways*, 7.

149. Nonetheless, for lack of a better solution, these furnishings remained in place until 1991. See Noah McClain, "Social Control, Object Interventions and Social-Material Recursivity" (unpublished manuscript, August 14, 2008), 13–17.

150. See Danielle Schwartz, "Modernism for the Masses: The Industrial Design of John Vassos," *Archives of American Art Journal* 46, no. 1–2 (2006): 11.

151. See Wilson, Pilgrim, and Tashjian, *The Machine Age in America*, 83.

152. One of the earliest patent applications for these turnstiles emphasizes how important such measures were for the system to function efficiently: "In addition, suitable barriers must be placed when the turnstile is installed to direct the flow of people through it." See the patent description for *G-E Automatic Electric Turnstile, Pre-Bulletin 44316*, General Electric Company, September 1923, 2, accessible in the NYTMA.

153. Saul Bellow, *Herzog* (New York: Penguin, 2003), 192–193.

154. Schwartz, "Modernism for the Masses," 11.

155. Quoted in Schwartz, "Modernism for the Masses," 11. Later versions, like model no. 97 (introduced in 1946) and model no. 107 (introduced in 1968), could be used in both directions, so that one and the same turnstile could operate as both an entrance and an exit gate.

156. Hood, *722 Miles*, 222.

157. Using a turnstile also requires that one carry nothing exceeding a certain volume. In limited circumstances, people were allowed to use a gate near the turnstile. This pertained particularly to baby strollers, which again demonstrates that in general the equipment was not scripted for use by women.

158. The turnstile as such can be considered a central infrastructure and cipher of modernity, as philosopher Giorgio Agamben has claimed with respect to refugee camps, prisoner camps, and concentration camps. See Giorgio Agamben, *Homo Sacer: Sovereign Power and Bare Life* (Stanford, CA: Stanford University Press, 1998). See also Anke Hagemann, "Filter, Ventile und Schleusen: Die Architektur der Zugangsregulierung," in *Kontrollierte Urbanität: Zur Neoliberalisierung städtischer Sicherheitspolitik*, ed. Volker Eick, Jens Sambale, and Eric Töpfer (Bielefeld: transcript, 2007), 301–328.

159. Anke Hagemann, "Drehkreuz," *archplus* 191–192 (March 2009): 38–39.

160. Bernhard Siegert, "Doors: On the Materiality of the Symbolic," *Grey Room* 47 (Spring 2012): 20.

161. See Giedion, *Mechanization Takes Command*, 242–252; Dorothee Brantz, "On the Nature of Urban Growth: Building Abattoirs in 19th-Century Paris and Chicago," *Cahiers Parisiens* 5 (2009): 17–30.

162. Perey Turnstiles, Inc. still exists today. See "Company Evolution and History," *Perey Turnstiles*, http://www.turnstile.com/history.

163. Schwartz, "Modernism for the Masses," 11.

164. For more, see Wolfgang Schivelbusch, *The Railway Journey*, 171–177.

165. Siegert, "Doors: On the Materiality of the Symbolic," 18.

166. See Hagemann, "Filter, Ventile und Schleusen: Die Architektur der Zugangsregulierung," 304.

167. On this distinction, see Carlo Caduff, "Anticipations of Biosecurity," in *Biosecurity Interventions: Global Health & Security in Question*, ed. Andrew Lakoff and Stephen J. Collier (New York: Columbia University Press, 2008), 257–277.

168. See Hagemann, "Drehkreuz."

169. Advertising by the Perey Manufacturing Company in *Electric Railway Journal* 58, no. 13 (November 1921): 169.

170. See Madeleine Akrich and Bruno Latour, "A Summary of a Convenient Vocabulary for the Semiotics of Human and Nonhuman Assemblies," in *Shaping Technology / Building Society: Studies in Sociotechnical Change*, ed. Wiebe J. Bijker and John Law (Cambridge, MA: MIT Press, 1992), 259–264.

171. For the early twentieth century, it is difficult to prove that passengers actually exhibited such behavior. Subway operators did not keep records of it and there were rarely witnesses. How to get past these barriers was considered common knowledge, passed on mostly by word of mouth. American sociologist Noah McClain has been able to gather recollections of such practices by interviewing retired subway employees. McClain, "Social Control, Object Interventions and Social-Material Recursivity," 14–19.

One can also find a few literary mentions of the practice, for example in William S. Burroughs's novel *Naked Lunch* (1959): "I can feel the heat closing in, feel them out there making their moves, setting up their devil doll stool pigeons, crooning over my spoon and dropper I throw away at Washington Square Station, vault a turnstile and two flights down the iron stairs, catch an uptown A train. . . ." William S. Burroughs, *Naked Lunch* (New York: Grove Atlantic, 1992), 1.

172. Although it contains five hundred pages of details, the subway company's annual report from the year 1921 does not mention fare evasion at all. See State of New York Transit Commission, "First Annual Report."

173. In 1953, when a new gate system was introduced in the newly consolidated subway system, tokens were produced for use at the turnstiles to prevent fare evasion. Chapter 5 sketches how that innovation led to a number of new and clever tactics.

174. For more on how the ideas of scientific management influenced the architecture of the times, see Mauro F. Guillén, *The Taylorized Beauty of the Mechanical: Scientific Management and the Rise of Modernist Architecture* (Princeton, NJ: Princeton University Press, 2009).

175. Squire J. Vickers, "Design of Subway and Elevated Stations," *Municipal Engineers Journal* 3, no. 9 (1917): 114–120, esp. 114. See also Squire J. Vickers, "The Architectural Treatment of Special Elevated Stations of the Dual System, New York City," *Journal of the American Institute of Architects* 3, no. 11 (1915).

176. See David J. Framberger, "Architectural Designs for New York's First Subway," *Historical American Engineering Record, Survey Number HAER NY-122* (New York, 1979), 365–412.

177. See New York Transit Museum, *Subway Style*, 110–116.

178. The very first vending machine on the American continent was supposedly set up on a platform at one of New York's elevated trains, selling Adams' Tutti-Frutti chewing gum. See New York Transit Museum, *Subway Style*, 106.

179. The scales were not removed until 1972, when the decision was made that they were too high-maintenance, took up valuable space, and presented an impediment to the flow of passengers. See New York Transit Museum, *Subway Style*, 106.

180. See State of New York Transit Commission, "First Annual Report," 132. The subway was a place where mass culture could spread quickly and passengers were addressed as consumers. This began with advertising boards that were brought in immediately after the subway opened. See chapter 3.

181. Vickers, "Design of Subway and Elevated Stations," 116–118.

182. Hood, *722 Miles*, 93.

183. For example, most stations had four stairways to the platforms, two of which were originally entrances and the other two exits. Station guards called out commands and occasionally used force to ensure that passengers used the right stairways. The aboveground stations rarely strayed from this layout, as upside-down versions of the underground stations. The BRT used the same layout for most stations. Engineers and architects found this construction efficient, modern, and functional. See Framberger, "Architectural Designs for New York's First Subway."

184. See Foucault, *Discipline and Punish*, 141–148.

185. Foucault, *Discipline and Punish*, 143.

186. See Patrick Joyce, *The Rule of Freedom: Liberalism and the Modern City* (London: Verso, 2003), 210–217.

187. One iconic example of the transfer of transit aesthetics to other buildings is Le Corbusier und Pierre Jeanneret's "transformable house" in the Weissenhof Estate near Stuttgart, Germany. Its interior design was made to resemble modern Pullman coaches and club cars—even the beds disappear into a wall closet during the day.

188. Siegert, "Doors: On the Materiality of the Symbolic," 16.

189. See Gene Sansone, *New York Subways: An Illustrated History of New York City's Transit Cars* (Baltimore: Johns Hopkins University Press, 1977), 179.

190. Chapter 4 deals with restructuring the visual routines of passengers.

191. Michel Serres, *The Five Senses: A Philosophy of Mingled Bodies* (London: Bloomsbury, 2008), 146.

192. As with most of the technical and architectural elements of the system, it would be possible to go on at length regarding controversies around the cars, the experiments used to test them, and the

innovations that followed. Here, we focus solely on the technological developments that made passengers subservient. Gene Sansone has collected details on all of the types of cars ever built for the subway; see Sansone, *New York Subways*. According to Sansone, from 1904 until the founding of the New York City Transit Authority in 1953, the system used several dozens of different coach types. Each of the three subway companies built and regularly rebuilt cars for their own lines, which led to constant changes in configuration, furnishings, material, and equipment. Many of these models rolled through the tunnels for decades when subway financing postponed the purchase of newer versions. Some models were only prototypes that added numerous technical complications and had to be promptly removed or refitted. August Belmont, the IRT's major investor, had an exclusive private train built for his pleasure, paneled inside with mahogany. It had a bathroom, kitchen, and specially designed porcelain dinnerware. Dubbed "Mineola," to this day it remains the only private subway car ever built. See E. J. Quinby, "Minnie Was a Lady," *Railroad Magazine—The Magazine of Adventurous Railroading* 67, no. 1 (1956): 54–74.

193. See Sansone, *New York Subways*, 57. The first cars were composite models (a mixture of wood and steel), but these were soon replaced by cars cased entirely in steel to guarantee stability and fire resistance.

194. Arnold, *Report No. 1–7 on the Subway*, No. 3, 7.

195. Sansone, *New York Subways*, 68.

196. Later models were equipped with roof ventilation, improved pneumatic brake systems, and other refinements. See Sansone, *New York Subways*, 64.

197. Sansone, *New York Subways*, 149.

198. Before the 1970s, almost no subway system in the world had air conditioning. Ventilators and air vents were used to cool passengers, particularly in the summer, but with mediocre results. Due to a very restricted budget, the BRT's first car models were quite austere, with only fifteen lights and no ventilators in the ceiling. In the 1940s, cars were refurnished to include additional lights above the seats and four ventilators. BRT coaches were also sturdier, wider, and longer than IRT coaches, and they could carry more passengers and reach greater speeds. See Sansone, *New York Subways*, 155. Models from the late 1930s reached speeds of more than 90 kmh. New York Transit Museum, *Subway Style*, 210. In preparation for the 1939 World's Fair, the subway operators introduced new models equipped with display systems that indicated lines and directions. See Sansone, *New York Subways*, 98.

199. Originally, the team for a normal subway train pulling ten cars consisted of one train conductor and five attendants responsible for opening and closing doors. In 1915 the Pullman Company introduced Multiple Unit Door Control (MUDC), which opened and closed all doors electrically from one control panel. This reduced the number of employees by two-thirds between 1919 and 1939. See Hood, *722 Miles*, 222.

200. In the year of Arnold's survey, the number of passengers had already surpassed 800,000; six years later, the system carried 1.2 million per year. Hood, "The Impact of the IRT on New York City," 147.

201. Arnold, *Report No. 1–7 on the Subway*, No. 2, 33.

202. In 1916 Daniel L. Turner determined the volume of air needed for each passenger as 18.5 cubic feet. See Statement of D. L. Turner, Deputy Engineer of Subway Construction, in *Report of Commission on Building Districts and Restriction,* ed. Commission on Building Districts and Restrictions (New York: Board of Estimate and Apportionment, 1916), 191. Reprinted in: The City Club of New York, *Subway Overcrowding* (New York, 1930), 5. See also Richard Levine, "Seeking Bearable Level of Subway Discomfort," *New York Times,* October 10, 1987.

203. For more on the history and function of black-boxing see Klose, *The Container Principle,* 216–219.

For the definition of the concept as used in actor-network theory, see Bruno Latour, *Pandora's Hope: Essays on the Reality of Science Studies* (Cambridge, MA: Harvard University Press, 1999), 183–193.

204. For a definition of the "container subject," see Alexander Klose, "Who do you want to be today? Annäherungen an eine Theorie des Container-Subjekts," in *Das Motiv der Kästchenwahl: Container in Psychoanalyse, Kunst, Kultur,* ed. Insa Härtel and Olaf Knellessen (Göttingen: Vandenhoeck & Ruprecht, 2012), 21–38.

205. This also facilitated a certain degree of freedom, as we will see in chapter 3.

206. According to Foucault, true desubjectivation is only possible by using literary techniques of the kind found in the works of Nietzsche and Bataille. Michel Foucault, *Remarks on Marx: Conversations with Duccio Trombadori,* trans. R. James Goldstein and James Cascaito (New York: Semiotext(e), 1991), 31–32.

Agamben was the first to identify phenomena of a desubjectified *bare life* outside of literary techniques. See Agamben, *Homo Sacer.*

207. John G. Blair, *Modular America: Cross-Cultural Perspectives on the Emergence of an American Way* (Westport, CT: Greenwood Press, 1988), 2. I am thankful to Alexander Klose for drawing my attention to this work.

208. This seating model followed the classical layout of railway cabins, which in turn had been based on stagecoaches and sedan chairs. See Schivelbusch, *The Railway Journey,* 70–72.

209. This was the case, for instance, in cars from the BU 900 series. See Sansone, *New York Subways,* 128–129.

210. Sansone, *New York Subways,* 144.

211. Schivelbusch, *The Railway Journey,* 73–77.

212. Schivelbusch, *The Railway Journey,* 74.

213. In chapter 3, we will discuss the threat posed by this situation, and new practices that evolved to cope with isolation and apathy.

214. Jim Dwyer, *Subway Lives: 24 Hours in the Life of the New York City Subway* (New York: Crown, 1991), 74–75.

215. As of the late 1920s, plans for new cars for the publicly operated IND included a number of possible seating arrangements and door combinations, tested using complex models. The resulting cars had four

doors on each side, a combination of seats running lengthwise and crosswise, and narrower seats in order to accommodate more passengers. Model R1 was also designed to handle maximum acceleration. In the true spirit of logistical rationality and calculable efficiency, engineers found that using four double doors per car reduced entering and exiting times by almost four seconds. Projected across the entire system, this meant an increase in the speed of circulation allowing for an entire additional train, in other words, moving an additional 3,000 people. See Sansone, *New York Subways*, 179–286.

216. See Sansone, *New York Subways*, 193–203. R11–R16, the first new models after the Second World War and the last models produced before the NYCTA was founded in 1953, mark the zenith of machine-age paradigms for design and technology in many respects. With their casings of gleaming rustproof steel, futuristic shape, porthole door windows, and new engine and brake systems, they were regarded by many contemporaries as "the subway car of tomorrow." New York Transit Museum, *Subway Style*, 210.

The incipient economic crisis that eventually led to the decline of the entire system prevented any large-scale production of these models. As of the 1950s, cars did include a few—much less spectacular—innovations. Nonetheless, the paradigm of rational, efficient circulation proved to be enduring imperatives for designing the subway's world of containers.

217. City Club of New York, *Subway Overcrowding*. The study does not name its authors, perhaps in order to underscore its presentation as objective, rational science.

218. City Club of New York, *Subway Overcrowding*, 5.

219. City Club of New York, *Subway Overcrowding*, 4, 18.

220. City Club of New York, *Subway Overcrowding*, 4.

221. City Club of New York, *Subway Overcrowding*, 7.

222. City Club of New York, *Subway Overcrowding*, 7.

223. City Club of New York, *Subway Overcrowding*, 5.

224. See Erving Goffman, *Relations in Public: Microstudies in the Public Order* (New York: HarperCollins Publishers, 1971).

225. Michel Foucault, *Society Must Be Defended: Lectures at the Collège de France, 1975–76* (Picador: New York, 2003), 253.

226. City Club of New York, *Subway Overcrowding*, 5.

227. Arline L. Bronzaft, Stephen B. Dobrow, and Timothy J. O'Hanlon, "Spatial Orientation in a Subway System," *Environment and Behavior* 8, no. 4 (December 1976): 575–594; Susan Saegert, "Crowding: Cognitive Overload and Behavioral Constraint," *Environmental Design Research* 2 (January 1973): 254–261.

228. City Club of New York, *Subway Overcrowding*, 10.

Chapter 3

1. Helmut Lethen, *Cool Conduct: The Culture of Distance in Weimar Germany*, trans. Don Reneu (Berkeley: University of California Press, 2002), 28.

2. Karl Marx, "Economic and Philosophic Manuscripts," in *Karl Marx: Early Writings*, trans. R. Livingstone (Harmondsworth: Penguin, 1975), 353.

3. Mark M. Smith, "Producing Sense, Consuming Sense, Making Sense: Perils and Prospects for Sensory History," *Journal of Social History* 40, no. 4 (Spring 2007): 841–858.

4. For more on the concept of the container and containment, which goes back to British psychoanalyst Wilfred R. Bion (1897–1979), see Rosemarie Kennel, "Bions Container-Contained-Modell—und die hieraus entwickelte Denktheorie," in *Das Motiv der Kästchenwahl: Container in Psychoanalyse, Kunst, Kultur*, ed. Insa Härtel and Olaf Knellessen (Göttingen: Vandenhoeck & Ruprecht, 2012), 69–85; Alexander Klose and Jörg Potthast, "Container/Containment: Zur Einleitung," *Tumult* 38 (2012): 8–12.

5. Lethen, *Cool Conduct*, 33.

6. Alain Corbin, *The Foul and the Fragrant: Odor and the French Social Imagination* (Leamington Spa: Berg), 1986, 5–8. See also Alain Corbin, "Zur Geschichte und Anthropologie der Sinneswahrnehmung," in *Wunde Sinne: Über die Begierde, den Schrecken und die Ordnung der Zeit im 19. Jahrhundert* (Stuttgart: Klett-Cotta, 1993), 197–211.

7. For a critique of the analysis of culture, society, and artefacts based solely on texts, see David Howes, "Introduction," in *Empire of the Senses: The Sensual Culture Reader*, ed. David Howes (Oxford: Berg, 2005), 1–20; Hans Peter Hahn, *Materielle Kultur: Eine Einführung* (Berlin: Dietrich Reimer Verlag, 2005), 137–142.

8. Mary Douglas, *Purity and Danger: An Analysis of Concepts of Pollution and Taboo* (London: Routledge, 2002).

9. Corbin, *The Foul and the Fragrant*, 89–110.

10. Jonathan Crary, *Techniques of the Observer: On Vision and Modernity in the Nineteenth Century* (Cambridge, MA: MIT Press, 1992); Alain Corbin, *The Foul and the Fragrant*, 1988; Michael Serres, *The Five Senses: A Philosophy of Mingled Bodies* (London: Bloomsbury, 2008).

11. Susan Stewart, "Remembering the Senses," in *Empire of the Senses: The Sensual Culture Reader*, ed. David Howes (Oxford: Berg, 2005), 59–69, esp. 60–61.

12. A number of authors have made this point: Alexander F. Cowan and Jill Steward, eds., *The City and the Senses: Urban Culture Since 1500* (Farnham: Ashgate, 2007); Jonathan Crary, *Suspensions of Perception: Attention, Spectacle, and Modern Culture* (Cambridge, MA: MIT Press, 2001); David Frisby, *Fragments of Modernity. Theories of Modernity in the Work of Simmel, Kracauer, and Benjamin* (London: Routledge, 2013); Wolfgang Kaschuba, *Die Überwindung der Distanz: Zeit und Raum in der europäischen Moderne* (Frankfurt am Main: Fischer Taschenbuch Verlag, 2004).

13. "The psychological basis of the metropolitan type of individuality consists in the intensification of nervous stimulation which results from the swift and uninterrupted change of outer and inner stimuli. Man is a differentiating creature. His mind is stimulated by the difference between a momentary impression and the one which preceded it. Lasting impressions, impressions which differ only slightly from one another, impressions which take a regular and habitual course and show regular and habitual contrasts— all these use up, so to speak, less consciousness than does the rapid crowding of changing images, the sharp discontinuity in the grasp of a single glance, and the unexpectedness of onrushing impressions." Georg Simmel, "The Metropolis and Mental Life," in *The Blackwell City Reader*, ed. Gary Bridge and Sophie Watson (Oxford: Wiley-Blackwell, 2002), 11.

14. Georg Simmel, *Sociology: Inquiries into the Construction of Social Forms*, ed. and trans. Anthony J. Blasi et al. (Leiden: Brill, 2009), 573.

15. Crary, *Suspensions of Perception*, 11–80.

16. Herbert A. Simon, "Designing Organizations for an Information-rich World," in *Computers, Communications, and the Public Interest*, ed. Martin Greenberger (Baltimore, MD: Johns Hopkins Press, 1971), 37–72.

For a discussion of the similar concept of "affect-economy," see Norbert Elias, *The Civilizing Process: Sociogenic and Psychogenetic Investigations* (Malden, MA: Blackwell, 2000), 171.

17. Ernst Cassirer presents a similar model of perception in "Mythic, Aesthetic and Theoretical Space," *Man and World* 2 (February 1969): 3–17; and *The Philosophy of Symbolic Forms, Volume 1: Language*, trans. Ralph Manheim (New Haven: Yale University Press, 1955).

18. Simon, "Designing Organizations for an Information-rich World," 40–41. In addition, see Fang Wu and Bernando A. Huberman, "Novelty and Collective Attention," *Proceedings of the National Academy of Sciences* 104, no. 45 (November 2007): 17599–17601.

19. Simmel, "The Metropolis and Mental Life," 11.

20. Wolfgang Schivelbusch, *The Railway Journey: The Industrialization of Time and Space in the Nineteenth Century* (New York: Urizen Books, 1979), 159.

21. Simmel, *Sociology*, 578.

22. Walter Benjamin, *The Arcades Project*, trans. Howard Eiland and Kevin McLaughlin (Cambridge, MA: Harvard University Press, 1999), 84. The English translation of the text actually reads "underworld of names." It is unclear why the connotations of the original word, "Hades," were not incorporated into the English translation. See Walter Benjamin, *Das Passagen-Werk*, vol. 1 (Frankfurt am Main: Suhrkamp, 1983), 135.

23. Benjamin, *The Arcades Project*, 84.

24. Quoted in Michael Brooks, *Subway City: Riding the Trains, Reading New York* (New Brunswick, NJ: Rutgers University Press, 1997), 1; David L. Pike, *Subterranean Cities: The World beneath Paris and London, 1800–1945* (Ithaca: Cornel University Press, 2005); David L. Pike, *Metropolis on the Styx: The Underworlds of Modern Urban Culture, 1800–2001* (Ithaca: Cornell University Press, 2007).

25. George A. Soper, *Air and Ventilation of Subways* (New York: John Wiley & Sons, 1908), 130.

26. James Blaine Walker, *Fifty Years of Rapid Transit, 1864–1917* (North Stratford, NH: Ayer Publishing, 1918), 263–268.

27. The cartoon is also reprinted and thoughtfully analyzed in Brooks, *Subway City*, 70–72.

28. Simmel, *Sociology*, 578.

29. Simmel, *Sociology*, 579.

30. George Orwell, *The Road to Wigan Pier* (London: Penguin, 1989), 119, emphasis in original.

31. Simmel, *Sociology*, 577.

32. For further detail, see chapter 5.

33. Simmel, *Sociology*, 577.

34. See Robert Jütte, *Geschichte der Sinne* (Munich: C. H. Beck, 2000), 290–291. For further reading on the social function of disgust, see Winfried Menninghaus, *Ekel: Theorie und Geschichte einer starken Empfindung* (Frankfurt am Main: Suhrkamp, 2002).

35. We will return to the topic of advertisements in the subway later in this chapter.

36. To this day, the New York subway contributes significantly to noise pollution. See Robyn R. M. Gershon et al., "Pilot Survey of Subway and Bus Stop Noise Levels," *Journal of Urban Health: Bulletin of the New York Academy of Medicine* 83, no. 5 (September 2006): 802–812.

The authors of study recorded an average level of noise on subway platforms at 86 decibels, with the highest levels above 106 decibels. This is about as loud as an industrial power drill. The first trains were much louder than those of today, but while some noise reduction was achieved over the years by insulating motors, the introduction of air conditioning added more noise once again as of the early 1970s.

37. J. R. Sedden, "Auritis—A Subway Disease," *New York Times*, October 29, 1904.

38. Schivelbusch, *The Railway Journey*, 74.

39. Simmel, *Sociology*, 590–591. Simmel's teacher, philosopher and psychologist Moritz Lazarus (1824–1903), felt similarly about "car conversations" in trains and horse-drawn coaches: "The most exquisite conversations, as almost all writers have proven, take place among total strangers, during travel, and with new acquaintances." Moritz Lazarus, *Über Gespräche* (Berlin: Hensel, 1986), 14, translated by the author.

40. Simmel, *Sociology*, 592.

41. Richard Sennett, *The Fall of Public Man* (New York: Alfred A. Knopf, 1977), 337–339.

42. Simmel, *Sociology*, 575.

43. For many years, city planners and architects neglected to consider the sounds of urban space. Recently this topic has gained more attention (by composer Murray Schafer, among others), and architects have begun to reflect on the soundscapes of the buildings they design. See Jean-Paul Thibaud, "The

Sonic Composition of the City," in *The Auditory Culture Reader*, ed. Michael Bull and Less Back (Oxford: Berg, 2003), 329–342. Daniel Morat points out that historical research on the act of hearing has also been neglected for a long time. Daniel Morat, "Geschichte des Hörens: Ein Forschungsbericht," *Archive für Sozialgeschichte* 51 (October 2011): 695–716.

44. Susie J. Tanenbaum, *Underground Harmonies: Music and Politics in the Subways of New York* (Ithaca: Cornell University Press, 1995).

45. The ban on music was regulated by the same paragraph that banned begging and peddling. Tanenbaum, *Underground Harmonies*, 40.

46. Tanenbaum, *Underground Harmonies*, 40.

47. Jim Dwyer, *Subway Lives: 24 Hours in the Life of the New York City Subway* (New York: Crown, 1991), 121.

48. Regula Valérie Burri, *Doing Images: Zur Praxis medizinischer Bilder* (Bielefeld: transcript, 2008), 17, translated by the author. See also the following inspiring essay: Barbara Duden and Ivan Illich, "Die skopische Vergangenheit Europas und die Ethik der Opsis. Plädoyer für eine Geschichte des Blickes und Blickens," *Historische Anthropologie* 3, no. 2 (1995): 203–221.

49. Simmel, *Sociology*, 570–579.

50. Simmel, *Sociology*, 571.

51. Simmel, *Sociology*, 572.

52. The ability to remain silent in the midst of huge crowds can be seen as a cultural practice that was particularly encouraged by urbanization. "It is natural to speak and an art to remain silent," according to Moritz Lazarus in *Über Gespräche*, 21, translated by the author.

53. Simmel, *Sociology*, 573. In his essay on the flaneur, Walter Benjamin also quotes this description by Simmel, juxtaposing it with the flaneur. Walter Benjamin, "The Paris of the Second Empire in Baudelaire," in *The Writer of Modern Life: Essays on Charles Baudelaire*, ed. Michael W. Jennings, trans. Howard Eiland et al. (Cambridge, MA: Belknap Press of Harvard University Press, 2006), 69.

54. Erving Goffman, *Relations in Public: Microstudies in the Public Order* (New York: HarperCollins, 1971), 125–126.

55. Erving Goffman, *Behavior in Public Places: Notes on the Social Organization of Public Gatherings*, 4th ed. (New York: The Free Press, 1969).

56. Henry Viscardi Jr., *A Man's Stature* (New York: John Day, 1952), 70, quoted in Goffman, *Behavior in Public Places*, 89.

57. Susan Saegert, "Cognitive Overload and Behavioral Constraint," *Environmental Design Research* 2 (1973): 254–261, esp. 256.

58. Schivelbusch, *The Railway Journey*, 64–68.

59. "Things Seen and Heard Along the Underground," *New York Times*, October 28, 1904.

60. Dwyer, *Subway Lives*, 153–154; Steve Rivo, "Daily News," in *Encyclopedia of New York City*, ed. Kenneth T. Jackson (New Haven: Yale University Press, 1995), 307–308.

61. Dwyer, *Subway Lives*, 153.

62. Goffman, *Relations in Public*, 322; 331–332.

63. Lyn Lofland calls this behavior "cooperative motility [. . .] the idea that strangers work together to traverse space without incident." Lyn H. Lofland, *The Public Realm: Exploring the City's Quintessential Social Territory* (Piscataway, NJ: Transaction Publishers, 1998), 29.

64. As Hartmut Böhme has pointed out, "in terms of phenomenology, the sense of touch works predominately in a flowing mode," which may explain why experiencing physical contact in a crowd of passengers seems unavoidable and therefore tolerable. Hartmut Böhme, "Der Tastsinn im Gefüge der Sinne: Anthropologische und historische Ansichten vorsprachlicher Aisthesis," in Kunst- und Ausstellungshalle der Bundesrepublik Deutschland, ed., *Tasten*, vol. 7, Schriftenreihe Forum (Göttingen: Steidl, 1996), 191, translated by the author.

65. Hannes Böhringer, *Orgel und Container* (Berlin: Merve, 1993), 12, translated by the author.

66. Lethen, *Cool Conduct*, 27.

67. Doris Bachmann-Medic, "Was heißt 'Iconic/Visual Turn'?," *Gegenworte* 20 (Autumn 2008): 10–15, esp. 12, translated by the author. In an analysis of viewing screens in public space, Nanna Verhoeff also uses the term "visual regime," albeit with slightly different connotations. See Nanna Verhoeff, *Mobile Screens: The Visual Regime of Navigation* (Amsterdam: Amsterdam University Press, 2012).

68. Jérôme Denis and David Pontille, "The Graphical Performation of a Public Space: The Subway Signs and their Scripts," in *Urban Plots, Organizing Cities*, ed. Giovanna Sonda, Claudio Coletta, and Francesco Gabbi (Farnham: Ashgate, 2010), 12.

69. Madeleine Akrich, "The De-Scription of Technical Objects," In *Shaping Technology / Building Society: Studies in Sociotechnical Change*, ed. Wiebe J. Bijker and John Law (Cambridge, MA: MIT Press, 1992), 205–224.

70. Literature on the logistical organization of sign systems presents a large number of possible ways to classify signs. The order used here is based in large part on Qingjie Zeng et al., "Performance Evaluation of Subway Signage: Part I—Methodology" (Transportation Research Board 90th Annual Meeting, Washington, DC, January 23–27, 2011).

71. Denis and Pontille, "The Graphical Performation of a Public Space: The Subway Signs and their Scripts," 5.

72. Gillian Fuller, "The Arrow-Directional Semiotics: Wayfinding in Transit," *Social Semiotics* 12, no. 3 (2002): 231–244, esp. 223. On the function of signage in airports, see Gillian Fuller, *Avipolis: A Book about Airports* (London: Black Dog, 2004).

73. Benson Bobrick, *Labyrinths of Iron: Subways in History, Myth, Art, Technology, and War* (New York: Henry Holt, 1994); Pike, *Metropolis on the Styx*.

74. Benjamin, *The Arcades Project*, 429.

75. "Things Seen and Heard Along the Underground."

76. New York Transit Museum, *Subway Style: 100 Years of Architecture and Design in the New York City Subway* (New York: Stewart, Tabori & Chang, 2004), 137–141.

77. Heins & LaFarge often added small ornaments to the signs in reference to the history of the area surrounding a particular station. For example, a beaver on the sign at Astor Place station played on the fact that the Astor family had made their millions in fur trading. For more on the ornaments in the subway, see the NYTMA's comprehensive compendium: Philip Ashforth Coppola, *Silver Connections: A Fresh Perspective on the New York Area Subway Systems*, vol. 3 (Maplewood, NJ: Four Oceans Press, 1988).

78. City Club of New York, "New York City Transit: A Memorandum Addressed to the Public Service Commission of the First District" (New York, 1907), 19.

79. Paul Shaw, *Helvetica and the New York City Subway System* (Cambridge, MA: MIT Press, 2011), 8–9.

80. New York Transit Museum, *Subway Style*, 159.

81. Sweeny Lithograph Co., *Municipal Railway-Broadway Line through the Heart of Manhattan, 1919* (Rapid Transit Company Collection. Part of the Lionel Pincus and Princess Firyal Map Division of the New York Public Library). Taken from New York Transit Museum, *Subway Style*, 164.

82. Mark Ovenden, *Transit Maps of the World* (London: Penguin, 2007), 33; Denis Wood, *The Power of Maps* (New York: Guilford Press, 1992).

83. Ovenden, *Transit Maps of the World*, 33; Wood, *The Power of Maps*.

84. This was admittedly easier to achieve in a well-organized system like the London Underground. In New York different technocrats, designers, and engineers worked for each of the autonomous subway companies. See Christian Wolman, *The Subterranean Railway: How the London Underground Was Built and How It Changed the City Forever* (London: Atlantic Books, 2005).

85. Janin Hadlaw, "The London Underground Map: Imagining Modern Time and Space," *Design Issues* 19, no. 1 (Winter 2003): 25–35, esp. 25. For further details on the importance of Beck's map, see David Leboff and Timothy Demuth, *No Need to Ask! Early Maps of London's Underground Railways* (Middlesex: Capital Transport Publishing, 1999); Ken Garland, *Mr. Beck's Underground Map* (Harrow Weald: Capital Transport, 1994); Karl Schlögel, *Im Raume lesen wir die Zeit: über Zivilisationsgeschichte und Geopolitik* (Munich: Carl Hanser, 2003), 96–100.

86. George Salomon, "Out of the Labyrinth: A Plea and a Plan for Improved Passenger Information on the New York Subways" (1957), unpublished manuscript, accessible in the NYTMA.

87. Salomon complaints in "Out of the Labyrinth": "There are four subway stations called 14th Street, five called 23rd Street. The Sixth Avenue line has a station called Seventh Avenue. The 207th Street station of the Broadway-Seventh Avenue line is on Tenth Avenue."

88. Images taken from Salomon, "Out of the Labyrinth." Unfortunately, these reproductions were made from black and white copies and fail to replicate the color and material quality of Salomon's work. It can be viewed at the New York City Transit Museum Archive.

89. Salomon, "Out of the Labyrinth," 4.

90. Salomon, "Out of the Labyrinth," 3.

91. Salomon, "Out of the Labyrinth, 6.

92. New York Transit Museum, *Subway Style*, 163.

93. See Clifton Hood, *722 Miles: The Building of the Subways and How They Transformed New York* (Baltimore: Johns Hopkins University Press, 2004), 100.

94. Hood, *722 Miles*, 116.

95. Donald F. Davis, "North American Urban Mass Transit, 1890–1959: What If We Thought about It as a Type of Technology?," *History and Technology* 12, no. 4 (1995): 309–326, esp. 315.

96. "One Real Danger of the Subway," *New York Times*, November 26, 1904.

97. Already in his 1939 magnum opus *The Civilizing Process*, Norbert Elias discussed the phenomenon of public spitting, showing that well into the nineteenth century it was seen not only as normal behavior, but as actually necessary for one's health. Over long periods of time this was apparently more than simply a custom; it was considered a need to spit as often as possible. About two hundred years ago, Western cultures began stigmatizing spitting as embarrassing and disgusting, and something to be done in private. Elias: "The modification of the manner of spitting, and finally the more or less complete elimination of the need for it, is a good example of the malleability of the psychic economy of humans." Elias, *The Civilizing Process*, 135.

98. Jeanne E. Abrams, "Spitting Is Dangerous, Indecent, and against the Law! Legislating Health Behavior during the American Tuberculosis Crusade," *Journal of the History of Medicine and Allied Sciences* 68, no. 3 (July 2013): 416–450.

99. Abrams, "Spitting Is Dangerous, Indecent, and against the Law!," 419–420.

100. Matthew Gandy, "Life without Germs: Contested Episodes in the History of Tuberculosis," in *The Return of the White Plague: Global Poverty and the "New" Tuberculosis,* ed. Matthew Gandy and Alimuddin Zumla (London: Verso, 2003), 15–38, esp. 29. Cited also in Abrams, "Spitting Is Dangerous, Indecent, and against the Law! Legislating Health Behavior during the American Tuberculosis Crusade," 6.

101. "In practice, many fines and arrests for spitting in America were leveled at members of the middle/upper classes, some of whom at least appeared to view spitting as an acceptable practice." Abrams, "Spitting Is Dangerous, Indecent, and against the Law! Legislating Behavior during the American Tuberculosis Crusade," 17.

102. In 1908 Hermann M. Biggs, head of the New York Public Health Department, reported: "All street cars, elevated and underground railways, ferryboats, public buildings, piers, etc., have been

placarded with large signs prohibiting spitting. The sanitary police of the Department have constantly made arrests of persons violating the law, and the newspapers have aided by giving the matter proper publicity." Department of Health, *Brief History of the Campaign against Tuberculosis in New York City* (New York, 1908), 13.

103. "Spitting in the Subway," *New York Tribune*, November 19, 1904.

104. See chapter 1.

105. Numerous physicians stressed that the lack of sunlight in the subway created perfect conditions for dangerous bacteria to thrive. See "Bacilli Invade the Subway," *New York Tribune*, November 19, 1904, and "Justice on the Trial of the Subway Germ," *New York Times*, February 17, 1905.

106. Soper, *Air and Ventilation of Subways*.

107. Soper, *Air and Ventilation of Subways*, 136.

108. Abrams, "Spitting Is Dangerous, Indecent, and against the Law!," 447–448.

109. Scott M. Cutlip, *Unseen Power: Public Relations—A History* (Hillsdale, NJ: Erlbaum Associates, 1994). Along with Sigmund Freud's nephew Edward Barnays (1891–1995), Ivy Lee is considered the father of public relations. Before his death in 1934, Lee also worked closely with the German chemistry firm IG Farben and interacted with the leading cadre of the NSDAP. For more information, see Ray Eldon Hiebert, *Courtier to the Crowd: The Story of Ivy Lee and the Development of Public Relations* (Ames: Iowa State University Press, 1966).

110. The New York City Transit Museum stores a collection of Subway Sun and Elevated Express editions. Princeton University Library has also digitalized parts of Ivy Lee's remaining papers and made them available online /http://pudl.princeton.edu/collections/pudl0036. The collection contains 385 posters from the Subway Sun and the Elevated Express.

111. Brooks, *Subway City*, 94–96.

112. Both Brooks and Fitzpatrick discuss the Subway Sun's history, but not the managerial appeals that these posters made to passengers. See Brooks, *Subway City*, 96 and Tracy Fitzpatrick, *Art and the Subway: New York Underground* (Piscataway, NJ: Rutgers University Press, 2009), 207–209.

113. "McClellan Motorman of First Subway Train," *New York Times*, October 28, 1904.

114. The *New York Times* reported that just three days after the subway opened, anger escalated at a meeting of the Transit Commission. Clavon Tomkins, president of the Municipal Art Society, threw a fistful of shards on the table, gathered from between the feet of men attaching billboards. He shouted: "Does the contract with the operating company permit the destruction of the city property, and the defacement of one of the most completely elaborate and beautiful bits of work ever finished by a public commission in this fashion? It is a public disgrace and an outrage that the men who are hanging these hideous works of art should be allowed to continue this destruction, and on behalf of the public generally and the taxpayers I ask that this outrage be stopped at once." Several complaints were filed the same day, including a petition from the Chamber of Commerce threatening to boycott the advertising firms,

should they not immediately remove the billboards. "Broken Tile Exhibit Halts Subway Signs," *New York Times*, November 4, 1904. See also Fitzpatrick, *Art and the Subway*, 208.

115. A statement made by Transit Commissioner Charles Stewart Smith on November 19, 1904, quoted in Hood, *722 Miles*, 96.

116. "Architectural League condemns Subway Ads," *New York Times*, November 2, 1904.

117. See David J. Framberger, "Architectural Designs for New York's First Subway," in *Historical American Engineering Record: Interborough Rapid Transit Subway (Original Line) NY-122* (New York, 1979), 365–412.

118. "Company has 48 Hours to move Subway Signs," *New York Times*, February 4, 1905.

119. Brooks, *Subway City*, 69–70.

120. It was not insignificant for the court's ruling that the profit from selling advertising space was calculated as covering 1.5 percent of the system's debt. See Brooks, *Subway City*, 70.

121. Kathleen Hulser, "Outdoor Advertising," in *Encyclopedia of New York City*, ed. Kenneth T. Jackson (New Haven: Yale University Press, 1995), 869–870. The world's first advertising agency, N. W. Ayer & Son, had been founded in Philadelphia in 1869, marking the beginning of the rise of an industry that already had a business volume of over ninety-five million dollars by the time the subway opened. See Lawrence B. Glickman's introduction, "Born to Shop? Consumer History and American History," in *Consumer Society in American History: A Reader*, ed. Lawrence B. Glickman (Ithaca: Cornell University Press, 1999), 1–16, esp. 3.

122. For more on the concept of the consumer, see Raymond Williams, *Keywords. A Vocabulary of Culture and Society,* revised ed. (New York: Oxford University Press, 1983), s.v. "Consumer."

123. Andreas Reckwitz, *Das hybride Subjekt: Eine Theorie der Subjektkulturen von der bürgerlichen Moderne zur Postmoderne* (Weilerswist: Velbrück Wissenschaft, 2010), 397–409.

124. Thorstein Veblen, *Theory of the Leisure Class* (New York: Dover, 1994).

125. Veblen, *Theory of the Leisure Class*, 36–52.

126. Reckwitz, *Das hybride Subjekt*, 400, translated by the author.

127. A comprehensive photo collection can be found in the New York Transit Museum Archives. The relevant section here is NYCTA Photo Collection 2005.4.

128. Benjamin also described billboards as, "Locust swarms of print, which already eclipse the sun of what city dwellers take for intellect, will grow thicker with each succeeding year." Walter Benjamin, "One-Way Street," in *One-Way Street and Other Writings* (London: Harcourt Brace Jovanovich, 1979), 62.

129. Janet Ward has shown this to be the case for metropolitan experience in the Weimar Republic. Janet Ward, *Weimar Surfaces: Urban Visual Culture in 1920s Germany* (Berkeley: University of California Press, 2001). See Susan Buck-Morss, *The Dialectics of Seeing: Walter Benjamin and the Arcades Project* (Cambridge, MA: MIT Press, 1991), esp. 110–125.

130. Reckwitz, *Das hybride Subjekt*, 399–400.

131. T. J. Jackson Lears, *Fables of Abundance: A Cultural History of Advertising in America* (New York: Basic Books, 1994), 12.

132. Elmo Roper, *Roper Surveys Subway Riders* (New York: New York Subways Advertising Company, 1941).

133. Roper, *Roper Surveys Subway Riders*, 3–5.

134. This poster, as well as many others, can be found in the wonderful book *Meet Miss Subways: New York's Beauty Queens, 1941–1976* (Brookline, ME: Seapoint Books and Media, 2014), featuring photography by Fiona Gardner, text by Amy Zimmer, and an excellent introduction by historian Kathy Peiss.

135. Nan Robertson, "Miss Subways Reigns: Persephone to 5 Million; Glamour Girls Out Once Picked Three," New York Times, February 18, 1957.

136. Melanie Bush, "Miss Subways, Subversive and Sublime," *New York Times*, October 24, 2004.

137. Enid Nemy, "Miss Subways of '41, Meet Miss Subways of '71," *New York Times*, December 8, 1971.

138. Melanie Bush, "Miss Subways, Subversive and Sublime."

139. Melanie Bush, "Miss Subways, Subversive and Sublime."

140. Maxine Leeds Craig, *Ain't I a Beauty Queen? Black Women, Beauty, and the Politics of Race* (New York: Oxford University Press, 2002), 68–69.

141. See Brooks, *Subway City*, 183.

142. William E. Geist, "Subway Queens of Old to Gather for Reunion," *New York Times*, October 15, 1983.

143. New York Transit Museum, *Subway Style*, 183–184.

144. Quoted in Thomas P. Hughes, *American Genesis: A Century of Invention and Technological Enthusiasm, 1870–1970* (Chicago: University of Chicago Press, 1989), 330.

CHAPTER 4

1. Wolfgang Sachs, "Herren über Raum und Zeit: Ein Rückblick in die Geschichte unserer Wünsche," in *Nahe Ferne—Fremde Nähe: Infrastrukturen und Alltag*, ed. Barbara Mettler-Meibom and Christine Bauhardt (Berlin: Edition Sigma, 1993), 59–68, esp. 66, translated by the author.

2. Luc Boltanski and Eve Chiapello, *The New Spirit of Capitalism* (London: Verso, 2018).

3. Boltanski and Chiapello, *The New Spirit of Capitalism*, 36–48. For a critical discussion of these concepts, see Maurizio Lazzerato, "Die Missgeschicke der 'Künstlerkritik' und der kulturellen Beschäftigung," in *Kritik der Kreativität*, ed. Gerald Raunig and Ulf Wuggenig (Vienna: Turia + Kant, 2007), 190–206.

4. Susan Sontag underscores this point in her essay *Regarding the Pain of Others* (New York: Farrar, Straus and Giroux, 2017).

5. Boltanski and Chiapello, *The New Spirit of Capitalism*, 36.

6. Martin Rubin, "The Crowd, the Collective, and the Chorus: Busby Berkeley and the New Deal," in *Movies and Mass Culture*, ed. John Belton (New Brunswick, NJ: Rutgers University Press, 1996), 59–94.

7. George Steinmetz, "Hot War, Cold War: The Structures of Sociological Action, 1940–1955," in *Sociology in America: A History*, ed. Craig Calhoun (Chicago: University of Chicago Press, 2007), 314–366.

8. Elwyn Brooks White, "Here Is New York," in *Essays of E. B. White* (New York: HarperCollins, 2006), 148–168, esp. 152.

9. Railroad companies offered to convert their employees' train tickets into "commutation tickets," which cost less than paying each fare individually. Nick Palmgarten, "There and Back Again: The Soul of the Commuter," *New Yorker*, April 16, 2007. For more on commuter culture see Rachel Bowbly, "Commuting," in *Restless Cities*, ed. Matthew Beaumont and Gregory Dart (London: Verso, 2010), 43–58.

10. Christopher Morley, "Thoughts in the Subway," in *Plum Pudding: A Literary Concoction,* ed. Christopher Morley (Maryland: Wildeside Press LLC, 2005), 133–135, esp. 134.

11. For example, on the occasion of the opening Berlin's underground between Neukölln and Gesundbrunnen, Siegfried Kracauer underscored the ambivalent nature of employees and workers being at the mercy of the "proletarian express trains." Siegfried Kracauer, "Proletarische Schnellbahn," in *Aufsätze 1927–1931* 5.2, ed. Inka Müller-Bach (Frankfurt am Main: Suhrkamp, 1990), 179–180.

12. White, "Here Is New York," 153.

13. White, "Here Is New York," 153.

14. Elwyn Brooks White, "The Commuter," in *The Lady Is Cold and Other Poems* (New York: Harper & Brother, 1929), 26.

15. For more on Depero's work while in New York, see Laura Chiesa, "Transnational Multimedia: Fortunato Depero's Impressions of New York City (1928–1930)," *California Italian Studies Journal* 1, no. 2 (2010): 1–33.

16. Tracy Fitzpatrick, *Art and the Subway: New York Underground* (New Brunswick, NJ: Rutgers University Press, 2009), 78.

17. As Fitzpatrick has shown, Evans was far from the only artist taking pictures in the subway with a hidden camera. Arthur Leipzig and Arthur Frank also clandestinely explored the system. For more detail, see Fitzpatrick, *Art and the Subway*, 117–131.

18. Michael Brooks, *Subway City: Riding the Trains, Reading New York* (New Brunswick, NJ: Rutgers University Press, 1997), 167–168.

19. Jeff L. Rosenheim, "Afterword," in *Many Are Called*, ed. Walker Evans, Jeff L. Rosenheim, and Luc Sante (New York: Yale University Press, 2004), 197–204; Mia Fineman, "Notes for Underground: The Subway Portraits," in *Walker Evans*, ed. Maria Morris Hambourg and Jeff L. Rosenheim (New York: Metropolitan Museum of Art, 2000), 106–119.

20. For a more detailed interpretation of Tooker's painting, especially in terms of gender identity, see Katherine Jane Hauser, "George Tooker, Surveillance, and Cold War Sexual Politics," *GLQ: A Journal of Lesbian and Gay Studies* 11, no. 3 (June 2005): 391–425.

21. Fitzpatrick, *Art and the Subway*, 87–97.

22. David Riesman, *The Lonely Crowd: A Study of the Changing American Character* (New Haven: Yale University Press, 2001).

23. Riesman, *The Lonely Crowd*, 24–31. Riesman's conceptions of types of character are sketched relatively simply, but they do overlap in some aspects with the concepts of type and forms of subjects. These are all abstractions (29), constructions, or types (31), that are necessary for the purpose of an analytical description of society.

24. Riesman's example for one such inner-directed model was the development of Protestant ethics as described by Max Weber. See Max Weber, *The Protestant Ethic and the Spirit of Capitalism* (London: Routledge, 2013).

25. Riesman, *The Lonely Crowd*, 32–33.

26. Riesman, *The Lonely Crowd*, 25 and 33. Lethen also notes this technical metaphor: Helmut Lethen, *Cool Conduct: The Culture of Distance in Weimar Germany*, trans. Don Reneu (Berkeley: University of California Press, 2002), 187.

27. Riesman, *The Lonely Crowd*, 33.

28. Riesman, *The Lonely Crowd*, 48.

29. Riesman, *The Lonely Crowd*, 260.

30. As Helmuth Lethen has emphasized, the strength and originality of Riesman's approach lies in combing other-directedness with personal autonomy. See Lethen, *Cool Conduct*, 193.

31. See Riesman, *The Lonely Crowd*, 206–224, for more detail.

32. Georg Simmel, "The Metropolis and Mental Life," in *The Blackwell City Reader*, ed. Gary Bridge and Sophie Watson (Oxford: Wiley-Blackwell, 2002), 11–19; See also Georg Simmel, *The Philosophy of Money* (New York: Routledge, 2011), 522–533.

33. Riesman, *The Lonely Crowd*, 244. Simmel's influence on Riesman is certainly also responsible for the fact that despite its subtitle (*A Study of the Changing American Character*), the book explicitly claims that the historical dynamics it describes are not restricted to the North American continent. As Riesman repeatedly stresses, the new type of character emerges globally, wherever capitalism, industrialization, and urbanization are on the rise; it would therefore sooner or later spread around the entire world.

34. See, for example, Robert H. Walker, "The Poet and the Rise of the City," *Mississippi Valley Historical Review* 39, no. 1 (June 1962), 85; Hosokawa, Shuhei, "The Walkman Effect." *Popular Music* 4 (1984): 165–180; Fitzpatrick, *Art and the Subway*, 90–92.

35. Wright C. Mills, *White Collar: The American Middle Classes* (New York: Oxford University Press, 1951).

36. Mills had lived for many years in New York and taught at Columbia University. See Marjorie Harrison, "Mills, Charles Wright," in *Encyclopedia of New York City*, ed. Kenneth T. Jackson (New Haven: Yale University Press, 1995), 763.

37. Mills, *White Collar*, 229–233.

38. Mills, *White Collar*, 252.

39. Mills, *White Collar*, 233.

40. See Steinmetz, "Hot War, Cold War: The Structures of Sociological Action, 1940–1955."

41. Thomas P. Hughes, *American Genesis: A Century of Invention and Technological Enthusiasm, 1870–1970* (Chicago: University of Chicago Press, 1989).

42. For instance, his book from 1922: Lewis Mumford, *The Story of Utopias* (Whitefish, MT: Kessinger Publishing, 1962).

43. Lewis Mumford, *The Myth of the Machine: Technics and Human Development* (Harcourt Brace Javanovich, 1967); Lewis Mumford, *The Myth of the Machine: The Pentagon of Power* (Harcourt Brace Javanovich, 1970).

44. Mumford, *The Myth of the Machine: Technics and Human Development*, 188–211.

45. The term machine is explicitly neither allegorical nor symbolic here; Mumford uses it in the strict sense of the word. The machine is initially a social formation, only secondarily connoted in terms of technology. This makes Mumford's concept of the machine a major source of inspiration for the work of Gilles Deleuze and Félix Guattari. See the introduction to this book, as well as Gilles Deleuze and Félix Guattari, *Anti-Oedipus: Capitalism and Schizophrenia* (Minneapolis: University of Minnesota Press, 1983), 141.

46. See Hartmut Böhme, Peter Matussek, and Lothar Müller, *Orientierung Kulturwissenschaft: Was sie kann, was sie will* (Reinbek bei Hamburg: Rowohlt, 2000), 172–173.

47. Mumford, *The Myth of the Machine: The Pentagon of Power*, 164–185.

48. Mumford, *The Myth of the Machine: The Pentagon of Power*, 300.

49. Lewis Mumford, *The Culture of Cities* (San Diego: Harvest/HBJ, 1970), 229–230.

50. Lewis Mumford, "The Metropolitan Milieu," in *America and Alfred Stieglitz: A Collective Portrait*, ed. Waldo Frank et al. (Garden City, NY: Doubleday, Doran & Co., 1934), 33–56, esp. 40. Quoted in Brooks, *Subway City*, 114.

51. Mumford, *The Culture of Cities*, 426.

52. Lewis Mumford, "Attacking the Housing Problem on Three Fronts," *Nation* 109, no. 2827 (June 1919): 332–333.

53. Mumford, "Attacking the Housing Problem on Three Fronts," 333.

54. Lewis Mumford, *Technics and Civilization* (New York: Harcourt, Brace, 1934), 333.

55. Lewis Mumford, "The Intolerable City: Must It Keep on Growing?," *Harper's* 152 (February 1926): 283–293, esp. 284. See also Brooks, *Subway City*, 115–116.

56. Mumford, "The Intolerable City: Must It Keep on Growing?," 285.

57. Lewis Mumford, *Art and Technics* (New York: Columbia University Press, 2000), 145.

58. See Thomas P. Hughes, *American Genesis: A Century of Invention and Technological Enthusiasm, 1870–1970* (Chicago: University of Chicago Press, 1989), 13.

59. Giedion, *Mechanization Takes Command*. Giedion wrote most of this work during his stay at Yale and Harvard Universities, and it was first published in the United States in 1948. It appeared in German translation only more thirty years later as *Die Herrschaft der Mechanisierung: Ein Beitrag zur anonymen Geschichte* (Frankfurt am Main: Europäische Verlagsanstalt, 1982).

60. Giedion, *Mechanization Takes Command*, 715.

61. Wilhelm Reich, *Character Analysis* (New York: Farrar, Straus and Giroux, 1990).

62. Erich Fromm, *Man for Himself: An Inquiry into the Psychology of Ethics* (New York: Routledge, 1999).

63. Theodor W. Adorno and Max Horkheimer, *Dialectic of Enlightenment*, trans. John Cumming (New York: Verso, 1997).

64. Boltanski and Chiapello, *The New Spirit of Capitalism*, 439.

65. Within this context, it is also necessary to mention Elias Canetti's *Crowds and Power* from 1960 (New York: Farrar, Straus and Giroux, 1984). This is more of a literary or lyrical text than a sociological analysis, but Canetti—like Riesman—understands the masses in terms of density and the leveling of differences.

66. Boltanski and Chiapello, *The New Spirit of Capitalism*, 439–441.

67. Herbert Marcuse, *One-Dimensional Man: Studies in the Ideology of Advanced Industrial Society* (New York: Routledge, 2002), 231. For more on the origins of this book, see Barry Katz, *Herbert Marcuse and the Art of Liberations: An Intellectual Biography* (New York: Verso, 1982).

68. For more on this transformation, see Georg Vrachliotis, *Geregelte Verhältnisse: Architektur und technisches Denken in der Epoche der Kybenetik* (Vienna: Springer, 2012).

69. First published in English in a small print run in 1960, the text later appeared in a revised edition as Reyner Banham, *Theory and Design in the First Machine Age* (New York: Praeger, 1967), esp. 9–12.

70. Banham, *Theory and Design in the First Machine Age*, 10.

71. For a history of these concepts, see Peter James Taylor, *Modernities: A Geohistorical Interpretation* (Minneapolis: University of Minnesota Press, 1999). See also Maria Kaika and Erik Swyngedouw, "Fetishizing the Modern City: The Phantasmagoria of Urban Technological Networks," *International Journal of Urban and Regional Research* 24, no.1 (March 2000): 120–138; Ulrich Beck, Scott Lash, and Anthony Giddens, *Reflexive Modernization: Politics, Tradition, and Aesthetics in the Modern Social Order* (Stanford, CA: Stanford University Press, 1994); James C. Scott, *Seeing Like a State: How Certain Schemes to Improve the Human Condition Have Failed* (New Haven, CT: Yale University Press, 1999).

72. See, for example, Jürgen Habermas, "What Does a Crisis Mean Today? Legitimation Problems in Late Capitalism," *Social Research* 40, no. 4 (1973): 643–667; Fredric Jameson, *Postmodernism, or The Cultural Logic of Late Capitalism* (Durham: Duke University Press, 1991); Ernest Mandel, Late Capitalism (London: Humanities Press, 1975).

73. Stephen B. Goddard, *Getting There: The Epic Struggle between Road and Rail in the American Century* (Chicago: University of Chicago Press, 1996), 164–165.

74. A definitive work on Robert Moses and the transformation of New York during his time is Robert A. Caro, *The Power Broker: Robert Moses and the Fall of New York* (New York: Vintage, 1975). For an assessment and interpretation in terms of cultural history and theory, see Marshall Berman, *All That Is Solid Melts into Air: The Experience of Modernity* (New York: Verso, 1991), 290–312.

75. James E. Vance, Jr., *The Continuing City: Urban Morphology in Western Civilization* (Baltimore: Johns Hopkins University Press, 1990), 434–445.

76. Quoted in Anthony Flint, *This Land: The Battle over Sprawl and the Future of America* (Baltimore: Johns Hopkins University Press, 2008), 31. Flint does not provide the source of this quotation.

77. Made in 1938/1939 for presentation at the New York World's Fair, the movie is about 41 minutes long. It was directed by Ralph Steiner and Willard Van Dyke, produced by the American Institute of Planners, and narrated by Lewis Mumford.

78. Peter Hall, *Cities of Tomorrow: An Intellectual History of Urban Planning and Design in the Twentieth Century* (Chichester: Wiley-Blackwell, 2002), 86–118.

79. Goddard, *Getting There*, 195–206.

80. Marshall McLuhan, *Mechanical Bride: Folklore of Industrial Man* (New York: Vanguard Press, 1951).

81. See Brian J. Cudahy, *Under the Sidewalks of New York: The Story of the Greatest Subway System in the World* (New York: Fordham University Press, 1995), 123.

82. Kenneth T. Jackson, *Crabgrass Frontier: The Suburbanization of the United States* (New York: Oxford University Press, 1987), 272–276.

83. David Harvey, *The Condition of Postmodernity* (Cambridge, MA: Blackwell, 1990), 129–130.

84. Cudahy, *Under the Sidewalks of New York*, xv.

85. While shutting down one line of the elevated train system after another, the subway system was expanded in 1956 to include a line to Rockaway Beach. Cudahy, *Under the Sidewalks of New York*, 130–132.

86. Clifton Hood, *722 Miles: The Building of the Subways and How They Transformed New York* (Baltimore: Johns Hopkins University Press, 2004), 254.

87. Fitzpatrick, *Art and the Subway*, 83.

88. Walter Benjamin, *The Arcades Project*, trans. Howard Eiland and Kevin McLaughlin (Cambridge, MA: Harvard University Press, 1999), 84.

CHAPTER 5

1. Cecile G. Cutler to the Transit Authority, November 29, 1964, NYTMA.
   The following letters, replies, and internal reports can be found in the New York Transit Museum Archives (NYTMA). Most of the documents are in eight boxes filed by date of arrival, but not further inventoried in any way.

2. Anonymous to Mayor Wagner, July 24, 1964, NYTMA.

3. Emanuel Perlmutter, "Major Crime Up 52% in Subways and 9% Citywide," *New York Times*, February 10, 1965.

4. "Negroes Slay Youth in Subway; One of Many New York Murders," *Gadsen Times*, March 14, 1965.

5. For example, see the Transit Police internal report titled "Attempted Fare Evasions" from February 11, 1965, NYTMA, or the report from the Transit Police to the Transit Authority, "Letter of Complaint—Mrs. Rhoda Barky—Relative Fare Evasion," February 11, 1965, NYTMA. Regarding spitting, see the Transit Authority's reply to a complaint made by Mr. Hoffmann, February 2, 1965, NYTMA.

6. Ellen Reiner to the Transit Authority, February 28, 1965, NYTMA.

7. Mrs. Schwalbe to the mayor and Transit Authority, February 17, 1965, NYTMA.

8. Transit Police report to the Transit Authority, "Arrest of 4 Males, Assault & Robbery—Felonious Assault (Knife)," February 18, 1965, NYTMA.

9. Marlene Connor to the Transit Authority, February 19, 1965, NYTMA.

10. Carl Heck to Mayor Wagner, February 20, 1965, NYTMA.

11. Paul Stone to the Transit Authority, February 8, 1965, NYTMA.

12. Transit Police report to the Transit Authority, "Investigation Re: Letter Complaining of Loiterers in Toilets," February 8, 1965, NYTMA.

13. Transit Police report to the Transit Authority, "Investigation Re: Anonymous Allegation of Loitering, Men's Toilet, Northbound Platform, Christopher St. 7 Ave. Linen, IRT Division," February 23, 1965, NYTMA.

14. Morris W. Watkins to the mayor and Transit Authority, February 13, 1965, NYTMA.

15. Transit Police report, "Investigation Re: Lost Property—Bag with Gun," February 11, 1965, NYTMA. This was not an isolated case. Officers repeatedly found guns on the subway, but in most cases could not find the owners.

16. Transit Police report to Transit Authority, "Investigation Re: Wallet found containing Threatening Notes," February 25, 1965, NYTMA.

17. Because they had only recently been discovered in the New York Transit Museum Archives, the contents of these boxes have not yet been properly archived or evaluated. The body of material was comprehensively screened by the author during months-long reading sessions in 2010, 2011, and 2014.

18. Michel Foucault, "The Life of Infamous Men," in *Foucault: Power, Truth, Strategy,* ed. Meaghan Morris and Paul Patton Michel (Sydney: Feral, 1979), 76–91.

19. A handful of sociological studies have been done on the forms and functions of complaining, but these do not deal with submitting complaints to institutions; instead, they focus on the collective moments of informally complaining about social grievances or precarious conditions in life and work. See Julian Baggini, *Complaint: From Minor Moans to Principled Protests* (London: Profile Books, 2010); Charles F. Hanna, "Complaint as a Form of Association," *Qualitative Sociology* 4, no. 4 (December 1981): 298–311; John Weeks, *Unpopular Culture: The Ritual of Complaint in a British Bank* (Chicago: University of Chicago Press, 2004).

20. Felix Mühlberg, *Bürger, Bitten und Behörden: Geschichte der Eingabe in der DDR* (Berlin: Dietz, 2004); Ina Merkels, *Wir sind doch nicht die Mecker-Ecke der Nation: Briefe an das DDR-Fernsehen* (Cologne: Böhlau Verlag, 1998); Peter Becker and Alf Lüdkte, eds., *Akten. Eingaben. Schaufenster. Die DDR und ihre Texte: Erkundungen zu Herrschaft und Alltag* (Berlin: Akademie Verlag, 1997); Steffen Elsner, "Flankierende Stabilisierungsmechanismen diktatorischer Herrschaft: Das Eingabewesen in der DDR," in *Repression und Wohlstandsversprechen: Zur Stabilisierung von Parteiherrschaft in der DDR und der CSSR,* ed. Christoph Boyer and Peter Skyba (Dresden: Hannah-Arendt-Institut für Totalitarismusforschung, 1999), 75–86; Ulf Rathje and Roswitha Schröder, "Bewertung von Eingaben der Bürger an den Präsidenten der DDR: Bestand DA 4 der Präsidialkanzlei," *Mitteilungen aus dem Bundesarchiv* 14 (2006), 65–70; Christiane Streubel, "Wir sind die geschädigte Generation: Lebensrückblicke von Rentnern in Eingaben an die Staatsführung der DDR," in *Graue Theorie: Die Kategorien Alter und Geschlecht im kulturellen Diskurs,* ed. Heike Hartung et al. (Cologne: Böhlau Verlag, 2007), 241–264.

21. See "The Topic of Denunciation" in Luc Boltanski, *Distant Suffering: Morality, Media and Politics,* trans. Graham D. Burchell (Cambridge: Cambridge University Press, 1999), 57–76; as well as Marie-Ange Schiltz, Yann Darré, and Luc Boltanski, "La Dénonciation," *Actes de la Recherche en Sciences Sociales* 51, no. 1 (March 1984): 3–40.

22. Arlette Farge and Michel Foucault, *Disorderly Families: Infamous Letters from the Bastille Archives* (Minneapolis: University of Minnesota Press, 2017).

23. Sheila Fitzpatrick and Robert Gellately, "Introduction to the Practices of Denunciation in Modern European History," in *Accusatory Practices: Denunciation in Modern European History, 1789–1989*, ed. Sheila Fitzpatrick and Robert Gellately (Chicago: University of Chicago Press, 1977), 1–21; Sheila Fitzpatrick, *Everyday Stalinism: Ordinary Life in Extraordinary Times: Soviet Russia in the 1930s* (New York: Oxford University Press, 1999); Robert Gellately, "Denunciation as a Subject of Historical Research," *Historical Social Research* 26, nos. 2–3 (2001): 16–29; Robert Gellately, *Backing Hitler* (New York: Oxford University Press, 2001).

24. Aside from studies that rely on letters of complaint and petitions as historical sources, beginning in the 1980s, the topic also became a focus of psychologically oriented management literature. This literature, however, primarily discusses various scenarios from behavioral economics in an attempt to make certain forms of customer communication fruitful for marketing. This makes them less valuable for our purposes; nonetheless, the following literature can be taken as representative for this kind of analysis: Jeffrey G. Blodgett, Kirk Wakefield, and James H. Barnes, "The Effects of Customer Service on Consumer Complaining Behavior," *Journal of Services Marketing* 9, no. 4 (October 1995): 31–42; Gary L. Clark, Peter F. Kaminski, and David R. Rink, "Consumer Complaints: Advice on How Companies Should Respond Based on an Empirical Study," *Journal of Consumer Marketing* 9, no. 3 (March 1992): 5–14; Robin M. Kowalski, "Complaints and Complaining. Functions, Antecedents, and Consequences," *Psychological Bulletin* 119, no. 2 (March 1996): 179–196; John P. Liefeld, Fred H. C. Edgecombe, and Linda Wolfe, "Demographic Characteristics of Canadian Consumer Complainers," *Journal of Consumer Affairs* 9, no. 1 (Summer 1975): 73–80; Marie Marquis and Pierre Filiatrault, "Understanding Complaining Responses through Consumers' Self-Consciousness Disposition," *Psychology and Marketing* 19, no. 3 (March 2002): 267–292; John Thøgersen, Hans Jørn Juhl, and Carsten Stig Poulsen, "Complaining: A Function of Attitude, Personality, and Situation," *Psychology and Marketing* 26, no. 8 (July 2009): 760–777; Heribert Meffert and Manfred Bruhn, "Beschwerdeverhalten und Zufriedenheit von Konsumenten," in *Marktorientierte Unternehmensführung im Wandel*, ed. Heribert Meffert (Wiesbaden: Gabler Verlag, 1999), 91–118.

25. Despite extensive research on the part of the archivist, it was not possible to trace who saved these documents and how they became part of the New York Transit Museum Archives.

26. Foucault, "The Life of Infamous Men," 79.

27. As mentioned above, the high number of letters received in March and April of 1965 reflects the murder of Andrew Mormile in March, which evoked a surge of complaints and demands for more security in the subway.

28. It is difficult to obtain reliable data on the history of the New York City Transit Police. The following information has been taken mostly from the introduction to a memoir by Officer Joseph Rivera, *Vandal Squad: Inside the New York City Transit Police Department, 1984–2004* (New York: Powerhouse, 2008).

29. Press release from the Director of Public Relations at the Transit Authority, March 6, 1955, NYTMA.

30. New York City Transit Police Department: Fiscal Report, July 1, 1967–June 30, 1968, NYTMA.

31. Transit Police Department, New York City Transit Authority, Manual of Procedure, 1975, NYMTA.

32. Brian J. Cudahy, *Under the Sidewalks of New York: The Story of the Greatest Subway System in the World* (New York: Fordham University Press, 1995), xv.

33. Michel Foucault, "The Confession of the Flesh," 195.

34. Furthermore, rarely does a complaint from these sources mention the overcrowding of the subway. While people continued to abandon the subway, it was still massively overloaded, and the relative absence of complaints on this subject suggests that these circumstances had come to be accepted as normal, with only extreme cases prompting expressions of outrage. Complaints were filed, however, about passengers who blocked doors and stairways or otherwise impeded circulation through the system. For example, Mildred Cain to the Transit Authority, November 13, 1963, NYTMA; or Morris Messing's letter to the Transit Authority, February 4, 1963, NYTMA.

35. Judith Butler, *Giving an Account of Oneself* (New York: Fordham University Press, 2005), 8–19.

36. Butler, *Giving an Account of Oneself*, 7–8.

37. Mary Fulbrook and Ulinka Rublack, "In Relation: The 'Social Self' and Ego-Documents," *German History* 28, no. 3 (August 2010): 263–272.

38. Judith Butler, *The Psychic Life of Power: Theories in Subjection* (Stanford, CA: Stanford University Press, 1997); Michel Foucault, *The Hermeneutics of the Subject: Lectures at the Collège de France, 1981–1982*, ed. Frederic Gros, trans. Graham Burchell (New York: Picador, 2006).

39. Fulbrook and Rublack, "In Relation: The 'Social Self' and Ego-Documents," 265–266.

40. Accessible in the NYTMA.

41. For example, Fred James Wiebelt to the Transit Authority, April 5, 1963, NYTMA; Glendale Taxpayers' Association Inc. to the Transit Authority, January 25, 1965, NYTMA.

42. K. Burke to the Transit Authority, July 31, 1965, NYTMA.

43. In 1957 the Transit Police counted 8,741 violations of the ban on smoking and 1,318 violations of the ban on spitting. They also reported 265 cases of littering and 669 fines issued for urinating in public. See Transit Police report, "Unsanitary Conditions," May 23, 1958, NYTMA.

In 1962 the number of violations of the ban on smoking soared to 13,767. See Transit Police report to the Transit Authority, "Investigation re: Smoking on the Transit System," 27 August 1963, NYTMA. Two years later, the count had reached 14,460, and in 1965 over 18,000 people were fined for smoking in the subway. See Transit Police report to the Transit Authority, "Investigation re: Smoking on the Train," September 4, 1964, NYTMA; Transit Police report to the Transit Authority, "Allegation re: Smoking and Spitting on the Transit System," July 26, 1966, NYTMA.

44. James Daly to the Transit Authority, July 17 1966, NYTMA.

45. Norman Zareko to the New York Department of Sanitation, May 29, 1958, NYTMA.

46. David Kloss to the Transit Authority, March 17, 1965, NYTMA.

47. For example, Richard Kawanagh to the Transit Authority, May 5, 1966, NYTMA.

48. For example, Transit Police report to the Transit Authority, "Campaign Literature Posted at Subway Entrances," July 25, 1966, NYTMA; Harry Schwart to the Transit Authority, August 1, NYTMA.

49. An enormous amount of literature has been written on New York's graffiti culture and the role it played for the subway in the 1970s. Particularly relevant are the following publications: Joe Austin, *Taking the Train: How Graffiti Art Became an Urban Crisis in New York City* (New York: Columbia University Press, 2002); Craig Castleman, *Getting Up: Subway Graffiti in New York* (Cambridge, MA: MIT Press, 1982); Maryalice Sloan-Howitt and George L. Kelling, "Subway Graffiti in New York City: 'Getting Up' vs. 'Meaning It and Cleaning It,'" *Security Journal* 1, no. 3 (1990): 131–136. A classic work of theoretical reflection on this phenomenon is Jean Baudrillard, "KOOL KILLER, or The Insurrection of Signs," in *Symbolic Exchange and Death*, trans. Iain Hamilton Grant (Sage: London, 1975), 76–86.

50. For a more precise discussion of scribbling, which cannot be clearly classified as pictures, writing, or symbols, see Christian Driesen, "Die Kritzelei als Ereignis des Formlosen," in *Über Kritzeln: Graphismen zwischen Schrift, Bild, Text und Zeichen*, ed. Christian Driesen et al. (Berlin: Diaphanes Verlag, 2012), 23–37.

51. For example, Rose Schwewe to the Transit Authority, February 21, 1955, NYTMA; Transit Authority report to the Transit Police, "Complaints about Man Drawing Obscene Pictures," February 23, 1955, NYTMA.

52. James J. Duff to the Transit Authority, April 12, 1958, NYTMA.

53. Ben Benowitz to the Transit Authority, June 27, 1957, NYMTA.

54. Chairman of the Jewish War Veterans of the USA to the Transit Authority, June 3, 1963, NYTMA.

55. For example, Samuel Kupfer to the Board of Transportation, May 5, 1956.

56. For example, the letter from the Anti-Defamation League to the Transit Authority, January 27, 1964, NYTMA.

57. Transit Police report to the Transit Authority, "Meeting at City Hall Re Section 484-H of the Penal Law," November 4, 1965, NYTMA.

58. Alfred E. Clark, "City Starts Drive on Subway Smut," *New York Times*, March 21, 1965.

59. Some passengers complained about that campaign, saying that it reminded them of George Orwell's novel *1984*. For example, Thomas E. Walsh to the mayor, March 24, 1965, NYMTA.

60. A. Witty to the Transit Authority, October 30, 1963, NYTMA.

61. Transit Police report, "Proselytism on Trains," June 18, 1957, NYTMA. There were also complaints of nuns collecting alms and doing missionary work, written up by a passenger named Mr. Fondiller who called himself a member of a secular culture. The chief of police remarks in the report: "Perusal

of department records revealed that it is a fetish with Mr. Fondiller to make this complaint, noting two identical complaints received in the prior year." Transit Police report to the general manager of the Transit Authority, "Investigation re: Violation of Transit Authority Rule, Nuns Soliciting Alms," August 14, 1967, NYTMA.

62. For more on these subversive practices and authorities' attempts to curtail them, see Stefan Höhne, "Tokens, Suckers und der Great New York Token War," *Zeitschrift für Medien- und Kulturforschung* 2, no. 1 (May 2011): 143–158.

63. For example, Marie Kuwashima to the Transit Authority, May 10, 1966, NYTMA; Joseph Calderon to the Transit Authority, August 18, 1966, NYTMA.

64. For example, Loretta Knudsen to the Transit Authority, October 17, 1963, NYTMA.

65. Agnes Hunter to the Transit Authority, June 6, 1957, NYTMA; Harry Hirsch to the Transit Authority, June 2, 1957. In 1955 the Transit Police arrested 4,169 sleeping passengers. See Transit Police report to the Transit Authority, May 1, 1956, NYTMA. Within a few years, that number had almost doubled.

66. For example, Alon Lemaco to Mayor Wagner, March 23, 1965, NYTMA.

67. Anonymous to the Transit Authority, February 14, 1966, NYTMA.

68. Matteo Mahanga to the Transit Authority, March 23, 1965, NYTMA. Capitalization and orthography as in Mahanga's letter.

69. Dorothy Bartel to the Transit Authority, August 6, 1957, NYTMA.

70. Transit Police report, August 18, 1957, NYTMA.

71. For example, Clara De Luisi to the Transit Authority, July 21, 1964, NYTMA.

72. Louis Halpern to the Transit Authority, March 30, 1955, NYTMA.

73. In addition, denunciations are typically made in writing and are normally delivered discreetly, rather than issued publicly. Denouncers are to be distinguished from informers, who are normally officially co-workers assigned with tasks of observation or spies with long-term relationships to authorities. See Fitzpatrick and Gellately, "Introduction to the Practices of Denunciation in Modern European History," 1–2.

74. The Transit Police was rigorous with those asking for alms. In 1954 they arrested 383 beggars. Two years later there were 1,314 arrests and fines issued for beggars and peddlers. Transit Police report, "Professional Beggars working the Lexington Ave. Subway," February 2, 1955, NYTMA.

75. For example, Transit Police report on the complaint submitted by George G. Gerson on December 14, 1956, NYTMA. Another woman angrily reports of a blind beggar who in her opinion was not blind at all. A protocol noting her highly emotional complaint made on the telephone remarks that "She was so aggravated over this incident, that when she arrived home, she sat right down and wrote a postal card to the mayor." Transit Authority report, "Post Card to Mayor's Office—Complaining about Beggars" regarding Beatrice Morris's complaint filed on February 15, 1956, NYTMA.

76. Frederik Allen Watson's complaint filed with the Transit Authority, June 11, 1965, NYTMA.

77. For example, John Weiss's letter written on behalf of the Ozone Tudor Civic Association to the Transit Authority, May 16, 1966, NYTMA; G. Cardsis from the Camellia Flower Shop to the Transit Authority, February 16, 1956, NYTMA.

78. Mrs. M. Roozebom to the mayor of New York, March 3, 1964, NYTMA.

79. J. Prissor to the Transit Authority, June 10, 1955, NYTMA.

80. Fitzpatrick and Gellately, "Introduction to the Practices of Denunciation in Modern European History," 759.

81. Gellately, "Denunciation as a Subject of Historical Research," 19–20.

82. See examples given by Janet Chan, "The New Lateral Surveillance and a Culture of Suspicion," in *Surveillance and Governance: Crime Control and Beyond (Sociology of Crime, Law, and Deviance 10)*, ed. Mathieu Deflem and Jeffrey T. Ulmer (Bingley: Emerald Group Publishing Limited, 2008), 223–239.

83. Leo Meyerowitz to the Transit Authority, February 17, 1955, NYTMA.

84. John L. Young to the Transit Authority, December 13, 1956, NYTMA.

85. George G. Gerson to the Transit Authority, November 29, 1956, NYTMA.

86. Transit Police report to the general manager of the Transit Authority, "Investigation re: Vagrants and Sleepers," November 20, 1963, NYTMA. In response to these complaints, the Transit Police took action: in 1962 a total of 7,282 homeless people were arrested, and a total of 8,102 homeless people were arrested in 1963. See Transit Police report to the general manager of the Transit Authority, "Investigation re: Undesirables on E and F trains," February 19, 1963, NYTMA; Transit Police report to the general manager of the Transit Authority, "Investigation re: Allegation of Drunks, beggars, and obscene writings," February 12, 1964, NYTMA.

87. For example, Grace M. Covey to Mayor Wagner, May 6, 1964, NYTMA.

88. For example, Taylor Mead to the Transit Authority, January 22, 1964, NYTMA.

89. James K. Burgart, for example, made this suggestion to Mayor Wagner on June 6, 1964, NYTMA.

90. John Berger to the Transit Authority, February 22, 1957, NYTMA.

91. Michael Brooks, *Subway City: Riding the Trains. Reading New York* (New Brunswick, NJ: Rutgers University Press, 1997), 183–189.

92. On attacks by passengers, see Anna Topoil to the Transit Authority, April 5, 1956, NYTMA. On attacks by subway employees, see Alexander Brooks to the Transit Authority, November 21, 1955, NYTMA.

93. Janet L. Abu-Lughod, *Race, Space, and Riots in Chicago, New York, and Los Angeles* (New York: Oxford University Press, 2007).

94. Anonymous to the mayor of New York, March 30, 1965, NYTMA.

95. For example, anonymous to the Transit Authority, April 3, 1955, NYTMA.

96. Transit Police report regarding a complaint by Mrs. Betz: "Rowdy Negroes—Lack of Police Protection," December 10, 1956, NYTMA.

97. Beatrice Kay to the Transit Authority, March 31, 1964, NYTMA.

98. Gellately, "Denunciation as a Subject of Historical Research," 24; Farge and Foucault, *Disorderly Families*, 255–261.

99. Michael Merryman to the Transit Authority, March 16, 1965, NYTMA.

100. Hattie S. Adams to the Transit Authority, November 21, 1954, NYTMA.

101. Natalie Raymonds to Mayor Lindsay, August 8, 1967, NYTMA.

102. Franziska Roller, "Flaneurinnen, Straßenmädchen, Bürgerinnen: Öffentlicher Raum und gesell-schaftliche Teilhabe von Frauen," in *Geschlechter-Räume: Konstruktionen von "gender" in Geschichte, Literatur und Alltag*, ed. Margarete Hubrath (Cologne: Böhlau Verlag, 2001), 253–254.

103. For the genesis and critical turn in the debate on "spaces of fear" in German discourse, see Thomas Bürk, *Gefahrenzone, Angstraum, Feindesland: Stadtkulturelle Erkundungen zu Fremdenfeindlichkeit und Rechtsradikalismus in ostdeutschen Kleinstädten* (Münster: Westfälisches Dampfboot, 2012), 264–278.

104. For example, I. D. Brown to Mayor Wagner, September 2, 1964, NYTMA.

105. Mrs. A. Schwalbe to Mayor Wagner, February 17, 1965, NYTMA. See also Edith Arenson to the Transit Authority, February 3, 1965, NYTMA.

106. Station restrooms, in particular, were considered dangerous and frightening. Transit Authority report, "Complaints of Drunken Man being in the Ladies Room on the Downtown Side of Delancey St. Station," February 20, 1956, NYTMA.

107. Lesley Goldberg to Mayor Lindsay, December 31, 1966, NYTMA.

108. For example, Mrs. A. Walter to the Transit Authority, June 13, 1955, NYTMA; Jacquelyn Lyles to the Transit Authority, August 8, 1963, NYTMA; Robin Baker to the mayor of New York, August 6, 1963, NYTMA; and many more.

109. Sally Klein to the Transit Authority, December 2, 1954, NYTMA.

110. In 1954 officials registered 257 incidents involving exhibitionists and other forms of sexual harassment; the number of unreported cases was probably much higher. Transit Police report, "Complaint of indecent Exposure," April 20, 1955, NYTMA.

The Transit Police did not begin to file these cases systematically in terms of severity or other criteria until 1975. In a 1977 analysis of 1,700 officially registered incidents, approximately half of them involved indecent exposure and obscenity, and the rest were serious harassment and sexual abuse. Attempted and actual rapes were relatively infrequent, in comparison, although more than sixty incidents were reported for the year 1977. See Anne Beller, Sanford Garelik, and Sydney Cooper, "Sex Crimes in the Subway," *Criminology* 18, no. 1 (May 1980): 37–38.

111. For example, R. W. Morgan (New York Telephone Company) to the Transit Authority, April 15, 1958, NYTMA.

112. For example, Grace Stryker to the Transit Authority, January 14, 1955, NYTMA.

113. Pearl Jacob to the Transit Authority, May 7, 1955, NYTMA.

114. In the state of New York, sexual abuse was not a category for crime until 1967. Until then, sexual violence was only illegal when it involved minors or people with intellectual disabilities. The law pursued sexual offenses among adults only in cases of rape or physical assault. Revision of the penal law was undertaken in part as a result of the insight that sexual violence was often difficult to prove in terms of these narrow categories. See Beller, Garelik, and Cooper, "Sex Crimes in the Subway," 39.

115. Beller, Garelik, and Cooper, "Sex Crimes in the Subway."

116. Beller, Garelik, and Cooper, "Sex Crimes in the Subway," 36.

117. Beller, Garelik, and Cooper, "Sex Crimes in the Subway," 51; Martin Gill, "Addressing the Security Needs of Women Passengers on Public Transport," *Security Journal* 21, nos. 1–2 (February 2008): 117–133.

118. Waltraud Ernst, "Möglichkeiten (in) der Stadt: Überlegungen zur Öffentlichkeit und Privatheit geschlechtlicher Raumordnungen," in *Street Harassment: Machtprozesse und Raumproduktion,* ed. Feministisches Frauenkollektiv (Vienna: Mandelbaum, 2008), 75–93.

119. Wanda A. Manfre to Mayor Lindsay, November 28, 1966, NYTMA.

120. Tracy Fitzpatrick, *Art and the Subway: New York Underground* (Piscataway, NJ: Rutgers University Press, 2009), 183.

121. Leonard Katz and John Cashman, "Terror in the Subways," *New York Post*, 1965. Quoted in Brooks, *Subway City*, 194.

122. One NYC Transit Authority assessment lists over 3,000 acts of vandalism performed by schoolchildren between September 1954 and March 1955: Transit Authority, "Vandalism by School Children," April 28, 1955, NYTMA. At about that time, the first newspaper articles also appeared on the topic. See "Incident in the Subway," *Daily Mirror*, October 8, 1955.

123. Transit Authority report, "Complaints of Hoodlums," November 17, 1954, NYTMA.
In 1963, damage in the subway caused by the vandalism of schoolchildren amounted to more than one hundred thousand dollars. It was an immense financial burden for the Transit Authority to repair and replace seats, lighting, fire extinguishers, and window glass, and to remove "obscene writings." See Walter L. Schlager Jr. (General Manager of the Transit Authority) to Mr. Diego Bernardete, April 3, 1964, NYTMA.

124. For example, Transit Authority report, 28 June 1955, NYTMA; G. Lawrence Marchand to the Transit Authority, June 26, 1957, NYTMA.

125. Mrs. M. G. Townsend to the Transit Authority, December 27, 1956, NYTMA.

126. For example, Transit Police memo sent to the Transit Authority, "Supplementary Report Re: Youth apprehended for Disorderly Acts on Transit System," March 12, 1956, NYTMA.

127. For example, Harold B. Schleifer to the Transit Authority, December 26, 1963, NYTMA.

128. Transit Authority report, "Letter from High School Rowdy," November 8, 1962, NYTMA.

129. Newspaper article titled "Subway Prank" from an unnamed publication attached to a letter from Loretta Chernin to the Transit Authority, April 7, 1965, NYTMA. See also "Teens Held in Subway Mugging," *Daily News*, April 7, 1965.

130. Lily B. Smith to Mr. Stark, President of the Brooklyn Council, April 26, 1963, NYTMA.

131. For example, G. Werlinsky to the Transit Authority, June 22, 1956, NYTMA; M. Curna to Mr. Bingham of the Transit Authority, October 31, 1954, NYTMA.

132. Transit Authority report, "Investigation Resulting in the Arrest of Six Female Students Responsible for Unprovoked Assault," May 14, 1964, NYTMA; Jaye M. Abenanty to the Transit Authority, July 20, 1964, NYTMA.

133. For example, Mrs. W. Murphy to the Transit Authority, November 22, 1965, NYTMA.

134. For example, Nathan Bernstein to Mayor Wagner, December 7, 1963, NYTMA.

135. For example, M. Ryan to the director of Bronx High School of Science, March 16, 1964, NYTMA.

136. Letter from Frank Woehr, Director of the Manhattan High School of Aviation Trades to the Transit Authority, March 6, 1957, NYTMA.

137. Transit Authority report, "Complaints that Transit Authority does not have enough Police Coverage of Subway Stations," June 26, 1956, NYTMA.

138. For example, report from the Director of Public Relations at the Transit Authority to the General Manager: "Radio Broadcast on Bus Vandalism Resulting from Meeting with Education Officials, Student Representatives and Transit Authority," April 30, 1963, NYTMA. Some letters also suggest that police officers not only admonish and arrest juvenile passengers, but also physically punish them on the spot. See Transit Authority report, "Letter suggesting Plainclothesmen use Rubber Hose on Youth in Subway," March 22, 1965, NYTMA.

139. Report from the Deputy Chief of the Transit Police to the General Manager of the Transit Authority, "28 Youth Arrested—Unlawful Assembly," December 13, 1965, NYTMA.

140. Jan Gabriel to the Transit Authority, April 5, 1965, NYTMA.

141. Larry Friedman, "Wagner Orders Policeman on Every Subway Train," *The Day*, April 6, 1965.

142. Transit Authority report, "Anti-Crime Program—Closing Down of Rear Cars," October 22, 1965, NYTMA.

143. "Address by Mayor Robert F. Wagner on Meeting the Problem of Crime in the Subways," Speech Manuscript, April 5, 1965, 7, NYTMA.

144. For example, Sylvia Zola to the Transit Authority, April 8, 1965, NYTMA.

145. For example, Rita Guttman to Mayor Wagner, December 15, 1965, NYTMA.

146. For example, Brian A. Carey to the chief of the Transit Police, August 12, 1966, NYTMA; Roy Jaffee to Mayor Lindsay, September 28, 1966, NYTMA.

147. For example, Leon Lieberthal to the Transit Police, April 16, 1963, NYTMA; and a telegram sent by L. L. Marks to Mayor Lindsay, September 12, 1966, NYTMA.

148. Transit Police report to the Transit Authority, November 22, 1966, NYTMA.

149. Transit Police report to the Transit Authority, September 21, 1965, NYTMA.

150. Accessible in the NYTMA.

151. For example, Phyllis Rollin to Mayor Lindsay, September 2, 1966, NYTMA; and Anne Foranoff, "Subway Alarm System to reduce Crime," *New York Times*, February 2, 1965.

152. For example, Paul T. Roger to Mayor Wagner, June 27, 1965, NYTMA; anonymous to Mayor Wagner, 28 January 1965, NYTMA; Teresa McCaffrey Kistler to Transit Authority, May 29, 1964, NYTMA.

153. For example, Nathan Bernstein to Mayor Wagner, December 7, 1963, NYTMA; Louis Halpern to the Transit Authority, March 30, 1955, NYTMA.

154. Irma M. Lasker to the Transit Authority, undated, received in July 1956, NYTMA.

155. For example, Rita Morais to the Transit Authority, June 19, 1963, NYTMA.

156. For example, Anne Goldberg to the Transit Authority, August 15, 1966, NYTMA. Goldberg writes: "This incident, having been previously victimized, frightened me to the extent that it affected my throat and caused me laryngitis and I could scarcely speak above a whisper the next day."

157. Raye Eilperin to Mayor Lindsay, June 23, 1966, NYTMA.

158. Seymour Epstein to the Transit Police, September 3, 1965, NYTMA.

159. Mrs. C. L. Dean to the Transit Authority, November 5, 1964, NYTMA.

160. Accessible in the NYTMA.

161. Nancy Arnstein to the Transit Authority, April 7, 1965, NYTMA.

162. Anonymous to Mayor Wagner, February 1, 1964, NYTMA.

163. Farge and Foucault, *Disorderly Families*, 258.

164. For example, Maude E. Kelheller to the Transit Authority, March 19, 1964, NYTMA.

165. For example, a petition from the residents of Long Island University to Mayor Wagner, April 5, 1963, NYTMA, or a complaint send on December 3, 1964, to the Transit Authority and signed by "Workers of the Esplanade Hotel," accompanied by a long list of signatures. NYTMA.

166. For example, D. E. Robinson to the Transit Authority, November 3, 1954, NYTMA.

167. For example, Irving Ostrofky to Noah Goldstein (Member of the Assembly), August 20, 1965, NYTMA.

168. For example, Transit Police report to the Transit Authority regarding a complaint from Mr. Silby, "Investigation re: Soliciting on the Transit System," January 23, 1964, NYTMA.

169. Roberta Michaels to the mayor, October 30, 1963, NYTMA.

170. For example, Berta Schwarz to the Transit Authority, April 11, 1964, NYTMA.

171. Fred Rose to the Transit Authority, June 3, 1964, NYTMA.

172. Jaqueline Bisagna to the chairman of the Transit Authority, June 2, 1964, NYTMA.

173. For example, Louis Halpern to the Transit Authority, March 30, 1955, NYTMA.

174. For example, anonymous to Mayor Wagner, February 16, 1963, NYTMA.

175. Transit Police report to the Transit Authority, "Anonymous Complaint," February 20, 1964, NYTMA.

176. Anonymous to the Transit Authority, April 8, 1965, NYTMA.

177. Anonymous to the Transit Authority, August 5, 1963, NYTMA.

178. Anonymous to the Transit Authority, May 22, 1967, NYTMA.

179. Anonymous to the Transit Authority, January 23, 1966, NYTMA.

180. Anonymous to the Mayor's Office, September 10, 1968, NYTMA.

181. Anonymous to the Transit Authority, January 6, 1968, NYTMA.

182. Anonymous to the Transit Authority, March 5, 1966, NYTMA.

183. Anonymous to Mayor Wagner, August 26, 1964, NYTMA.

184. Transit Authority report: "Anonymous Complaint," August 20, 1964, NYTMA.

185. Franz Kafka, *The Man who Disappeared (America)* (New York: Oxford University Press, 2012), 16.

186. Transit Police report, "Investigation re: Officer Assaulting Passenger," July 12, 1966, NYTMA.

187. Transit Authority department memo regarding a letter from Mrs. Cohen, July 8, 1955, NYTMA.

188. For example, Transit Police report, "Investigation Re. Letter Complaining of Loiterers in Toilets," February 8, 1965, NYTMA.

189. Because the source material has gaps, it cannot be ascertained beyond a doubt that this procedure was actually followed in every single case, but the files that have survived give no indication that there were deviations from this rule.

190. Transit Police report, "Suggestions to Curb Crime," July 13, 1964, NYTMA.

191. For example, the Transit Authority report "Investigation Re: Bums Sleeping on Trains," August 25, 1966. The file includes the remark: "Unable to furnish any further information concerning the subject, Mrs. Murphy launched into a lengthy discourse on other matters and displayed copies of such correspondences from various other agencies."

192. Luc Boltanski and his colleagues discovered a similar mechanism used for evaluating letters to the editor of the French newspaper *Le Monde*. See Marie-Ange Schiltz, Yann Darré, and Luc Boltanski, "La Dénonciation," *Actes de la Recherche en Sciences Sociales* 51, no. 1 (March 1984): 3–40.

193. Butler, *Giving an Account of Oneself*, 22–23.

194. For example, Transit Police report, "Investigation re: Concern for Citizen Riders," February 16, 1965, NYTMA.

195. Transit Police report to the Transit Authority, "Investigation Re: Alleged Biased Attitude of Patrolman," May 31, 1967, NYTMA; Transit Police report to the Transit Authority, "Investigation re: Alleged Improper Action of Patrolman," May 23, 1967, NYTMA.

196. Transit Police report to the Transit Authority, "Investigation re: Unreasonable Detention of Passenger," February 2, 1967, NYTMA.

197. Transit Police report: "Civilian Complaint," May 10, 1968, NYTMA.

198. Farge and Foucault, *Disorderly Families*, 282.

199. Jan Waxman to Mayor Wagner, June 23, 1964, NYTMA.

200. Transit Police report, "Investigation re: Female Afraid to Ride Subway," July 13, 1964, NYTMA.

201. Louis Althusser, *On the Reproduction of Capitalism: Ideology and Ideological State Apparatuses* (London: Verso, 2014), 191.

202. Alois Hahn and Volker Kapp, "Ideologie und Selbstzeugnis: Bekenntnis und Geständnis," in *Selbstthematisierung und Selbstzeugnis: Bekenntnis und Geständnis*, ed. Alois Hahn and Volker Kapp (Frankfurt am Main: Suhrkamp, 1987), 7–8.

203. Bernhard Siegert, *Passagiere und Papiere: Schreibakte auf der Schwelle zwischen Spanien und Amerika* (Munich: Wilhelm Fink Verlag, 2006), 158, translated by the author.

CHAPTER 6

1. As most chroniclers have focused on the first decades of the system, there are few works on the history of the subway after the 1960s. For a good survey of the system's technical and administrative development, see essays by Mark S. Feinman, *The New York Transit Authority in the 1970s* and Mark S. Feinman, *The New York Transit Authority in the 1980s*. These can be found at http://nycsubway.org.

2. See *Subway Ridership Reporting*, MTA report from 22 March 2012, compiled by Rob Hickey (Unit Chief, Revenue Analysis) and Bill Amarosa (Manager, Revenue & Ridership Analysis). These reports

can be found at the NYTMA, and at http://www.pcac.org/wp-content/uploads/2012/04/subway-ridership-032212.pdf.

3. "Subway Ridership Hits 65 Year Low," *New York Daily News*, October 1, 1982.

4. On the strikes of the Transport Workers Union, see the detailed epilogue to Joshua B. Freeman, *In Transit: The Transport Workers Union in New York City, 1933–1966*, 2nd ed. (Philadelphia: Temple University Press, 1992), 337–346. On the economic situation, see Joshua B. Freeman, *Working-class New York Life and Labor since World War II* (New York: New Press, 2000), 156; Eric Lichten, *Class, Power & Austerity: The New York City Fiscal Crisis* (South Hadley, MA: Bergin & Garvey Publishers, 1986); Wolfgang Quante, *The Exodus of Corporate Headquarters from New York City* (New York: Praeger, 1976), 56–69; Sharon Zukin, "Space and Symbols in an Age of Decline," in *Re-Presenting the City, Ethnicity, Capital and Culture in the 21st Century Metropolis*, ed. Anthony D. King (New York: New York University Press, 1996), 43–59.

5. See Feinman, *The New York Transit Authority in the 1980s*; Feinman, *The New York Transit Authority in the 1970s*.

6. Paul Theroux, "Subway Odyssey," *New York Times*, January 31, 1982.

7. Despite much discussion in the media, it is difficult to come by murder statistics for the subway. Based on information provided by the New York Transit Police, the *New York Times* has published some numbers. See Theroux, "Subway Odyssey."

8. Feinman, *The New York Transit Authority in the 1970s*.

9. "Subway Ridership Hits 65 Year Low."

10. Fred Brathwait and Henry Geldzahler, ed., *Bruce Davidson: Subway* (New York: Aperture, 2011), 32.

11. Richard Conway, "Grit, Grime, and Graffiti: Christopher Morris on the New York Subway, 1981," *Time*, January 22, 2014.

12. This applied also to construction work on a new line on the east shore of Manhattan, the so-called Second Avenue Subway. Plans for it had been drawn up in 1929 and from time to time smaller segments of it had been built, but never completed. In 2007 work was once again taken up to extend what had meanwhile become the world's largest underground infrastructure project. The first part of the line, costing 5.4 billion dollars with three new stations on Manhattan's Upper East Side, was opened in 2017. For more detail, see Verya Nasri, "Design of Second Avenue Subway in New York" (conference paper, World Tunnel Congress: Underground Facilities for Better Environment and Safety, India, 2008), 1372–1382, http://ctta.org/FileUpload/ita/2008/data/pdf/141.PDF.

13. Paul Shaw, *Helvetica and the New York City Subway System* (Cambridge, MA: MIT Press, 2001), 38. For more on the conflict involved in introducing this new visual regime see Stefan Höhne, "How to Make a Map for the Hades of Names—Transit Maps and the Representation of the City," in *Cultural Histories of Sociabilities, Spaces and Mobilities* 7, ed. Colin Divall (New York: Routledge, 2015), 83–98.

14. For more detail on the history of New York during that phase, see contributions to John H. Mollen-kopf and Manuel Castells, ed., *Dual City: Restructuring New York* (New York: Russell Sage Foundation, 1992); Miriam Greenberg, *Branding New York: How a City in Crisis was Sold to the World* (New York: Routledge, 2008); John H. Mollenkopf, *New York City in the 1980s: A Social, Economic, and Political Atlas* (New York: Simon & Schuster, 1993).

15. On the history of the Guardian Angels see Dennis Jay Kenney, "Crime on The Subways: Measuring the Effectiveness of the Guardian Angels," *Justice Quarterly* 3, no. 4 (1986): 481–496.

16. "The Subway Savages," *New York Daily News*, January 18, 1980.

17. On London, see Barry Webb and Gloria Laycock, *Reducing Crime on the London Underground: An Evaluation of Three Pilot Projects* (London: Home Office, 1992). http://www.popcenter.org/library/scp/pdf/189-Webb_and_Laycock.pdf.

18. David Clark, *Urban Decline* (New York: Routledge, 2013).

19. See "65 Cent Fare considered in Talks on Coping with Subway Crime," *New York Times*, September 27, 1980. Quoted in Feinman, *The New York Transit Authority in the 1980s*.

20. On Bernhard Goetz, see George P. Fletcher, *A Crime of Self-Defense: Bernhard Goetz and the Law on Trial* (Chicago: University of Chicago Press, 1990). For a broader historical investigation of this case, see also Paul J. Grayson, "Vigilantism in Canada and the United States," *Legal Studies Forum* 16 (1992): 21–41.

21. For debate on these dynamics, see Gunter Dreher and Thomas Feltes, eds., *Das Modell New York: Kriminalprävention durch 'Zero Tolerance'? Beiträge zur aktuellen kriminalpolitischen Diskussion* (Holzkirchen: Felix-Verlag, 1998); Bernard E. Harcourt, *Illusion of Order: The False Promise of Broken Windows Policing* (Cambridge, MA: Harvard University Press, 2004); Neil Smith, "Guiliani Time: The Revanchist 1990s," *Social Text* 57 (Winter 1998): 1–20; Neil Smith, *New Urban Frontier: Gentrification and the Revanchist City* (New York: Routledge, 1996).

22. Janet L. Abu-Lughod, *New York, Chicago, Los Angeles: America's Global Cities* (Minneapolis: University of Minnesota Press, 1999), 285–320.

23. Greenberg, *Branding New York: How a City in Crisis Was Sold to the World*, 146.

24. See "MTA Cites Cool, Cleaner Cars—Oh What a Feelin', Subways!," *New York Daily News*, 13, 1989, quoted in Feinman, *The New York Transit Authority in the 1980s*.

25. Harcourt, *Illusion of Order*, 154–159.

26. Harvey Molotch, *Against Security: How We Go Wrong at Airports, Subways, and Other Sites of Ambiguous Danger* (Princeton, NJ: Princeton University Press, 2012), 50–84.

27. Susanne Krasmann, "Gouvernementalität der Oberfläche: Aggressivität (ab-)trainieren beispiels-weise," in *Gouvernementalität der Gegenwart: Studien zur Ökonomisierung des Sozialen,* ed. Ulrich Bröckling, Susanne Krasmann, and Thomas Lemke (Frankfurt am Main: Suhrkamp, 2000), 198, translated by the author.

28. On the efficacy of the campaign, see Molotch, *Against Security*, 53–58.

29. Manny Fernandez, "A Phrase for Safety after 9/11 Goes Global," *New York Times*, May 10, 2010.

30. Stefan Höhne and Bill Boyer, "Subway," in *Encyclopedia of Urban Studies*, ed. Ray Hutchinson (London: SAGE, 2009), 784–786.

31. David Armitage, "What's the Big Idea? Intellectual History and the *Longue Durée*," *History of European Ideas* 38, no. 4 (December 2012): 493–507.

32. Andreas Reckwitz, *Das hybride Subjekt: Eine Theorie der Subjektkulturen von der bürgerlichen Moderne zur Postmoderne* (Weilerswist: Velbrück Wissenschaft, 2010), 14, translated by the author.

33. Wolfgang Sachs, "Herren über Raum und Zeit: Ein Rückblick in die Geschichte unserer Wünsche," in *Nahe Ferne—Fremde Nähe: Infrastrukturen und Alltag*, ed. Barbara Mettler-Meiborn and Christine Bauhardt (Berlin: Edition Sigma, 1993), 59–68.

34. Jean Baudrillard, "KOOL KILLER, or The Insurrection of Signs," in *Symbolic Exchange and Death*, trans. Iain Hamilton Grant (London: Sage, 1975), 76–86.

35. Alexander Klose and Jörg Potthast, "Container/Containment: Zur Einleitung," *Tumult* 38 (Spring 2011): 8–12; Alexander Klose, *The Container Principle: How a Box Changes the Way We Think* (Cambridge, MA: MIT Press, 2015).

36. Lieven De Cauter, "The Capsule and the Network: Preliminary Notes for a General Theory," *OASE* 54 (2001): 131.

37. See Alexander Klose, "Who Do You Want to Be Today? Annäherung an eine Theorie des Container-Subjekts," in *Das Motiv der Kästchenwahl: Container in Psychoanalyse, Kunst, Kultur*, ed. Insa Härtel and Olaf Knellessen (Göttingen: Vandenhoeck & Ruprecht, 2012), 21–38.

38. Michel de Certeau, *The Practice of Everyday Life* (Berkeley: University of California Press, 1984), 111–114.

39. De Certeau, *The Practice of Everyday Life*, 113.

40. De Certeau, *The Practice of Everyday Life*, 111.

41. Michel Foucault, "Of Other Spaces," trans. Jay Miskowiec, *Diacritics* 16, no. 1 (Spring 1986): 27.

42. De Certeau, *The Practice of Everyday Life*, 111.

43. De Certeau, *The Practice of Everyday Life*, 114.

44. Alexander Klose, "Die Containerisierung des Passagiers," in *Blätter für Technikgeschichte* 75–76, ed. Technisches Museum für Industrie und Gewerbe (Vienna: Technisches Museum Wien, 2013–2014), 72.

45. Michel Foucault, *The Order of Things* (New York: Pantheon Books, 1971), 386.

BIBLIOGRAPHY

Abrams, Jeanne E. "'Spitting Is Dangerous, Indecent, and against the Law!' Legislating Health Behavior during the American Tuberculosis Crusade." *Journal of the History of Medicine and Allied Sciences* 68, no. 3 (July 2013): 416–450.

Abu-Lughod, Janet L. *New York, Chicago, Los Angeles: America's Global Cities*, Minneapolis: University of Minnesota Press, 1999.

Abu-Lughod, Janet L. *Race, Space, and Riots in Chicago, New York, and Los Angeles*. New York: Oxford University Press, 2007.

Adas, Michael. *Machines as the Measure of Men: Science, Technology, and Ideologies of Western Dominance*. Ithaca: Cornell University Press, 1989.

Adorno, Theodor W., and Max Horkheimer. *Dialectic of Enlightenment*. Translated by John Cumming. New York: Verso, 1997.

Agamben, Giorgio. *Homo Sacer: Sovereign Power and Bare Life*. Stanford: Stanford University Press, 1998.

Agamben, Giorgio. "What Is an Apparatus?" In *What Is an Apparatus? And Other Essays*, 1–24. Stanford: Stanford University Press, 2009.

Akrich, Madeleine. "The De-Scription of Technical Objects." In *Shaping Technology/Building Society: Studies in Sociotechnical Change*, edited by Wiebe J. Bijker and John Law, 205–224. Cambridge, MA: MIT Press, 1992.

Akrich, Madeleine, and Bruno Latour. "A Summary of a Convenient Vocabulary for the Semiotics of Human and Nonhuman Assemblies." In *Shaping Technology/Building Society: Studies in Sociotechnical Change*, edited by Wiebe J. Bijker and John Law, 259–264. Cambridge, MA: MIT Press, 1992.

Alexander, Jennifer Karns. *The Mantra of Efficiency: From Waterwheel to Social Control.* Baltimore: Johns Hopkins University Press, 2008.

Alland, Alexander. *Jacob A. Riis: Photographer and Citizen.* New York: Aperture, 1993.

Allen, Irving L. *City in Slang: New York Life and Popular Speech.* New York: Oxford University Press, 1995.

Allerkamp, Andrea. *Anruf, Adresse, Appell: Figurationen der Kommunikation in Philosophie und Literatur.* Bielefeld: transcript, 2005.

Althusser, Louis. *On the Reproduction of Capitalism: Ideology and Ideological State Apparatuses.* London: Verso Books, 2014.

Amin, Ash, and Nigel Thrift. *Cities: Reimagining the Urban.* Cambridge: Polity Press, 2002.

Anbinder, Tyler. *Five Points: The 19th-Century New York City Neighborhood that Invented Tap Dance, Stole Elections, and Became the World's Most Notorious Slum.* New York: Blume, 2002.

Anderson, Ake E. "The Four Logistical Revolutions." *Papers in Regional Science* 59, no. 1 (January 1986): 1–12.

Armitage, David. "What's the Big Idea? Intellectual History and the Longue Durée." *History of European Ideas* 38, no. 4 (December 2012): 493–507.

Arnold, Bion J. *Report no. 1–7 on the Subway of the Interborough Rapid Transit Company of New York City.* New York: Public Service Commission, 1909.

Asendorf, Christoph. *Batteries of Life: On the History of Things and Their Perception in Modernity.* Berkeley: University of California Press, 1993.

Augé, Marc. *In the Metro.* Minneapolis: University of Minnesota Press, 2002.

Augé, Marc. *Non-Places: An Introduction to Supermodernity.* London: Verso, 2008.

Austin, Joe. *Taking the Train: How Graffiti Art Became an Urban Crisis in New York City.* New York: Columbia University Press, 2002.

Bachelard, Gaston. *The Poetics of Space.* Translated by Maria Jolas. Boston: Beacon Press, 1994.

Baggini, Julian. *Complaint: From Minor Moans to Principled Protests.* London: Profile Books, 2010.

Banham, Reyner. *Theory and Design in the First Machine Age.* New York: Praeger, 1967.

Barles, Sabine. "Urban Metabolism of Paris and Its Region." *Journal of Industrial Ecology* 13, no. 6 (December 2009): 896–913.

Barnes, Catherine A. *Journey from Jim Crow: Desegregation of Southern Transit.* New York: Columbia University Press, 1983.

Barrows, Susanna. *Distorting Mirrors: Visions of the Crowd in Late Nineteenth-Century France.* New Haven: Yale University Press 1981.

Bartz, Christina, and Marcus Krause. "Einleitung: Spektakel der Normalisierung." In *Spektakel der Normalisierung*, edited by Christina Bartz and Marcus Krause, 7–24. Paderborn: Fink, 2007.

Bascomb, Neal. *Higher: A Historic Race to the Sky and the Making of a City*. New York: Broadway Books, 2004.

Baudelaire, Charles. "The Painter of Modern Life." In *The Painter of Modern Life and Other Essays*, edited and translated by Jonathan Mayne, 1–47. London: Phaidon Press, 1995.

Baudrillard, Jean. "KOOL KILLER, or The Insurrection of Signs." In *Symbolic Exchange and Death*, translated by Iain Hamilton Grant, 76–86. London: Sage, 1975.

Baumann, Zygmunt. *Liquid Modernity*. Oxford: Polity Press, 2000.

Beck, Ulrich, Scott Lash, and Anthony Giddens. *Reflexive Modernization: Politics, Tradition and Aesthetics in the Modern Social Order*. Stanford, CA: Stanford University Press, 1994.

Becker, Peter, and Alf Lüdtke, eds. *Akten. Eingaben. Schaufenster. Die DDR und ihre Texte: Erkundungen zu Herrschaft und Alltag*. Berlin: Akademie Verlag 1997.

Beller, Anne, Sanford Garelik, and Sydney Cooper. "Sex Crimes in the Subway." *Criminology* 18, no. 1 (May 1980): 35–52.

Belliger, Andréa, and David J. Krieger. *ANThology: Ein einführendes Handbuch zur Akteur-Netzwerk-Theorie*. Bielefeld: transcript, 2006.

Bellow, Saul. *Herzog*. New York: Penguin, 2003.

Beniger, James. *The Control Revolution: Technological and Economic Origins of the Information Society*. Cambridge, MA: Harvard University Press, 2009.

Benjamin, Walter. *The Arcades Project*. Translated by Howard Eiland and Kevin McLaughlin. Cambridge, MA: Harvard University Press, 1999.

Benjamin, Walter. *Das Passagen-Werk*, vol. 1, Frankfurt am Main: Suhrkamp, 1983.

Benjamin, Walter. "One-Way Street." In *One-Way Street and Other Writings*, translated by Edmund Jephcott and Kingsley Shorter, 45–106. London: Harcourt Brace Jovanovich, 1979, 62.

Benjamin, Walter. "Paris, the Capital of the Nineteenth Century." In *The Writer of Modern Life: Essays on Charles Baudelaire*, edited by Michael W. Jennings, translated by Howard Eiland et al., 30–45. Cambridge, MA: Belknap Press of Harvard University Press, 2006.

Benjamin, Walter. "The Paris of the Second Empire in Baudelaire." In *The Writer of Modern Life: Essays on Charles Baudelaire*, edited by Michael W. Jennings, translated by Howard Eiland et al., 46–133. Cambridge, MA: Belknap Press of Harvard University Press, 2006.

Berman, Marshall. *All That Is Solid Melts into Air: The Experience of Modernity*. New York: Verso, 1991.

Bernard, Andreas. *Lifted: A Cultural History of the Elevator*. New York: NYU Press, 2014.

Biggs, Lindy. *Rational Factory: Architecture, Technology, and Work in America's Age of Mass Production*. Baltimore: Johns Hopkins University Press, 1996.

Blair, John G. *Modular America: Cross-Cultural Perspectives on the Emergence of an American Way*. Westport, CT: Greenwood Press, 1988.

Blodgett, Jeffrey G., Kirk L. Wakefield, and James H. Barnes. "The Effects of Customer Service on Consumer Complaining Behavior." *Journal of Services Marketing* 9, no. 4 (October 1995): 31–42.

Bobrick, Benson. *Labyrinths of Iron: Subways in History, Myth, Art, Technology, and War*. New York: Henry Holt, 1994.

Bogart, Michele Helene. *Public Sculpture and the Civic Ideal in New York City, 1890–1930*. Washington, DC: Smithsonian Institution Press, 1997.

Böhme, Hartmut. "Der Tastsinn im Gefüge der Sinne: Anthropologische und historische Ansichten vorsprachlicher Aisthesis." In *Tasten* 7, edited by Kunst und Ausstellungshalle der Bundesrepublik Deutschland, 185–210. Göttingen: Steidl, 1996.

Böhme, Hartmut, Peter Matussek, and Lothar Müller. *Orientierung Kulturwissenschaft: Was sie kann, was sie will*. Reinbek bei Hamburg: Rowohlt, 2000.

Böhringer, Hannes. *Orgel und Container*. Berlin: Merve, 1993.

Boltanski, Luc. *Distant Suffering: Morality, Media and Politics*. Translated by Graham D. Burchell. Cambridge: Cambridge University Press, 1999.

Boltanski, Luc, and Ève Chiapello. *The New Spirit of Capitalism*. London: Verso, 2018.

Boone, Christoper, and Ali Modarres. *City and Environment*. Philadelphia: Temple University Press, 2006.

Boutros, Alexandra, and Will Straw. "Introduction." In *Circulation and the City*, edited by Alexandra Boutros and Will Straw, 3–22. Montreal: McGill-Queen's University Press, 2010.

Bowlby, Rachel. "Commuting." In *Restless Cities*, edited by Mathew Beaumont and Gregory Dard, 43–58. London: Verso, 2010.

Boyer, Paul S. *Urban Masses and Moral Order in America, 1820–1920*. Cambridge, MA: Harvard University Press, 1992.

Braithwaite, Fred, and Henry Geldzahler, eds. *Bruce Davidson: Subway*. New York: Aperture, 2011.

Brand, Dana. *The Spectator and the City in Nineteenth-Century American Literature*. Cambridge: Cambridge University Press, 1991.

Brantz, Dorothee. "On the Nature of Urban Growth: Building Abattoirs in 19th-Century Paris and Chicago." *Cahiers Parisiens* 5 (2009): 17–30.

Brantz, Dorothee. *Slaughterhouse City: Paris, Berlin, and Chicago, 1780–1914*. Baltimore: Johns Hopkins University Press, forthcoming.

Braudel, Fernand. "Geschichte und Sozialwissenschaften. Die longue durée." In *Schrift und Materie der Geschichte: Vorschläge zu einer systematischen Aneignung historischer Prozesse*, edited by Fernand Braudel, Marc Bloch and Lucien Febvre, 47–85. Frankfurt am Main: Suhrkamp, 1977.

Brecher, Charles. "City Council." In *Encyclopedia of New York City*, edited by Kenneth T. Jackson, 229–230. New Haven: Yale University Press, 1995.

Bröckling, Ulrich. *The Entrepreneurial Self: Fabricating a New Type of Subject*. London: Sage, 2015.

Bronzaft, Arline L., Stephen B. Dobrow, and Timothy J. O'Hanlon. "Spatial Orientation in a Subway System." *Environment and Behavior* 8, no. 4 (December 1976): 575–594.

Brooks, Michael. *Subway City: Riding the Trains, Reading New York*. New Brunswick, NJ: Rutgers University Press, 1997.

Brouwer, Norman J. "Port of New York." In *Encyclopedia of New York City*, edited by Kenneth T. Jackson, 927–929. New Haven: Yale University Press, 1995.

Buck-Morss, Susan. *The Dialectics of Seeing: Walter Benjamin and the Arcades Project*. Cambridge, MA: MIT Press, 1991.

Buck-Morss, Susan. "The Flaneur, the Sandwichman, and the Whore: The Politics of Loitering." *New German Crtique* 39 (Autumn 1986): 99–140.

Buhr, Walter. "What Is Infrastructure?" Volkswirtschaftliche Diskussionsbeiträge—Discussion Paper no. 107-03, Department for Economic Science, University of Siegen, Siegen, 2003. https://EconPapers .repec.org/RePEc:sie:siegen:107-03.

Bührmann, Andrea D., and Werner Scheider. *Vom Diskurs zum Dispositiv: Eine Einführung in die Dispositivanalyse*. Bielefeld: transcript, 2008.

Bürk, Thomas. *Gefahrenzone, Angstraum, Feindesland: Stadtkulturelle Erkundungen zu Fremdenfeindlichkeit und Rechtsradikalismus in ostdeutschen Kleinstädten*. Münster: Westfälisches Dampfboot, 2012.

Burri, Regula Valérie. *Doing Images: Zur Praxis medizinischer Bilder*. Bielefeld: transcript, 2008.

Butler, Judith. *Giving an Account of Oneself*. New York: Fordham University Press, 2005.

Butler, Judith. *The Psychic Life of Power: Theories in Subjection*. Stanford, CA: Stanford University Press, 1997.

Caduff, Carlo. "Anticipations of Biosecurity." In *Biosecurity Interventions: Global Health and Security in Question*, edited by Andrew Lakoff and Stephen J. Collier, 257–277. New York: Columbia University Press, 2008.

Capra, Fritjof. *The Web of Life: A New Scientific Understanding of Living Systems*. New York: Anchor, 1997.

Caro, Robert A. *The Power Broker: Robert Moses and the Fall of New York*. New York: Vintage, 1975.

Cassirer, Ernst. "Mythic, Aesthetic and Theoretical Space." *Man and World* 2 (February 1969): 3–17.

Cassirer, Ernst. *Philosophy of Symbolic Forms*, volume 1: *Language*. Translated by Ralph Manheim. New Haven: Yale University Press, 1955.

Castleman, Craig. *Getting Up: Subway Graffiti in New York*. Cambridge, MA: MIT Press, 1982.

Chambers, John Whiteclay. *The Tyranny of Change: America in the Progressive Era, 1890–1920*. New Brunswick, NJ: Rutgers University Press, 2000.

Chan, Janet. "New Lateral Surveillance and a Culture of Suspicion." In *Surveillance and Governance: Crime Control and Beyond (Sociology of Crime, Law, and Deviance 10)*, edited by Mathieu Deflem and Jeffrey T. Ulmer, 223–239. Bingley: Emerald Group Publishing, 2008.

Chiesa, Laura. "Transnational Multimedia: Fortunato Depero's Impressions of New York City (1928–1930)." *California Italian Studies Journal* 1, no. 2 (2010): 1–33.

City Club of New York, The. *New York City Transit: A Memorandum Addressed to the Public Service Commission of the First District*. New York, 1907.

City Club of New York, The. *Subway Overcrowding*. New York, 1930.

Clark, David. *Urban Decline*. New York: Routledge, 2013.

Clark, Gary L., Peter F. Kaminski, and David R. Rink. "Consumer Complaints: Advice on How Companies Should Respond Based on an Empirical Study." *Journal of Consumer Marketing* 9, no. 3 (March 1992): 5–14.

Conley, Verena A. "The Passenger: Paul Virilio and Feminism." In *Paul Virilio: From Modernism to Hypermodernism and Beyond (Theory, Culture and Society)*, edited by John Armitage, 201–215. London: Sage, 2000.

Conway, Richard. "Grit, Grime, and Graffiti: Christopher Morris on the New York Subway, 1981." *Time*, January 22, 2014.

Coppola, Philip Ashforth. *Silver Connections*, volume 3: *A Fresh Perspective on the New York Area Subway Systems*. Maplewood, NJ: Four Oceans Press, 1988.

Corbin, Alain. *The Foul and the Fragrant: Odor and the French Social Imagination*. Leamington Spa: Berg, 1986.

Corbin, Alain. "Zur Geschichte und Anthropologie der Sinneswahrnehmung." In *Wunde Sinne: Über die Begierde, den Schrecken und die Ordnung der Zeit im 19. Jahrhundert*, 197–211. Stuttgart: Klett-Cotta, 1993.

Corey, Steven H. "Sanitation." In *Encyclopedia of New York City*, edited by Kenneth T. Jackson, 1041–1043. New Haven: Yale University Press, 1995.

Cowan, Alexander F., and Jill Steward, eds. *The City and the Senses: Urban Culture since 1500*. Farnham: Ashgate, 2007.

Cowen, Deborah. *The Deadly Life of Logistics: Mapping Violence in Global Trade*. Minneapolis: University of Minnesota Press, 2014.

Craig, Maxine Leeds. *Ain't I a Beauty Queen? Black Women, Beauty, and the Politics of Race*. New York, Oxford University Press, 2002.

Crary, Jonathan. *Suspensions of Perception: Attention, Spectacle, and Modern Culture*. Cambridge, MA: MIT Press, 2001.

Crary, Jonathan. *Techniques of the Observer: On Vision and Modernity in the Nineteenth Century*. Cambridge, MA: MIT Press, 1992.

Cudahy, Brian J. *Cash, Tokens, and Transfers: A History of Urban Mass Transit in North America*. New York: Fordham University Press, 1990.

Cudahy, Brian J. *A Century of Subways: Celebrating 100 Years of New York's Underground Railways*. New York: Fordham University Press, 2003.

Cudahy, Brian J. *Under the Sidewalks of New York: The Story of the Greatest Subway System in the World*. New York: Fordham University Press, 1995.

Cumo, Christopher. *Science and Technology in Twentieth Century American Life*. Westport, CT: Greenwood Publishing Group, 2007.

Cutlip, Scott M. *The Unseen Power: Public Relations—A History*. Hillsdale, NJ: Erlbaum, 1994.

Daston, Lorraine, and Peter Galison. *Objectivity*. New York: Zone Books, 2007.

Davis, Donald F. "North American Urban Mass Transit, 1890–1950: What if We Thought about It as a Type of Technology?" *History and Technology* 12, no. 4 (1995): 309–326.

De Cauter, Lieven. "The Capsule and the Network. Preliminary Notes for a General Theory." *OASE* 54 (2001): 122–134.

de Certeau, Michel. *Heterologies: Discourse on the Other*. Minneapolis: University of Minnesota Press, 1986.

de Certeau, Michel. *The Practice of Everyday Life*. Berkeley: University of California Press, 1984.

DeLanda, Manuel. *A New Philosophy of Society: Assemblage Theory and Social Complexity*. London/New York: Continuum, 2006.

Deleuze, Gilles. *Foucault*. Minnesota: University of Minnesota Press, 1988.

Deleuze, Gilles. "What Is a Dispositif?" In *Michel Foucault Philosopher*, translated by Timothy J. Armstrong, 159–168. New York: Routledge, 1992.

Deleuze, Gilles, and Félix Guattari. *Anti-Oedipus: Capitalism and Schizophrenia*. Minneapolis: University of Minnesota Press, 1983.

Deleuze, Gilles, and Félix Guattari. *A Thousand Plateaus: Capitalism and Schizophrenia*. New York: Continuum, 1988.

Delitz, Heike. *Gebaute Gesellschaft: Architektur als Medium des Sozialen*. Frankfurt am Main: Campus, 2010.

Denis, Jérôme, and David Pontille. "The Graphical Performance of a Public Space: The Subway Signs and Their Scripts." In *Urban Plots, Organizing Cities*, edited by Giovanna Sonda, Claudio Coletta, and Francesco Gabbi, 11–22. Farnham: Ashgate, 2010.

Department of Health. *Brief History of the Campaign against Tuberculosis in New York City*. New York, 1908.

Derrick, Peter. *Tunneling the Future: The Story of the Great Subway Expansion That Saved New York*. New York: NYU Press, 2001.

Diehl, Lorraine B. *Subways: The Tracks That Built New York City*. New York: Clarkson Potter, 2004.

Diogo, Maria Paula, and Dirk van Laak. *Europeans Globalizing: Mapping, Exploiting, Exchanging*. London: Palgrave Macmillan, 2016.

Dommann, Monika. "Handlung, Flowcharts, Logistik. Zur Wissensgeschichte und Materialkultur von Warenflüssen." In *Nach Feierabend: Zürcher Jahrbuch für Wissensgeschichte: Zirkulationen 7*, edited by David Gugerli et al., 75–103. Zurich: diaphanes, 2011.

Dommann, Monika. "Material Manövrieren: Eine Begriffsgeschichte der Logistik." *Via Storia*, no. 2 (2009): 13–27.

Douglas, George H. *Skyscrapers: A Social History of the Very Tall Building in America*. Jefferson, NC: McFarland, 2004.

Douglas, Mary. *Purity and Danger: An Analysis of Concepts of Pollution and Taboo*. London: Routledge, 2002.

Dreher, Gunther, and Thomas Felters, eds. *Das Modell New York: Kriminalprävention durch "zero Tolerance"? Beiträge zur aktuellen kriminalpolitischen Diskussion*. Holzkirchen: Felix-Verlag, 1998.

Driesen, Christian. "Die Kritzelei als Ereignis des Formlosen." In *Über Kritzeln: Graphismen zwischen Schrift, Bild, Text und Zeichen*, edited by Christian Driesen et al., 23–37. Berlin: Diaphanes Verlag, 2012.

DuBois, Ellen Carol. "Equality League of Self-Supporting Women." In *Encyclopedia of New York City*, edited by Kenneth T. Jackson, 281. New Haven: Yale University Press, 1995.

Duden, Barbara, and Ivan Illich. "Die skopische Vergangenheit Europas und die Ethik der Opsis: Plädoyer für eine Geschichte des Blickes und Blickens." *Historische Anthropologie* 3, no. 2 (1995): 203–221.

Duffy, John. *A History of Public Health in New York City, 1625–1866*. New York: Russell Sage Foundation, 1968.

Dwyer, Jim. *Subway Lives: 24 Hours in the Life of the New York City Subway*. New York: Crown, 1991.

Elias, Norbert. *The Civilizing Process: Sociogenic and Psychogenetic Investigations*. Translated by Edmund Jephcott with notes and corrections by the author. Edited by Eric Dunning, Johan Goudsblom, and Stephen Menell. Malden, MA: Blackwell, 2000.

Elsner, Steffen H. "Flankierende Stabilisierungsmechanismen diktatorischer Herrschaft: Das Eingabewesen in der DDR." In *Repression und Wohlstandsversprechen: Zur Stabilisierung von Parteiherrschaft in der DDR und der CSSR*, edited by Christoph Boyer and Peter Skyba, 75–86. Dresden: Hannah-Arendt-Institut für Totalitarismusforschung, 1999.

Ernst, Robert. *Immigrant Life in New York City: 1825–1863.* Syracuse, NY: Syracuse University Press, 1994.

Ernst, Waltraud. "Möglichkeiten (in) der Stadt: Überlegungen zur Öffentlichkeit und Privatheit geschlechtlicher Raumordnungen." In *Street Harassment: Machtprozesse und Raumproduktion,* edited by Feministisches Frauenkollektiv, 75–93. Vienna: Mandelbaum, 2008.

Essbach, Wolfgang. "Antitechnische und antiästhetische Haltungen in der soziologischen Theorie." In *Technologien als Diskurse,* edited by Andreas Lösch, 123–136. Heidelberg: Synchron Wissenschaftsverlag der Autoren, 2001.

Esteve, Mary. *The Aesthetics and Politics of the Crowd in American Literature.* Cambridge: Cambridge University Press, 2003.

Farge, Arlette, and Michel Foucault. *Disorderly Families: Infamous Letters from the Bastille Archives.* Minneapolis: University of Minnesota Press, 2017.

Farias, Ignacio. "Introduction: Decentering the Object of Urban Studies." In *Urban Assemblages: How Actor-Network Theory Changes Urban Studies,* edited by Thomas Bender and Ignacio Farias, 1–24. New York: Routledge, 2010.

Fawcett, Paul. *Managing Passenger Logistics: The Comprehensive Guide to People and Transport.* London: Kogan Page, 2000.

Felsch, Philipp. "Merves Lachen." *Zeitschrift für Ideengeschichte* 2, no. 4 (November 2008): 11–30.

Feinman, Mark S. *New York Transit Authority in the 1970s.* http://www.nycsubway.org.

Feinman, Mark S. *New York Transit Authority in the 1980s.* http://www.nycsubway.org.

Feinman, Mia. "Notes for Underground: The Subway Portraits." In *Walker Evans,* edited by Maria Morris Hambourg and Jeff L. Rosenheim, 106–119. New York: Metropolitan Museum of Art, 2000.

Fischler, Stan. *Uptown Downtown: A Trip through Time on New York's Subways.* New York: Hawthorn Books, 1976.

Fishman, Alfred P., and Dickinson W. Richards. *Circulation of the Blood: Men and Ideas.* New York: Oxford University Press, 1964.

Fitzpatrick, Sheila. *Everyday Stalinism: Ordinary Life in Extraordinary Times: Soviet Russia in the 1930s.* New York: Oxford University Press, 1999.

Fitzpatrick, Sheila, and Robert Gellately. "Introduction to the Practices of Denunciation in Modern European History." In *Accusatory Practices: Denunciation in Modern European History, 1789–1989,* edited by Sheila Fitzpatrick and Robert Gellately, 1–21. Chicago: University of Chicago Press, 1977.

Fitzpatrick, Tracy. *Art and the Subway: New York Underground.* New Brunswick, NJ: Rutgers University Press, 2009.

Fletcher, George P. *A Crime of Self Defense: Bernard Goetz and the Law on Trial.* Chicago: University of Chicago Press, 1990.

Flint, Anthony. *This Land: The Battle over Sprawl and the Future of America*. Baltimore: Johns Hopkins University Press, 2008.

Ford, Henry. *My Life and Work*. New York: Doubleday, Page, 1922.

Foucault, Michel. *The Birth of Biopolitics: Lectures at the Collège de France, 1978–1979*. Edited by Michel Senellart. Translated by Graham Burchell. New York: Palgrave Macmillan, 2008.

Foucault, Michel. "The Confession of the Flesh." In *Foucault: Power/Knowledge: Selected Interviews and other Writings 1972–1977*, edited by Colin Gordon, 194–228. New York: Pantheon Books, 1980.

Foucault, Michel. *Discipline and Punish: The Birth of the Prison*. New York: Random House, 1995.

Foucault, Michel. *The Hermeneutics of the Subject: Lectures at the Collège de France, 1981–1982*. Edited by Frederic Gros. Translated by Graham Burchell. New York: Picador, 2006.

Foucault, Michel. *The History of Sexuality*, volume 1: *The Will to Knowledge*. New York: Pantheon. 1978.

Foucault, Michel. "The Life of Infamous Men." In *Foucault: Power, Truth, Strategy*, edited by Meaghan Morris and Paul Patton Michel, 76–91. Sydney: Feral, 1979.

Foucault, Michel. *Madness and Civilization: A History of Insanity in the Age of Reason*. New York: Random House, 1988.

Foucault, Michel. "The Meshes of Power." In *Space, Knowledge, and Power: Foucault and Geography*, edited by Jeremy W. Crampton and Stuart Elden, 153–162. Aldershot: Ashgate, 2007.

Foucault, Michel. "Of Other Spaces." Translated by Jay Miskowiec, *Diacritics* 16, no. 1 (Spring 1986): 22–27.

Foucault, Michel. *The Order of Things*. New York: Pantheon Books, 1971.

Foucault, Michel. *Remarks on Marx: Conversations with Duccio Trombadori*. Translated by R. James Goldstein and James Cascaito. New York: Semiotext(e), 1991.

Foucault, Michel. *Security, Territory, Population: Lectures at the Collège de France, 1977–78*. Edited by Michel Senellart. Translated by Graham Burchell. Basingstoke: Palgrave Macmillan, 2014.

Foucault, Michel. "Self Writing." In *Ethics: Subjectivity and Truth*, edited by Paul Rabinow, 207–223. New York: The New Press, 1997.

Foucault, Michel. *Society Must Be Defended: Lectures at the Collège de France, 1975–76*. Edited by Mauro Bertani and Alessandro Fontana. Translated by David Macey. Picador: New York, 2003.

Foucault, Michel. "The Subject and Power." *Critical Inquiry* 8, no. 4 (Summer 1982): 777–795.

Foucault, Michel. "Technologies of the Self." In *Technologies of the Self: A Seminar with Michel Foucault*, edited by Luther H. Martin, Huck Gutman and Patrick H. Hutton, 16–49. Amherst, MA: University of Massachusetts Press, 1988.

Framberger, David J. "Architectural Designs for New York's First Subway." In *Historical American Engineering Record: Interborough Rapid Transit Subway (Original Line) NY-122*, 365–412. New York, 1979.

Frank, Susanne. *Stadtplanung im Geschlechterkampf: Stadt und Geschlecht in der Großstadtentwicklung des 19. und 20. Jahrhunderts.* Opladen: Leske + Budrich, 2003.

Freeman, Joshua B. *In Transit: The Transport Workers Union in New York City, 1933–1966.* 2nd ed. Philadelphia: Temple University Press, 1992.

Freeman, Joshua B. *Working-Class New York Life and Labor since World War II.* New York: New Press, 2000.

Freud, Sigmund. "Group Psychology and the Analysis of the Ego." In *The Standard Edition of the Complete Psychological Works of Sigmund Freud,* volume 18 (1920–1922): *Beyond the Pleasure Principle, Group Psychology and Other Works,* 65–144. London: Hogarth Press, 1955.

Freud, Sigmund. "Massenpsychologie und Ich-Analyse." In *Massenpsychologie und Ich-Analyse/Die Zukunft einer Illusion,* 31–106. Frankfurt am Main: Fischer, 1993.

Freud, Sigmund. "Three Essays on Sexuality." In *The Standard Edition of the Complete Psychological Works of Sigmund Freud,* volume 7 (1901–1905): *A Case of Hysteria, Three Essays on Sexuality and Other Works,* edited by James Strachney, 135–243. London: Hogarth Press, 1953.

Frisby, David. *Fragments of Modernity. Theories of Modernity in the Work of Simmel, Kracauer, and Benjamin.* London: Routledge, 2013.

Fromm, Erich. *Den Menschen verstehen: Psychoanalyse und Ethik.* Munich: Deutscher Taschenbuch Verlag, 2004.

Fromm, Erich. *Man for Himself: An Inquiry into the Psychology of Ethics.* New York: Routledge, 1999.

Fulbrook, Mary, and Ulinka Rublack. "In Relation: The 'Social Self' and Ego-Documents." *German History* 2, no. 3 (August 2010): 263–272.

Fuller, Gillian. "Arrow-Directional Semiotics: Wayfinding in Transit." *Social Semiotics* 12, no. 3 (2002): 231–244.

Fuller, Gillian. *Aviopolis: A Book about Airports.* London: Black Dog, 2004.

Gamper, Michael. *Masse lesen, Masse schreiben: eine Diskurs- und Imaginationsgeschichte der Menschenmenge 1765–1930.* Munich: Fink, 2007.

Gandy, Matthew. "Cyborg Urbanization: Complexity and Monstrosity in the Contemporary City." *International Journal of Urban and Regional Research* 29, no. 1 (March 2005): 26–49.

Gandy, Matthew. "Life without Germs: Contested Episodes in the History of Tuberculosis." In *Return of the White Plague: Global Poverty and the "New" Tuberculosis,* edited by Matthew Gandy and Alimuddin Zunla, 15–38. London: Verso, 2003.

Garland, Ken. *Mr. Beck's Underground Map.* Harrow Weald: Capital Transport, 1994.

Gellately, Robert. *Backing Hitler.* New York: Oxford University Press, 2001.

Gellately, Robert. "Denunciation as a Subject of Historical Research." *Historical Social Research* 26, no. 2/3 (2001): 16–29.

Gershon, Robyn R. M., Richard Neitzel, Marissa A. Barrera, and Muhammad Akram. "Pilot Survey of Subway and Bus Stop Noise Levels." *Journal of Urban Health: Bulletin of the New York Academy of Medicine* 83, no. 5 (September 2006): 802–812.

Giedion, Sigfried. *Mechanization Takes Command: A Contribution to Anonymous History.* New York: Oxford University Press, 1948.

Gill, Jonathan. *Harlem: The Four Hundred Year History from Dutch Village to Capital of Black America.* New York: Grove Press, 2011.

Glickman, Lawrence B. "Introduction: Born to Shop? Consumer History and American History." In *Consumer Society in American History: A Reader*, edited by Lawrence B. Glickman, 1–16. Ithaca: Cornell University Press, 1999.

Goddard, Stephen B. *Getting There: The Epic Struggle between Road and Rail in the American Century.* Chicago: University of Chicago Press, 1996.

Goffman, Erving. *Relations in Public: Microstudies in the Public Order.* New York: HarperCollins, 1972.

Graham, Stephen. *Vertical: The City from Satellites to Bunkers.* London: Verso, 2017.

Graham, Stephen, and Simon Marvin. *Splintering Urbanism: Networked Infrastructures, Technological Mobilities and the Urban Condition.* London: Routledge, 2001.

Grayson, J. Paul. "Vigilantism in Canada and the United States." *Legal Studies Forum* 16 (1992): 21–41.

Greenberg, Miriam. *Branding New York: How a City in Crisis was sold to the World.* New York: Routledge, 2008.

Greene, Asa. *A Glance at New York.* New York: Craighead & Allen, 1937.

Groce, Nancy. *New York, Songs of the City.* New York: Billboard Books, 1999.

Groneman, Carol, and David M. Reimers. "Immigration." In *Encyclopedia of New York City*, edited by Kenneth T. Jackson, 581–589. New Haven: Yale University Press, 1995.

Guattari, Félix. *Chaosmosis: An Ethico-Aesthetic Paradigm.* Bloomington: Indiana University Press, 1995.

Guillén, Mauro F. *The Taylorized Beauty of the Mechanical: Scientific Management and the Rise of Modernist Architecture.* Princeton, NJ: Princeton University Press, 2009.

Guldi, Jo. *Roads to Power: Britain Invents the Infrastructure State.* Cambridge, MA: Harvard University Press, 2012.

Habermas, Jürgen. *The Inclusion of the Other: Studies in Political Theory.* Edited by Ciaran P. Cronin and Pablo De Greiff. Cambridge, MA: MIT Press, 1998.

Habermas, Jürgen. "What Does a Crisis Mean Today? Legitimation Problems in Late Capitalism." *Social Research* 40, no. 4 (1973): 643–667.

Hacking, Ian. "Making Up People." In *Reconstructing Individualism: Autonomy, Individuality, and the Self in Western Thought,* edited by Thomas C. Heller, Morton Sosna, and David E. Wellbery, 222–236. Stanford: Stanford University Press, 1986.

Hadlaw, Janin. "The London Underground Map: Imaging Modern Time and Space." *Design Issues* 19, no. 1 (Winter 2003): 25–35.

Hagemann, Anke. "Drehkreuz." *archplus* 191–192 (March 2009): 38–39.

Hagemann, Anke. "Filter, Ventile und Schleusen: Die Architektur der Zugangsregulierung." In *Kontrollierte Urbanität: Zur Neoliberalisierung städtischer Sicherheitspolitik,* edited by Volker Eick, Jens Sambale, and Eric Töpfer, 301–328. Bielefeld: transcript, 2007.

Hahn, Alois, and Volker Kapp. "Selbstthematisierung und Selbstzeugnis: Bekenntnis und Geständnis." In *Selbstthematisierung und Selbstzeugnis: Bekenntnis und Geständnis,* edited by Alois Hahn and Volker Kapp, 7–8. Frankfurt am Main: Suhrkamp, 1987.

Hahn, Hans Peter. *Materielle Kultur: Eine Einführung.* Berlin: Dietrich Reimer Verlag, 2005.

Hall, Peter. *Cities of Tomorrow: An Intellectual History of Urban Planning and Design in the Twentieth Century.* Chichester: Wiley-Blackwell, 2002.

Hanna, Charles F. "Complaint as a Form of Association." *Qualitative Sociology* 4, no. 4 (December 1981): 298–311.

Harcourt, Bernard E. *Illusion of Order: The False Promise of Broken Windows Policing.* Cambridge, MA: Harvard University Press, 2004.

Harrison, Jarjorie. "Mills Charles Wright." In *Encyclopedia of New York City,* edited by Kenneth T. Jackson, 763. New Haven: Yale University Press, 1995.

Harvey, David. *The Condition of Postmodernity.* Cambridge, MA: Blackwell, 1990.

Hauser, Katherine Jane. "George Tooker, Surveillance, and Cold War Sexual Politics." *GLQ: A Journal of Lesbian and Gay Studies* 11, no. 3 (June 2005): 391–425.

Hessler, Martina. "Ansätze und Methoden der Technikgeschichtsschreibung (Zusatztexte im Internet)." In *Kulturgeschichte der Technik,* 1–65. Frankfurt am Main: Campus Verlag, 2010. http://studium.campus .de/sixcms/media.php/274/Hessler_Zusatzkapitel_Internet.pdf.

Hiebert, Ray Eldon. *Courtier to the Crowd: The Story of Ivy Lee and the Development of Public Relations in America.* Ames: Iowa State University Press, 1966.

Hillmann, Karl-Heinz. *Wörterbuch der Soziologie.* 5th ed. Stuttgart: Kröner, 2007.

Hofstadter, Richard. *Age of Reform: From Bryan to F.D.R.* New York: Vintage Books, 1971.

Höhne, Stefan. "How to Make a Map for the Hades of Names—Transit Maps and the Representation of the City." In *Cultural Histories of Sociabilities, Spaces and Mobilities* 7, edited by Colin Divall, 83–98. New York: Routledge, 2015.

Höhne, Stefan. "Tokens, Suckers, and the Great New York Token War." *Zeitschrift für Medien- und Kultur-forschung* 1 (May 2011): 143–158.

Höhne, Stefan. "Vereinzelungsanlagen: Die Genese des Drehkreuzes aus dem Geist automatischer Kontrolle." *Technikgeschichte* 83, no. 2 (2016): 103–124.

Höhne, Stefan, and Bill Boyer. "Subway." In *Encyclopedia of Urban Studies*, edited by Ray Hutchinson, 784–786. London: Sage, 2009.

Höhne, Stefan, and Rene Umlauf. "Die Akteur-Netzwerk Theorie. Zur Vernetzung und Entgrenzung des Sozialen." In *Theorien in der Raum- und Stadtforschung: Einführungen*, edited by Jörgen Ossenbrügge and Anne Vogelpohl, 195–214. Münster: Westfälisches Dampfboot, 2015.

Holt, Glen E. "The Changing Perception of Urban Pathology: An Essay on the Development of Mass Transit in the United States." In *Cities in American History*, edited by Kenneth T. Jackson and Stanley K. Schultz, 324–343. New York: Alfred A. Knopf, 1972.

Hood, Clifton. "Changing Perceptions of Public Space on the New York Rapid Transit System." *Journal of Urban History* 22, no. 3 (March 1996): 308–331.

Hood, Clifton. "The Impact of the IRT on New York City." In *Historical American Engineering Record: Interborough Rapid Transit Subway (Original Line) NY-122*, 145–206. New York, 1979.

Hood, Clifton. *722 Miles: The Building of the Subways and How They Transformed New York*. Baltimore: Johns Hopkins University Press, 2004.

Hosokawa, Shuhei. "The Walkman Effect." *Popular Music* 4 (1984): 165–180.

Hounshell, David. *From the American System to Mass Production, 1800–1932: The Development of Manufacturing Technology in the United States*. Baltimore: Johns Hopkins University Press, 1985.

Howes, David. "Introduction." In *Empire of the Senses: The Sensual Culture Reader*, edited by David Howes, 1–20. Oxford: Berg, 2005.

Hughes, Thomas P. *American Genesis: A Century of Invention and Technological Enthusiasm, 1870–1970*. Chicago: University of Chicago Press, 1989.

Hughes, Thomas P. "Electrification of America: System Builders." *Technology and Culture* 20, no. 1 (January 1979): 124.

Hughes, Thomas P. "Evolution of Large Technological Systems." In *Social Construction of Technological Systems: New Directions in the Sociology and History of Technology*, edited by Wiebe E. Bijker, Thomas P. Hughes, and Trevor J. Pinch, 51–82. Cambridge, MA: MIT Press, 1989.

Hulser, Kathleen. "Outdoor Advertising." In *Encyclopedia of New York City*, edited by Kenneth T. Jackson, 869–870. New Haven: Yale University Press, 1995.

Interborough Rapid Transit Company. *The New York Subway: Its Construction and Equipment*. New York: McGraw, 1904.

Jackson, Kenneth T. *Crabgrass Frontier: Suburbanization of the United States*. New York: Oxford University Press, 1987.

Jahns, Christopher, and Christine Schüffler. *Logistik: Von der Seidenstrasse bis heute*. Wiesbaden: Springer Gabler Verlag, 2008.

Jameson, Fredric. *Postmodernism or the Cultural Logic of Late Capitalism*. Durham: Duke University Press, 1991.

Jeffrey, Richard, and John M. MacKenzie. *The Railway Station: A Social History*. London: Faber & Faber, 2010.

Johnson, Steven. *Invention of Air: A Story of Science, Faith, Revolution, and the Birth of America*. New York: Riverhead Books, 2008.

Jordan, John M. *Machine-Age Ideology: Social Engineering and American Liberalism, 1911–1939*. Chapel Hill, NC: University of North Carolina Press, 1994.

Joyce, Patrick. *Rule of Freedom: Liberalism and the Modern City*. London: Verso, 2003.

Judge, Erica, Vincent Seyfried, and Andrew Sparberg. "Elevated Railways." In *Encyclopedia of New York City*, edited by Kenneth T. Jackson, 368–370. New Haven: Yale University Press, 1995.

Jütte, Robert. *Geschichte der Sinne*. Munich: C. H. Beck, 2000.

Kafka, Franz. *The Man Who Disappeared*. Translated by Ritchie Robertson. New York: Oxford University Press, 2012.

Kaika, Maria, and Erik Swyngedouw. "Fetishizing the Modern City: Phantasmagoria of Urban Technological Networks." *International Journal of Urban and Regional Research* 24, no. 1 (March 2000): 120–138.

Kantrowitz, Nathan. "Population." In *Encyclopedia of New York City*, edited by Kenneth T. Jackson, 920–923. New Haven: Yale University Press, 1995.

Kaschuba, Wolfgang. *Die Überwindung der Distanz: Zeit und Raum in der europäischen Moderne*. Frankfurt am Main: Fischer Taschenbuch Verlag, 2004.

Katz, Barry. *Herbert Marcuse and the Art of Liberation: An Intellectual Biography*. New York: Verso, 1982.

Kennel, Rosemarie. "Bions Container-Contained-Modell—und die hieraus entwickelte Denktheorie." In *Das Motiv der Kästchenwahl: Container in Psychoanalyse, Kunst, Kultur*, edited by Insa Härtel and Olaf Knellessen, 69–85. Göttingen: Vandenhoeck & Ruprecht, 2012.

Kenney, Dennis Jay. "Crime on the Subways: Measuring the Effectiveness of the Guardian Angels." *Justice Quarterly* 3, no. 4 (1986): 481–496.

Kern, Stephen. *Culture of Time and Space, 1880–1918*. Cambridge, MA: Harvard University Press, 2003.

Kible, Brigitte. "Subjekt." In *Historisches Wörterbuch der Philosophie* 10, edited by Joachim Ritter and Karlfried Gründer, 373–400. Stuttgart: Schwabe & Co. AG, 1998.

Klose, Alexander. *The Container Principle: How a Box Changes the Way We Think*. Cambridge, MA: MIT Press, 2015.

Klose, Alexander. "Die Containerisierung des Passagiers." In *Blätter für Technikgeschichte* 75–76, edited by Technisches Museum für Industrie und Gewerbe, 69–86. Vienna: Springer, 2013–2014.

Klose, Alexander. "Who Do You Want to Be Today? Annäherungen an eine Theorie des Container-Subjekts." In *Das Motiv der Kästchenwahl: Container in Psychoanalyse, Kunst, Kultur*, edited by Insa Härtel and Olaf Knellessen, 21–38. Göttingen: Vandenhoeck & Ruprecht, 2012.

Klose, Alexander, and Jörg Potthast. "Container/Containment: Zur Einleitung." *Tumult* 38 (2012): 8–12.

Knoblauch, Herbert A. "Erving Goffmans Reich der Interaktion—Einführung." In *Erving Goffman: Interaktion und Geschlecht*, 7–49. Frankfurt am Main: Campus Verlag, 2001.

Kopper, Christopher. "Mobile Exceptionalism? Passenger Transport in Interwar Germany." *Transfers* 3, no. 2 (June 2013): 89–107.

Koschorke, Albrecht. *Körperströme und Schriftverkehr: Mediologie des 18. Jahrhunderts*. Munich: Fink, 2003.

Kowalski, Robin M. "Complaints and Complaining: Functions, Antecedents, and Consequences." *Psychological Bulletin* 119, no. 2 (March 1996): 179–196.

Kracauer, Siegfried. "The Mass Ornament." In *The Mass Ornament: Weimar Essays*, translated, edited, and with an introduction by Thomas Y. Levi, 75–88. Cambridge, MA: Harvard University Press, 1995.

Kracauer, Siegfried. "Proletarische Schnellbahn." In *Aufsätze 1927–1931* 5.2, edited by Inka Müller-Bach, 179–180. Frankfurt am Main: Suhrkamp, 1990.

Krasmann, Susanne. "Gouvernementalität der Oberfläche. Aggressivität (ab-)trainieren beispielsweise." In *Gouvernementalität der Gegenwart: Studien zur Ökonomisierung des Sozialen*, edited by Ulrich Bröckling, Susanne Krasmann, and Thomas Lemke, 194–226. Frankfurt am Main: Suhrkamp, 2000.

Latour, Bruno. "The Berlin Key, or How to Do Things with Words." In *Matter, Materiality and Modern Culture*, edited by P. M. Graves-Brown, 10–21. London: Routledge, 2000.

Latour, Bruno. *Pandora's Hope: Essays on the Reality of Science Studies*. Cambridge, MA: Harvard University Press, 1999.

Latour, Bruno. "Technology Is Society Made Durable." *Sociological Review* 38, no. 1 suppl. (May 1990): 103–131.

Latour, Bruno. *We Have Never Been Modern*. Cambridge, MA: Harvard University Press, 1993.

Lazarus, Moritz. *Über Gespräche*. Berlin: Hensel, 1986.

Lazzerato, Maurizio. "Die Missgeschicke der 'Künstlerkrikik' und der kulturellen Beschäftigung." In *Kritik der Kreativität*, edited by Gerald Raunig and Ulf Wuggenig, 190–206. Vienna: Turia + Kant, 2007.

Lears, T. J. Jackson. *Fables of Abundance: A Cultural History of Advertising in America*. New York: Basic Books, 1994.

Leboff, David, and Timothy Demuth. *No Need to Ask! Early Maps of London's Underground Railways*. Middlesex: Capital Transport Publishing, 1999.

Le Bon, Gustave. *The Crowd: A Study of the Popular Mind*. Kitchener, ON: Batoche, 2001.

Le Corbusier. *La Ville radieuse: Eléments d'une doctrine d'urbanisme pour l'équipement de la civilisation machiniste*. Boulogne-Sur-Seine: Editions de l'Architecture d'aujourd'hui, 1935.

Lee, Benjamin, and Edward Li Puma. "Cultures of Circulation. The Imaginations of Modernity." *Public Culture* 14, no. 1 (January 2002): 191–213.

Lethen, Helmut. *Cool Conduct: The Culture of Distance in Weimar Germany*. Translated by Don Reneu. Berkeley: University of California Press, 2002.

Levinson, Marc. *The Box: How the Shipping Container Made the World Smaller and the World Economy Bigger*. Princeton, NJ: Princeton University Press, 2006.

Lichten, Eric. *Class, Power, and Austerity: The New York City Fiscal Crisis*. South Hadley, MA: Bergin & Garvey, 1986.

Liefeld, John P., Fred H. C. Edgecombe, and Linda Wolfe. "Demographic Characteristics of Canadian Consumer Complainers." *Journal of Consumer Affairs* 9, no. 1 (Summer 1975): 73–80.

Link, Jürgen. "'Normative' oder 'Normal'? Diskursgeschichtliches zur Sonderstellung der Industrienorm im Normalismus, mit einem Blick auf Walter Cannon." In *Normalität und Abweichung: Studien Zur Theorie und Geschichte der Normalisierungsgesellschaft*, edited by Werner Sohn and Herbert Mertens, 30–44. Opladen: Westdeutscher Verlag, 1999.

Link, Jürgen. *Versuch über den Normalismus: Wie Normalität produziert wird*. Opladen: Westdeutscher Verlag, 1999.

Lofland, Lyn H. *The Public Realm: Exploring the City's Quintessential Social Territory*. Piscataway, NJ: Transaction, 1998.

Lölfgren, Orvar. "Motion and Emotion: Learning to Be a Railway Traveler." *Mobilities* 3, no. 3 (October 2008): 331–351.

Lubove, Roy. *The Progressives and the Slums: Tenement House Reform in New York City, 1890–1917*. Pittsburgh: University of Pittsburgh Press, 1974.

Luhmann, Niklas. "Die Tücke des Subjekts und die Frage nach dem Menschen." In *Der Mensch, das Medium der Gesellschaft?*, edited by Peter Fuchs and Andrea Göbel, 40–56. Frankfurt am Main: Suhrkamp, 1994.

Makropoulos, Michael. *Theorie der Massenkultur*. Paderborn: Fink, 2008.

Mandel, Ernest. *Late Capitalism*. London: Humanities Press, 1975.

Marcus, George E., and Erkan Saka. "Assemblage." *Theory, Culture & Society* 23, no. 2–3 (May 2006): 101–106.

Marcuse, Herbert. *One-Dimensional Man: Studies in the Ideology of Advanced Industrial Society*. New York: Routledge 2002.

Marguis, Marie, and Pierre Filiatrault. "Understanding Complaining Responses through Consumers' Self-Consciousness Disposition." *Psychology and Marketing* 19, no. 3 (March 2002): 267–292.

Marshall, Alex. *Beneath the Metropolis: The Secret Lives of Cities*. New York: Caroll & Graf, 2006.

Marx, Karl. *Capital: A Critique of Political Economy*, volume 2: *The Process of Circulation of Capital*. Translated from the 2nd German edition by Ernest Untermann. Chicago: Charles H. Kerr, 1910.

Marx, Karl. "Economic and Philosophic Manuscripts." In *Karl Marx: Early Writings*, translated by Rodney Livingstone, 279–400. Harmondsworth: Penguin, 1975.

Marx, Karl. *Grundrisse: Foundations of the Critique of Political Economy*. London: Penguin UK, 2005.

Marx, Karl. "The Poverty of Philosophy." In *Marx–Engels Collected Works*, volume 6: *Marx and Engels, 1845–1848*, 105–212. New York: International Publishers, 1976.

Marx, Leo. *The Pilot and the Passenger: Essays on Literature, Technology, and Culture in the United States*. New York: Oxford University Press, 1988.

McClain, Noah. "Social Control, Object Interventions and Social-Material Recursivity." Unpublished manuscript, last modified August 14, 2008.

McGerr, Michael E. *A Fierce Discontent: The Rise and Fall of the Progressive Movement in America, 1870–1920*. New York: Oxford University Press, 2005.

McLuhan, Marshall. *Mechanical Bride, Folklore of Industrial Man*. New York: Vanguard Press, 1951.

McPhail, Clark. "Crowd Behavior." In *Blackwell Encyclopedia of Sociology*, edited by George Ritzer, 880–883. London: Blackwell, 2007.

McShane, Clay. *Down the Asphalt Path: The Automobile and the American City*. New York: Columbia University Press, 1995.

Meffert, Heribert, and Manfred Bruhn. "Beschwerdeverhalten und Zufriedenheit von Konsumenten." In *Marktorientierte Unternehmensführung im Wandel*, edited by Heribert Meffert, 91–118. Wiesbaden: Gabler Verlag, 1999.

Melosi, Martin. *Sanitary City: Urban Infrastructure in America from Colonial Times to the Present*. Baltimore: Johns Hopkins University Press, 2000.

Menninghaus, Winfried. *Disgust: Theory and History of a Strong Sensation*. Albany, NY: SUNY Press, 2003.

Merkel, Ina. *Wir sind noch nicht die Mecker-Ecke der Nation: Briefe an das DDR-Fernsehen*. Cologne: Böhlau, 1998.

Merz-Benz, Peter-Ulrich, and Gerhard Wagner, eds. *Der Fremde als sozialer Typus*. Constance: UVK Verlagsgesellschaft, 2007.

Mills, C. Wright. *White Collar: The American Middle Classes.* New York: Oxford University Press, 1951.

Mills, Nicolaus. *The Crowd in American Literature.* Baton Rouge: Louisiana State University Press, 1986.

Mollenkopf, John H. *New York City in the 1980s: A Social, Economic, and Political Atlas.* New York: Simon & Schuster, 1993.

Mollenkopf, John H., and Manuel Castells, eds. *Dual City: Restructuring New York.* New York: Russell Sage Foundation, 1992.

Molotch, Harvey. *Against Security: How We Go Wrong at Airports, Subways, and Other Sites of Ambiguous Danger.* Princeton, NJ: Princeton University Press, 2012.

Mooney, James E. "Sage, Russell." In *Encyclopedia of New York City*, edited by Kenneth T. Jackson, 1032. New Haven: Yale University Press, 1995.

Morat, Daniel. "Geschichte des Hörens. Ein Forschungsbericht." *Archiv für Sozialgeschichte* 51 (October 2011): 695–716.

Morehouse, Ward III. *Waldorf-Astoria: America's Gilded Dream.* New York: 1991.

Morgenstern, Oskar. "Note on the Formulation of the Theory of Logistics." *Naval Research Logistics Quarterly* 2, no. 3 (March 1955): 129–136.

Morley, Christopher. *Christopher Morley's New York.* New York: Fordham University Press, 1988.

Morley, Christopher. "Thoughts in the Subway." In *Plum Pudding: A Literary Concoction*, 133–135. Maryland: Wildside Press LLC, 2005.

Mühlberg, Felix. *Bürger, Bitten und Behörden: Geschichte der Eingabe in der DDR.* Berlin: Dietz, 2004.

Mújica, Francisco. *History of the Skyscraper.* Cambridge: Da Capo Press, 1977.

Mumford, Lewis. *Art and Technics.* New York: Columbia University Press, 2000.

Mumford, Lewis. "Attacking the Housing Problem on Three Fronts." *Nation* 109, no. 2827 (June 1919): 332–333.

Mumford, Lewis. *The Culture of Cities.* San Diego: Harvest/HBJ, 1970.

Mumford, Lewis. "The Intolerable City: Must It Keep on Growing?" *Harper's* 152 (February 1926): 283–293.

Mumford, Lewis. "The Metropolitan Milieu." In *America and Alfred Stieglitz: A Collective Portrait*, edited by Waldo Frank et al., 33–58. Garden City, NY: Doubleday, Doran, 1934.

Mumford, Lewis. *The Myth of the Machine: The Pentagon of Power.* Harcourt Brace Javanovich, 1970.

Mumford, Lewis. *The Myth of the Machine: Technics and Human Development.* Harcourt Brace Javanovich, 1967.

Mumford, Lewis. *The Story of Utopias.* Whitefish, MT: Kessinger Publishing, 1962.

Mumford, Lewis. *Technics and Civilization.* New York: Harcourt, Brace, 1934.

Nasri, Verya. "Design of Second Avenue Subway in New York." Conference paper at the World Tunnel Congress, "Underground Facilities for Better Environment and Safety," India, 2008.

Nelson, Daniel. *Managers and Workers: Origins of the Twentieth-Century Factory System in the United States, 1880–1920*. Madison: University of Wisconsin Press, 1996.

New York City Transit Authority. *The New York City Transit Police Department: History and Organization*. Internal Report, December 1990.

New York Transit Museum. *Subway Style: 100 Years of Architecture and Design in the New York City Subway*. New York: Stewart, Tabori & Chang, 2004.

Ortega y Gasset, José. *The Revolt of the Masses*. New York: W. W. Norton, 1932.

Orwell, George. *The Road to Wigan Pier*. London: Penguin, 1989.

Osterhammel, Jürgen. *Transformation of the World: A Global History of the Nineteenth Century*. Princeton, NJ: Princeton University Press, 2015.

Oudshoorn, Nelley, and Trevor Pinch. "Introduction: How Users and Non-Users Matter." In *How Users Matter: The Co-Construction of Users and Technology*, edited by Nelly Oudshoorn and Trevor Pinch, 1–25. Cambridge, MA: MIT Press, 2005.

Ovenden, Mark. *Transit Maps of the World*. London: Penguin, 2007.

Park, Robert Ezra. *The Crowd and the Public and Other Essays*. Chicago: University of Chicago Press, 1972.

Pearson, Marjorie. "Heins and La Farge." In *Encyclopedia of New York City*, edited by Kenneth T. Jackson, 537. New Haven: Yale University Press, 1995.

Perry, Clarence Arthur. *Housing for the Machine Age*. New York: Russell Sage Foundation, 1929.

Pike, David L. *Metropolis on the Styx: The Underworlds of Modern Urban Culture, 1800–2001*. Ithaca: Cornell University Press, 2007.

Pike, David L. *Subterranean Cities: The World beneath Paris and London, 1800–1945*. Ithaca: Cornell University Press, 2005.

Platt, Harold L. "Planning Modernism: Growing the Organic City in the 20th Century." In *Thick Space: Approaches to Metropolitanism*, edited by Dorothee Brantz, Sasha Disko, and Georg Wagner-Kyora, 165–212. Bielefeld: transcript, 2012.

Pluntz, Richar A. "On the Uses and Abuses of Air: Perfecting the New York Tenement, 1850–1901." In *Like and Unlike: Essays on Architecture and Art from 1870 to the Present*, edited by Josef P. Kleibues and Christina Rathgeber, 159–179. New York: Rizzoli, 1993.

Quante, Wolfgang. *The Exodus of Corporate Headquarters from New York City*. New York: Praeger, 1976.

Rabinbach, Anson. *The Human Motor: Energy, Fatigue, and the Origins of Modernity*. Berkeley: University of California Press, 1992.

Rathje, Ulf, and Roswitha Schröder. "Bewertung von Eingaben der Bürger an den Präsidenten der DDR: Bestand DA 4 der Präsidialkanzlei." *Mitteilungen aus dem Bundesarchiv* 14 (2006): 65–70.

Reckwitz, Andreas. *Das hybride Subjekt: Eine Theorie der Subjektkulturen von der bürgerlichen Moderne zur Postmoderne.* Weilerswist: Velbrück Wissenschaft, 2010.

Reckwitz, Andreas. *Subjekt.* Bielefeld: transcript, 2008.

Reich, Wilhelm. *Character Analysis.* New York: Farrar, Straus and Giroux, 1990.

Reid, Donald. *Paris Sewer and Sewermen: Realities and Representations.* Cambridge, MA: Harvard University Press, 1991.

Riesman, David. *The Lonely Crowd: A Study of the Changing American Character.* New Haven: Yale University Press, 2001.

Riis, Jacob A. *How the Other Half Lives.* New York: Barnes & Noble, 2004.

Rivera, Joseph. *Vandal Squad: Inside the New York City Transit Police Department, 1984–2004.* New York: Powerhouse, 2008.

Rivo, Steve. "Daily News." In *Encyclopedia of New York City*, edited by Kenneth T. Jackson, 307–308. New Haven: Yale University Press, 1995.

Rodgers, Daniel T. *Atlantic Crossings: Social Politics in a Progressive Age.* Cambridge, MA: Harvard University Press, 1998.

Roller, Franziska. "Flaneurinnen, Strassenmädchen, Bürgerinnen: Öffentlicher Raum und gesellschaftliche Teilhabe von Frauen." In *Geschlechterräume: Konstruktionen von "Gender" in Geschichte, Literatur und Alltag*, edited by Margarete Hubrath, 251–265. Cologne: Böhlau, 2001.

Roper, Elmo. *Roper Surveys Subway Riders.* New York: New York Subways Advertising Company, 1941.

Rose, Mark H., and Vincent Seyfried. "Streetcars." In *Encyclopedia of New York City*, edited by Kenneth T. Jackson, 1127–1128. New Haven: Yale University Press, 1995.

Rose, Nikolas S. *Governing the Soul: The Shaping of the Private Self.* London: Free Association Books, 1999.

Rose, Nikolas S., and Peter Miller. *Governing the Present: Administering Economic, Social and Personal Life.* New York: John Wiley & Sons, 2013.

Rosenheim, Jeff L. "Afterword." In *Walker Evans: Many Are Called*, 197–204. New York: Yale University Press, 2004.

Rubin, Martin. "The Crowd, the Collective, and the Chorus: Busby Berkeley and the New Deal." In *Movies and Mass Culture*, edited by John Belton, 59–94. New Brunswick, NJ: Rutgers University Press, 1996.

Sachs, Wolfgang. "Herren über Raum und Zeit. Ein Rückblick in die Geschichte unserer Wünsche." In *Nahe Ferne—Fremde Nähe: Infrastrukturen und Alltag*, edited by Barbara Mettler-Meibom and Christine Bauhardt, 59–68. Berlin: Edition Sigma, 1993.

Saegert, Susan. "Crowding: Cognitive Overload and Behavioral Constraint." *Environmental Design Research* 2 (1973): 254–261.

Salomon, Georg. "Out of the Labyrinth: A Plea and a Plan for Improved Passenger Information on the New York Subways." Unpublished manuscript. New York, 1957.

Sansone, Gene. *New York Subways: An Illustrated History of New York City's Transit Cars*. Baltimore: Johns Hopkins University Press, 1997.

Sarasin, Philipp, and Andreas Kilcher. "Editorial." In *Nach Feierabend. Zürcher Jahrbuch für Wissensgeschichte: Zirkulationen* 7, edited by David Gugerli et al., 8–10. Zurich: diaphanes, 2011.

Sassen, Saskia. *Deciphering the Global: Its Scales, Spaces and Subjects*. London: Routledge, 2007.

Sassen, Saskia. *The Global City: New York, London, Tokyo*. Princeton, NJ: Princeton University Press, 2001.

Schabacher, Gabriele. "Raum-Zeit-Regime: Logistikgeschichte als Wissenszirkulation zwischen Medien, Verkehr und Ökonomie." *Archive für Mediengeschichte* 8 (2008): 135–148.

Schiltz, Marie-Ange, Yann Darré, and Luc Boltanski. "La Dénonciation." *Actes de la Recherche en Sciences Sociales* 51, no. 1 (March 1984): 3–40.

Schivelbusch, Wolfgang. *Disenchanted Night: The Industrialization of Light in the Nineteenth Century*. Berkeley: University of California Press, 1998.

Schivelbusch, Wolfgang. *The Railway Journey: The Industrialization of Time and Space in the Nineteenth Century*. New York: Urizen Books, 1979.

Schlögel, Karl. *Im Raume lesen wir die Zeit: Über Zivilisationsgeschichte und Geopolitik*. Munich: Carl Hanser, 2003.

Schmucki, Barbara. "On the Trams: Women, Men, and Urban Pubic Transport in Germany." *Journal of Transport History* 23, no. 1 (March 2002): 60–72.

Schneider, Ulrich Johannes. *Michel Foucault*. Darmstadt: Primus, 2006.

Schroer, Markus, and Stephan Moebius. *Diven, Hacker, Spekulanten: Sozialfiguren der Gegenwart*. Frankfurt am Main: Suhrkamp, 2010.

Schwartz, Danielle. "Modernism for the Masses: The Industrial Design of John Vassos." *Archives of American Art Journal* 46, nos. 1–2 (2006): 4–23.

Scott, James C. *Seeing Like a State: How Certain Schemes to Improve the Human Condition Have Failed*. New Haven: Yale University Press, 1999.

Sennett, Richard. *The Corrosion of Character: The Personal Consequences of Work in the New Capitalism*. New York: W. W. Norton, 1998.

Sennett, Richard. *The Fall of Public Man*. New York: Alfred A. Knopf, 1977.

Sennett, Richard. *Flesh and Stone: The Body and the City in Western Civilization*. New York: W. W. Norton, 1996.

Serres, Michel. *The Five Senses: A Philosophy of Mingled Bodies*. London: Bloomsbury, 2008.

Seyfried, Vincent. "Blizzard of 1888." In *Encyclopedia of New York City*, edited by Kenneth T. Jackson, 118. New Haven: Yale University Press, 1995.

Shanor, Rebecca Read. *The City That Never Was: Two Hundred Years of Fantastic and Fascinating Plans that Might Have Changed the Face of New York City*. New York: Viking, 1988.

Shaw, Paul. *Helvetica and the New York City Subway System*. Cambridge, MA: MIT Press, 2011.

Shepherd, Roger, eds. *Skyscraper: The Search for an American Style, 1891–1941*. New York: McGraw-Hill, 2003.

Siegert, Bernhard. "Doors: On the Materiality of the Symbolic." *Grey Room* 47 (Spring 2012): 6–23.

Siegert, Bernhard. *Passagiere und Papiere: Schreibakte auf der Schwelle zwischen Spanien und Amerika*, Munich: Wilhelm Fink, 2006.

Siegert, Bernhard. *Relais: Geschicke der Literatur als Epoche der Post 1751–1913*. Berlin: Brinkmann u. Bose, 1993.

Simmel, Georg. "The Metropolis and Mental Life." In *The Blackwell City Reader*, edited by Gary Bridge and Sophie Watson, 11–19. Oxford: Wiley-Blackwell, 2002.

Simmel, Georg. *The Philosophy of Money*. New York: Routledge, 2011.

Simmel, Georg. *Sociology: Inquiries into the Construction of Social Forms (1908)*. Translated and edited by Anthony J. Blasi, Anton K. Jacobs, and Mathew Kanjirathinkal. Leiden: Brill, 2009.

Simon, Herbert A. "Designing Organizations for an Information-rich World," in *Computers, Communications, and the Public Interest*, edited by Martin Greenberger, 37–72. Baltimore, MD: Johns Hopkins Press, 1971.

Sloan-Howitt, Maryalice, and George L. Kelling. "Subway Graffiti in New York City: 'Getting Up' vs. 'Meaning It and Cleaning It.'" *Security Journal* 1, no. 3 (1990): 131–136.

Smith, Mark M. "Producing Sense, Consuming Sense, Making Sense: Perils and Prospects for Sensory History." *Journal of Social History* 40, no. 4 (Spring 2007): 841–858.

Smith, Neil. "Giuliani Time: The Revanchist 1990s." *Social Text* 57 (Winter 1998): 1–20.

Sohn, Werner. "Bio-Macht und Normalisierungsgesellschaft: Versuch einer Annäherung." In *Normalität und Abweichung: Studien zur Theorie und Geschichte der Normalisierungsgesellschaft*, edited by Werner Sohn and Herbert Mehrtens, 9–29. Opladen: Westdeutscher Verlag, 1999.

Sohn, Werner, and Herbert Mertens. *Normalität und Abweichung: Studien zur Theorie und Geschichte der Normalisierungsgesellschaft*. Opladen: Westdeutscher Verlag 1999.

Sollohub, Darius. "The Machine in Society." *Journal of Urban History* 34, no. 3 (May 2008): 532–40.

Sontag, Susan. *Regarding the Pain of Others*. New York: Farrar, Straus and Giroux, 2017.

Soper, George A. *The Air and Ventilation of Subways*. New York: John Wiley & Sons, 1908.

Spann, Edward K. "Grid Plan." In *Encyclopedia of New York City*, edited by Kenneth T. Jackson, 510. New Haven: Yale University Press, 1995.

Stabile, Donald R. "New York Chamber of Commerce and Industry." In *Encyclopedia of New York City*, edited by Kenneth T. Jackson, 825–826. New Haven: Yale University Press, 1995.

Stalter, Sunny. "The Subway Crush. Making Contact in New York City Subway Songs, 1904–1915." *Journal of American Culture* 34, no. 4 (December 2011): 321–331.

Star, Susan Leigh. "The Ethnography of Infrastructure." *American Behavioral Scientist* 43, no. 3 (November 1999): 377–391.

Steinmetz, George. "Hot War, Cold War: The Structures of Sociological Action, 1940–1955." In *Sociology in America: A History*, edited by Craig Calhoun, 314–366. Chicago: University of Chicago Press, 2007.

Stewart, Susan. "Remembering the Senses." In *Empire of the Senses: The Sensual Culture Reader*, edited by David Howes, 59–69. Oxford: Berg, 2005.

Streubel, Christiane. "Wir sind die geschädigte Generation: Lebensrückblicke von Rentnern in Eingaben an die Staatsführung der DDR." In *Graue Theorie: Die Kategorien Alter und Geschlecht im kulturellen Diskurs*, edited by Heike Hartung et al., 241–264. Cologne: Böhlau Verlag, 2007.

Tanenbaum, Susie J. *Underground Harmonies: Music and Politics in the Subways of New York*. Ithaca: Cornell University Press, 1995.

Taylor, Peter James. *Modernities: A Geohistorical Interpretation*. Minneapolis: University of Minnesota Press, 1999.

Thibaud, Jean-Paul. "The Sonic Composition of the City." In *Auditory Culture Reader*, edited by Michael Bull and Less Back, 329–342. Oxford: Berg, 2003.

Thøgersen, John, Hans Jørn, and Carsten Stig Poulsen. "Complaining: A Function Attitude, Personality, and Situation." *Psychology and Marketing* 26, no. 8 (July 2009): 760–777.

Töpfer, Georg. *Historisches Wörterbuch der Biologie: Geschichte und Theorie der biologischen Grundbegriffe* 2. Stuttgart: Metzler, 2011.

Trachtenberg, Alan. *The Incorporation of America: Culture and Society in the Gilded Age*. New York: Hill and Wang, 2007.

Turner, Victor. *The Ritual Process: Structure and Anti-Structure*. Chicago: Aldine, 1995.

Twaddell, Elizabeth. "The American Tract Society, 1814–1860." *Church History* 15, no. 2 (June 1946): 116–132.

Urry, John. *Mobilities*. Cambridge: Polity Press, 2007.

Urry, John. *Tourist Gaze*. London: Sage, 1990.

Vahrenkamp, Richard. *Die logistische Revolution: Der Aufstieg der Logistik in der Massenkonsumgesellschaft.* Frankfurt am Main: Campus Verlag, 2001.

Vance, James E., Jr. *The Continuing City: Urban Morphology in Western Civilization.* Baltimore: Johns Hopkins University Press, 1990.

Van Creveld, Martin L. *Supplying War: Logistics from Wallenstein to Patton.* Cambridge, MA: Cambridge University Press, 1980.

Van Gennep, Arnold. *The Rites of Passage.* London: Routledge, 2010.

Van Laak, Dirk. "Der Begriff 'Infrastruktur' und was er vor seiner Erfindung besagte." *Archive für Begriffsgeschichte* 41 (1999): 280–299.

Van Laak, Dirk. *Imperiale Infrastruktur: Deutsche Planungen für die Erschließung Afrikas 1880–1960.* Paderborn: Schöningh, 2004.

Van Laak, Dirk. "Infra-Strukturgeschichte." *Geschichte und Gesellschaft* 27, no. 3 (September 2001): 367–393.

Van Laak, Dirk. "Infrastruktur und Macht." In *Umwelt und Herrschaft in der Geschichte,* edited by François Duceppe-Lamarre and Jens Ivo Engels, 106–114. Munich: R. Oldenbourg, 2008.

Van Laak, Dirk. *Weiße Elefanten: Anspruch und Scheitern technischer Großprojekte im 20. Jahrhundert.* Munich: Deutsche Verlags-Anstalt, 1999.

Veblen, Thorstein. *Theory of the Leisure Class.* New York: Dover, 1994.

Verhoeff, Nanna. *Mobile Screens: The Visual Regime of Navigation.* Amsterdam: Amsterdam University Press, 2012.

Vickers, Squire J. "The Architectural Treatment of Special Elevated Stations of the Dual System, New York City." *Journal of the American Institute of Architects* 3 no. 11 (1915): 501–502.

Vickers, Squire J. "Design of Subway and Elevated Stations." *Municipal Engineers Journal* 3 no. 9 (1917): 114–120.

Virilio, Paul. "Perception, Politics and the Intellectual." Interview with Niels Brügger in *Virilio Live: Selected Interviews,* edited by John Armitage, 82–96. London: Sage, 2001.

Virilio, Paul. *Logistics of Perception.* New York: Verso, 1989.

Virilio, Paul. *Negative Horizon: An Essay in Dromoscopy.* London: Continuum, 2005.

Virilio, Paul. *Speed and Politics: An Essay on Dromology.* New York: Semiotext(e), 1977.

Von Suttner, Bertha. *Das Maschinenzeitalter: Zukunftsvorlesungen über unsere Zeit.* Dresden: E. Pierson, 1899.

Vormann, Boris. *Global Port Cities in North America: Urbanization Processes and Global Production Networks.* London: Routledge, 2015.

Vrachliotis, Georg. *Geregelte Verhältnisse: Architektur und technisches Denken in der Epoche der Kybernetik.* Vienna: Springer-Verlag, 2012.

Wachsmuth, David. "Three Ecologies: Urban Metabolism and the Society-Nature Opposition." *Sociological Quarterly* 53, no. 4 (Autumn 2012): 506–523.

Walker, James Blaine. *Fifty Years of Rapid Transit, 1864–1917.* North Stratford, NH: Ayer Publishing, 1918.

Walker, Robert H. "The Poet and the Rise of the City." *Mississippi Valley Historical Review* 49, no. 1 (June 1962): 85.

Ward, David, and Oliver Zunz. "Between Rationalism and Pluralism: Creating the Modern City." In *Landscape of Modernity, New York City 1900–1940,* edited by David Ward and Oliver Zunz, 3–18. Baltimore: Johns Hopkins University Press, 1992.

Ward, Janet. *Weimar Surfaces: Urban Visual Culture in 1920s Germany.* Berkeley: University of California Press, 2001.

Ward, Simon. "The Passenger as Flaneur? Railway Networks in German-language Fiction since 1945." *Modern Language Review* 100 (April 2005): 412–428.

Wasik, John F. *The Merchant of Power: Sam Insull, Thomas Edison, and the Creation of the Modern Metropolis.* New York: Palgrave Macmillan, 2008.

Webb, Barry, and Gloria Laycock. *Reducing Crime on the London Underground: An Evaluation of Three Pilot Projects.* London: Home Office, 1992. http://www.popcenter.org/library/scp/pdf/189-Webb_and_Laycock.pdf.

Weber, Adna F. *The Growth of Cities in the Nineteenth Century: A Study in Statistics.* New York: Macmillan, 1899.

Weber, Adna F. "Rapid Transit and the Housing Problem." *Municipal Affairs* 6 (1902): 408–417.

Weber, Heike. *Das Versprechen mobiler Freiheit: Zur Kultur- und Technikgeschichte von Kofferradio, Walkman und Handy.* Bielefeld: transcript, 2008.

Weber, Max. *The Protestant Ethic and the Spirit of Capitalism.* London: Routledge, 2013.

Weeks, John. *Unpopular Culture: The Ritual of Complaint in a British Bank.* Chicago and London: University of Chicago Press, 2004.

Wesser, Robert F. "McClellan, George Brinton." In *Encyclopedia of New York City,* edited by Kenneth T. Jackson, 704. New Haven: Yale University Press, 1995.

White, Elwyn Brooks. "The Commuter." In *The Lady Is Cold and Other Poems,* 26. New York: Harper & Brother, 1929.

White, Elwyn Brooks. "Here Is New York." In *Essays of E. B. White,* 148–168. New York: HarperCollins, 2006.

Whitman, Walt. "A Broadway Pageant." In *Leaves of Grass*, 193. New York: William E. Chapin Printers, 1867.

Whyte, William Foote. *Street Corner Society: The Social Structure of an Italian Slum*. 4th ed. Chicago: University of Chicago Press, 1993.

Wiebe, Robert H. *The Search for Order, 1877–1920*. New York: Hill and Wang, 1967.

Williams, Raymond. *Keywords: A Vocabulary of Culture and Society*. Rev. ed. New York: Oxford University Press, 1983.

Williams, Rosalind. *Notes on the Underground: An Essay on Technology, Society, and the Imagination*. Cambridge, MA: MIT Press, 2008.

Wilson, Richard Guy, Dianne H. Pilgrim, and Dickran Tashijan. *The Machine Age in America, 1918–1941*. New York: Brooklyn Museum/Harry N. Abrams, 1986.

Wise, J. Macgregor. "Assemblage." In *Gilles Deleuze: Key Concepts*, edited by Charles J. Strivale, 77–86. Montreal: McGill-Queen's University Press, 2005.

Wolmar, Christian. *The Subterranean Railway: How the London Underground Was Built and How It Changed the City Forever*. London: Atlantic Books, 2005.

Wood, Denis. *The Power of Maps*. New York: Guilford Press, 1992.

Woolgar, Steve. "Configuring the User: The Case of Usability Trials." In *A Sociology of Monsters: Essays on Power, Technology, and Domination*, edited by John Law, 57–99. London: Routledge, 1991.

Wu, Fang, and Bernardo A. Hubermann. "Novelty and Collective Attention." *Proceedings of the National Academy of Sciences* 104, no. 45 (November 2007): 17599–17601.

Zeng, Qingjie, Ciprian Alecsandru, Kuo Cheng Huang, Behzad Rouhieh, Ali Raza Khan, and Martine Ouellet. "Performance Evaluation of Subway Signage: Part I—Methodology." Transportation Research Board 90th Annual Meeting, Washington, DC, January 23–27, 2011. ftp://ftp.hsrcunc.edu/pub/TRB2011/data/papers/11-0588.pdf.

Zimmer, Amy. *Meet Miss Subways: New York's Beauty Queens 1941–1976*. Brookline, ME: Seapoint Books and Media, 2014.

Žižek, Slavoj. *The Ticklish Subject: The Absent Centre of Political Ontology*. London: Verso, 1999.

Zukin, Sharon. "Space and Symbols in an Age of Decline." In *Re-Presenting the City, Ethnicity, Capital and Culture in the 21st Century Metropolis*, edited by Anthony D. King, 43–59. New York: NYU Press, 1996.